COMPUTATIONAL SCIENCE

COMPUTATIONAL SCIENCE

Editor

Giorgino Mondo

scitus
academics

Computational Science

Edited by **Giorgino Mondo**

This Book is Printed on Acid-Free Paper. ♾

ISBN: 9781681179582

Library of Congress Control Number: 2017949103

SCITUS Academics LLC,

616, Corporate Way, Suite 2, 4766,

Valley Cottage, NY 10989

www.scitusacademics.com

Notice

Printed in United States of America.

Preface

Computational Science is a swiftly growing multi- and interdisciplinary field that uses advanced computing and data analysis to understand and solve complex problems. Computational science refers to the use of computers, networks, storage devices, software, and algorithms to solve problems, the development of models, systems, algorithms, do simulations, or create new knowledge. Since the advent of the computer, its latent to advance scientific investigation has been recognized. From the design and application of early computers to aid in the cracking of codes during WWII and the simulations leading to the hydrogen bomb in 1952, the use of computers and the development of algorithms and systems in the service of computational science were born. Since that time, the computer has become a crucial appliance of business and an essential facet of science. Computational simulation is being accepted as a third methodology in scientific research, complementing the traditional approaches of theory and experiment. Computer simulations provide both qualitative and quantitative insights into many phenomena that are too complex to be dealt with by analytical methods or too expensive or dangerous to study by experiments. Many experiments and investigations that have traditionally been performed in a laboratory, a wind tunnel, or the field are being augmented or replaced by computer simulations. Some studies, such as nuclear repository integrity and global climate change, involve time scales that preclude the use of realistic physical experiments. The availability of high performance computers, graphic workstations, and high speed networks, coupled with major advances in algorithms and software, has brought about a revolution in the way scientific and engineering investigations are carried out.

Computational Science is aimed to provide important research in different areas covering all aspects of mathematical and computational science, such as numerical analysis, optimization, linear and nonlinear programming, and theory of computation, theory of algorithms, computational logic, and applied combinatorics. It combines computational thinking, modern computational methods, devices and collateral technologies to address problems far beyond the scope of traditional numerical methods. It also presents original research in simulation based science across all scientific disciplines and advanced innovative, interdisciplinary research where complex multi-scale, multi-domain problems in science and engineering are solved, integrating sophisticated numerical methods and computation.

Table of Contents

CHAPTER 1

Droplet Dynamics in Confinement

N. Ioannou[a], H. Liu[b], Y.H. Zhang[a]

[a]James Weir Fluids Laboratory, Department of Mechanical & Aerospace Engineering, University of Strathclyde, Glasgow G1 1XJ, UK

[b]School of Energy and Power Engineering, Xi'an Jiaotong University, Xi'an 710049, China

ABSTRACT

This study is to understand confinement effect on the dynamical behaviour of a droplet immersed in an immiscible liquid subjected to a simple shear flow. The lattice Boltzmann method, which uses a forcing term and a recolouring algorithm to realize the interfacial tension effect and phase separation respectively, is adopted to systematically study droplet deformation and breakup in confined conditions. The effects of capillary number, viscosity ratio of the droplet to the carrier liquid, and confinement ratio are studied. The simulation results are compared against the theoretical predictions, experimental and numerical data available in literature. We find that increasing confinement ratio will enhance deformation, and the maximum deformation occurs at the viscosity ratio of unity. The droplet is found to orient more towards the flow direction with increasing viscosity ratio or confinement ratio. Also, it is noticed that the wall effect becomes more significant for the confinement ratios larger than 0.4. Finally, the critical capillary number, above which the droplet breakup occurs, is found to be mildly affected by the confinement for the viscosity ratio of unity. Upon increasing the confinement ratio, the critical capillary number increases for the viscosity ratios less than unity, but decreases for the viscosity ratios more than unity.

Keywords: Droplet dynamics Lattice Boltzmann method Deformation Breakup Confinement

1. INTRODUCTION

Emulsions consist of immiscible fluids commonly found in production processes in food, chemical, and pharmaceutical industries. Since the droplet size and shape determine important emulsion properties such as stability, rheology, and particle morphology, it is important to understand the mechanism of droplet deformation and breakup during emulsification. In addition, the study of droplet deformation and breakup can provide valuable insights into immiscible fluid displacement in porous media, which plays an important role in enhanced oil recovery, geologic CO_2 sequestration, and remediation of nonaqueous-phase liquids. In recent years, the droplet deformation and breakup have received more attention because of the growing interest in microfluidic technologies, where droplets are circulated in channels and their size is often comparable with or even smaller than channel dimension. A significant number of theoretical, experimental and numerical studies have been reported regarding droplet deformation and breakup in a shear flow since the pioneering work of Taylor [1]. The main feature of shear flow is its relative simplicity, while it contains rich physics [2].

As sketched in Fig. 1, a spherical droplet with radius R is initially placed halfway between two parallel plates that are separated by a distance H. Shear flow is introduced to this system by the movement of the plates with equal but opposite velocities U, and the resulting shear rate is $\dot{\gamma}=2U/H$. The

droplet and the carrier fluid are assumed to have equal densities ρ, a constant interfacial tension σ, and their dynamic viscosities are μ_d and μ_m. In the simple shear flow, two most important forces exerting on the droplet are: the viscous forces ($\mu_m\dot{\gamma}R^2$), leading the droplet to deform and turn

towards the flow direction, and the capillary forces (σR), resisting the droplet deformation and retaining the spherical shape of the droplet. Consequently, the droplet undergoes an elongation in the direction of the L-axis, and a compression in the direction of the B-axis (see the right panel of Fig. 1). The capillary number, which is defined as $Ca=\dot{\gamma}R\mu_m/\sigma$, is used to

measure the relative magnitude of viscous and capillary forces. The inertial force can also influence the dynamical behaviour of the droplet, and its importance is described by the Reynolds number, defined as the ratio of inertial to viscous forces i.e. $Re=\rho\dot{\gamma}R^2/\mu_m$. In addition to the capillary and

Reynolds numbers, the viscosity ratio ($\lambda = \mu_d/\mu_m$) and the proximity of droplet to the walls are found to affect strongly the deformation and breakup of the droplet. The proximity is evaluated by the confinement ratio, defined as $2R/H$. Under consideration of low capillary number and confinement ratio, the droplet will eventually become ellipsoidal in the Stokes flow regime ($Re < 1$). The deformation parameter D can be defined by the lengths of the major (L) and minor (B) axes of the deformed droplet, i.e.

$$D = \frac{L - B}{L + B} \tag{1}$$

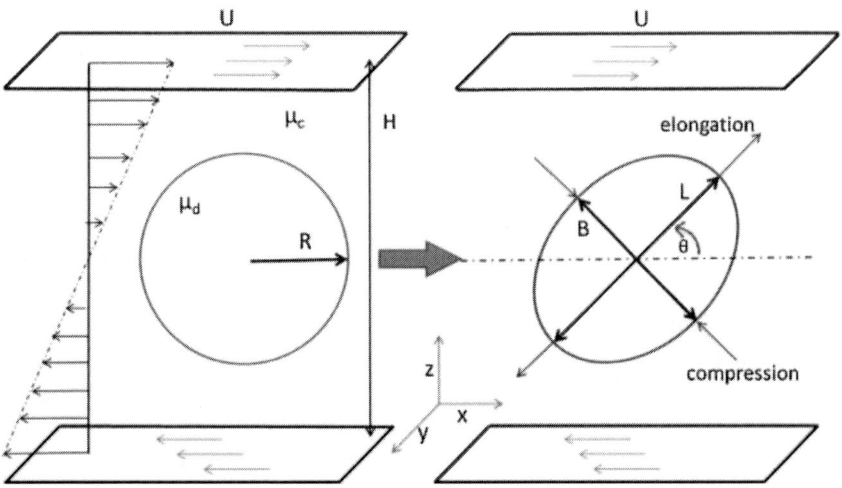

Figure. 1. Schematic illustration of droplet deformation in a simple shear flow.

With increasing Ca, the droplet can reach a steady state but deviate from the ellipsoidal shape. This causes difficulty in obtaining the correct values of deformation parameter and orientation angle (θ), which is defined as the angle between the droplet major axis and the horizontal plane. As the capillary number is further increased above the critical value, the capillary forces can no longer retain the shape of droplet; the dominant viscous forces lead the droplet to form a long thin neck and finally break up into two or more fragments. The critical capillary number (Ca_{cr}), above which the droplet breakup will occur, is influenced by both the viscosity and confinement ratios [3].

The interest in droplet deformation dates back to the work of Taylor (1934) [1], who derived a theoretical expression to describe small deformations in the bulk shear flow in terms of the viscosity ratio and the capillary number:

$$D_T = \frac{19\lambda + 16}{16\lambda + 16} Ca \quad (2)$$

This expression has been demonstrated to predict experimental results well in a variety of cases, where the value of confinement ratio was around 0.2. However, the influence of wall confinement is not taken into account in Eq. (2). It was reported that the presence of walls has a negligible contribution to the droplet deformation for the confinement ratio $2R/H \leq 0.4$ [4,5]. When the confinement ratio is higher than 0.4, the deformation cannot be predicted accurately by Eq. (2). Moreover, Eq. (2) is not able to describe the droplet deformation for very large viscosity ratios [1,6]. More discussion about this equation can be found in these review papers [7–10]. Taylor derived his model using small deformation perturbation procedure to the first order, with Ca as the expansion parameter, so he obtained a constant orientation angle of 45°. The perturbation procedure was later to be extended to the second order in Ca to yield an expression for the orientation angle [11–14]: $\theta = (\pi/4) - ((16 + 19\lambda)(3 + 2\lambda)/80(1 + \lambda))Ca$[15]. The orientation angle was also formulated differently in the phenomenological models of Maffetone and Minale [16] and Minale [17].

To address the wall confinement effect on droplet behaviour, Shapira and Haber [18] solved the Stokes flows around a droplet using the method of reflection, which takes into account the relative position of the droplet to the wall. The resulting SH model combines the Taylor deformation and an additional term accounting for the influence of walls on the deformation:

$$D_{SH} = D_T \left[1 + C_S \frac{1 + 2.5\lambda}{1 + \lambda} \left(\frac{R}{H}\right)^3\right], \quad (3)$$

where C_S is referred to as a shape parameter and its value depends on the relative position of the droplet to the walls. For a droplet positioned halfway between the two walls C_S is taken as 5.6996.

To predict the droplet deformation under transient conditions, Maffettone and Minale [16] proposed a phenomenological model, i.e. MM model, in which the droplet shape is assumed to remain ellipsoidal. A second order tensor S, whose eigenvalues reflect the squares of the semi-axes of an ellipsoid, is used to describe the droplet shape. Based on this assumption, they derived an evolution equation for the tensor S, which consists of a co-rotational derivative, the contributions of the viscous stress and the capillary force. The values of the semi-axes for several typical flows were predicted, including the simple shear flow, the uniaxial extensional flow and the planar hyperbolic flow. In particular, for the simple shear flow, the deformation parameter in a steady state is calculated as

$$D_{MM} = \frac{\sqrt{m_1^2 + Ca^2} - \sqrt{m_1^2 + (1 - m_2^2)Ca^2}}{m_2 Ca}, \qquad (4)$$

where

$$m_1 = \frac{40(\lambda + 1)}{(2\lambda + 3)(19\lambda + 16)}, \qquad (5)$$

and

$$m_2 = \frac{5}{2\lambda + 3} + \frac{3Ca^2}{2 + 6Ca^2} \qquad (6)$$

The droplet deformation was also investigated experimentally by Sibillo et al. [19] for the viscosity ratio of unity. Three different capillary numbers, i.e. 0.1, 0.2, and 0.3, were studied for the confinement ratios ranging from 0.14 to 1.0. For the same capillary number, the droplet could obtain a more elongated shape for larger confinement ratios. Also, the droplet shape deviates from ellipsoid at very high confinement ratios, i.e., $2R/H \geq 0.8$. A similar study, for a range of viscosity ratios, was performed by Vananroye et al. [6]. The difference in the droplet deformation is not significant for different viscosity ratios at low confinement ratios. However, this difference grows considerably at large confinement ratios, and is more significant at large capillary numbers. In an attempt to match the experimental results regarding large deformation at large capillary numbers (e.g., $Ca = 0.3$) with an analytical expression, two models were proposed. One is the orientation angle scaled SH model. However, it suffers from the difficulty of estimating the orientation angle at high confinement ratios [7]. The other is a combination of the aforementioned two models (i.e., SH and MM). Specifically, the term D_T in the SH model is replaced by D_{MM} in the MM model, leading to a new model, called as the MMSH model:

$$D_{MMSH} = D_{MM}\left[1 + C_S \frac{1 + 2.5\lambda}{1 + \lambda}\left(\frac{R}{H}\right)^3\right], \qquad (7)$$

Since the MMSH model does not need to estimate the droplet orientation angle and its predictions agree well with the experimental data [6], it will be used in Section 3 for validation of our numerical results. In addition, Minale [17] refined the phenomenological MM model, by forcing the model to recover the analytical limits of the SH model for moderate confinement ratios, i.e. $2R/H \leq 0.5$. For large confinement ratios, the experimental data from Sibillo et al. [19] and Vananroye et al. [6], were used to re-evaluate the coefficients m_1 and m_2 in the MM model.

Complementing theoretical and experimental studies, numerical simulations have been extensively used to investigate the droplet deformation and breakup in a simple shear flow [3,4,20–24]. Among these works, Janssen and his co-workers [3,23,24] used the boundary integral method (BIM) to study the effect of viscosity and confinement ratios on droplet deformation and the critical capillary number. It was found that the simulated shape of the droplet and the critical capillary number are both in excellent agreement with the experimental results. However, the BIM is usually for the solution of Stokes flow, in which the fluid inertia is totally neglected. Also, it suffers from the difficulties of accessing the low viscosity ratio regime [3], and of dealing with topological changes such as breakup and coalescence of the interfaces [7,25,26]. These disadvantages can be overcome by the lattice Boltzmann method (LBM), which tracks evolution of the distribution function of an assembly of molecules and is built upon microscopic models and mesoscopic kinetic equations. The LBM can capture multiphase interface and describe interfacial dynamics such as phase segregation and interfacial tension through the incorporation of intermolecular interactions. A number of multiphase and multicomponent models have been proposed, which can be divided into four main types: colour gradient model [27], free-energy model [28], interparticle-potential model [29], and mean-field theory model [30]. Although the droplet dynamics in a shear flow was successfully simulated by the former three models or their variants [31–36], none of them has systematically examined the impact of viscosity and confinement ratios on the droplet deformation and breakup, which is the focus of this work. In particular, the colour gradient LBM is used here because of its advantages such as low spurious currents, high numerical accuracy and strict mass conservation for each fluid which have been demonstrated in our recent works [37,38].

2. NUMERICAL METHOD

We consider isothermal motion of two immiscible fluids and adopt a continuum surface force (CSF) model [39] to describe the system. The CSF formulation allows us to consider the two phases as a single fluid with space-dependent properties and replaces the jump condition at the interface with an additional force that acts only in the interface region. Unlike a two-fluid formulation, it does not need to solve a moving boundary problem, which is computationally expensive and extremely difficult to implement especially when large topological changes occur around the interface. In the CSF formulation, the flow of two immiscible fluids is governed by a single set of Navier–Stokes equations (NSEs):

$$\nabla \cdot \vec{u} = 0 \quad , \tag{8}$$

$$\frac{\partial(\rho\vec{u})}{\partial t} + \nabla \cdot \left(\rho\vec{u}\vec{u}\right) = -\nabla p + \nabla \cdot [\mu(\nabla\vec{u} + \nabla\vec{u}^T)] + \vec{F}_s \quad , \tag{9}$$

where t is the time, $\vec{u}$ is the fluid velocity, ρ is the total density, p is the pressure, and μ is the dynamic viscosity. The last term in Eq. (9) represents the interfacial tension force acting at the interface between the two fluids and is defined as

$$\vec{F}_s = \sigma\kappa\vec{n}\delta_\Gamma \quad , \tag{10}$$

where σ is the interfacial tension; κ, the local interface curvature; $\vec{n}$, the unit normal to the interface; and δ_Γ, the Dirac delta function used to localize the force explicitly at the interface, which should satisfy

$$\int_{-\infty}^{\infty} \delta_\Gamma dz = 1 \tag{11}$$

in order to recover properly the stress jump condition in the sharp-interface limit. Here z is the spatial location normal to the interface.

Although Eqs. (8) and (9) allow computing the velocity field, an advection equation has to be additionally solved to capture the evolution of interface in traditional multiphase solvers, e.g., the volume-of-fluid (VOF) and level-set methods, where the sophisticated interface reconstruction algorithm or unphysical re-initialization process is often required. In order to avoid these issues, we use the lattice Boltzmann colour gradient model, first proposed by Gunstensen et al. [27] and later improved by Halliday and his coworkers [40], to simulate the dynamical behaviour of a droplet subjected to a simple shear flow. The colour gradient model possesses many advantages in simulating immiscible two-phase flows, including the ability of capturing interface automatically, low spurious velocities, high numerical accuracy and strict mass conservation for each fluid. In this model, two sets of distribution functions f_i^R and f_i^B are introduced to represent the "red" and "blue" fluids. The total distribution function is defined by $f_i = f_i^R + f_i^B$, which undergoes a Bhatnagar–Gross–Krook (BGK) collision step as

$$f_i^{\dagger}(\vec{x},t) = f_i(\vec{x},t) - \frac{1}{\tau}(f_i(\vec{x},t) - f_i^{eq}(\vec{x},t)) + \Phi_i \quad ,(12)$$

where $f_i(\vec{x},t)$ is the total distribution function in the ith velocity direction at the position $\vec{x}$ and time t, f_i^{eq} is the equilibrium distribution function of f_i, $f_i^{\dagger}$ is the post-collision distribution function, τ is the dimensionless relaxation time, and Φ_i is the forcing term.

The equilibrium distribution function is obtained by a second-order Taylor expansion of Maxwell–Boltzmann distribution with respect to the local velocity $\vec{u}$:

$$f_i^{eq} = \rho w_i \left[1 + \frac{\vec{e}_i \cdot \vec{u}}{c_s^2} + \frac{(\vec{e}_i \cdot \vec{u})^2}{2c_s^4} - \frac{\vec{u}^2}{2c_s^2}\right], \quad (13)$$

where ρ is calculated by $\rho = \rho_R + \rho_B$ with the subscripts 'R' and 'B' referring to the red and blue fluids respectively, c_s is the speed of sound, $\vec{e}_i$ is the lattice velocity in the ith direction, and wi is the weight factor. For the three-dimensional 19-velocity (D3Q19) model, the lattice velocity $\vec{e}_i$ and the weight factors are given by

$$\vec{e}_i = \begin{cases} (0,0,0)c, & i=0; \\ (\pm1,0,0)c, (0,\pm1,0)c, (0,0,\pm1)c, & i=1,2,\ldots,6; \\ (\pm1,\pm1,0)c, (0,\pm1,\pm1)c, (\pm1,0,\pm1)c, & i=7,8,\ldots,18, \end{cases} \quad (14)$$

$$w_i = \begin{cases} 1/3, & i=0; \\ 1/18, & i=1,2,\ldots,6; \\ 1/36, & i=7,8,\ldots,18 \end{cases}, \quad (15)$$

and the speed of sound $c_s = c/\sqrt{3} = \delta_x/\sqrt{3}\delta_t$ with δ_x and δ_t being the lattice length and time step, respectively.

The spatial distribution of the two fluids is described using a colour function (or phase-field function), which is defined as

$$\rho^N(\vec{x},t) = \frac{\rho_R(\vec{x},t) - \rho_B(\vec{x},t)}{\rho_R(\vec{x},t) + \rho_B(\vec{x},t)}, \quad -1 \le \rho^N \le 1 \quad . \quad (16)$$

By defining $\vec{n} = -\nabla\rho^N/|\nabla\rho^N|$, and $\delta_\Gamma = |\nabla\rho^N|/2$, Eq. (10) can be further written as

$$\vec{F}_s = -\frac{1}{2}\sigma\kappa\nabla\rho^N \quad (17)$$

where the local interface curvature κ is related to the unit normal to the interface by

$$\kappa = -\left[(\mathbf{I} - \vec{n}\otimes\vec{n})\cdot\nabla\right]\cdot\vec{n} = -\nabla\cdot\vec{n} \quad (18)$$

and **I** is the second-order identity tensor. With the interfacial tension force given by Eq. (17), the forcing term Φ_i that is applied to realize the interfacial tension effect, reads as [41]

$$\Phi_i = \left(1 - \frac{1}{2\tau}\right) w_i \left(\frac{\vec{e}_i - \vec{u}}{c_s^2} + \frac{\vec{e}_i\cdot\vec{u}}{c_s^4}\vec{e}_i\right)\cdot\vec{F}_s\delta_t \quad (19)$$

According to Guo et al. [42], the local fluid velocity should be defined to incorporate the spatially varying interfacial tension force, i.e.,

$$\rho\vec{u}(\vec{x},t) = \sum_i f_i(\vec{x},t)\vec{e}_i + \frac{1}{2}\vec{F}_s(\vec{x},t)\delta_t \quad (20)$$

Using the Chapman–Enskog multiscale expansion, Eq. (12) can reduce to the NSEs in the low frequency, long wavelength limit with the pressure and the fluid viscosity defined by

$$p = \rho c_s^2 \quad (21)$$

$$\mu = \rho c_s^2\left(\tau - \frac{1}{2}\right)\delta_t \quad (22)$$

To allow for unequal viscosities of the two fluids, we determine the viscosity of the fluid mixture by a harmonic mean, i.e.

$$\frac{1}{\mu\left(\rho^N\right)} = \frac{1+\rho^N}{2\mu_R} + \frac{1-\rho^N}{2\mu_B} \quad (23)$$

where μ_k (k = R or B) is the dynamic viscosity of fluid k. It has been shown that the choice of Eq. (23) can ensure a constant viscosity stress across the interface, which provides higher accuracy than other choices [43].

The partial derivatives required for the curvature and normal vector calculations are obtained using the 19-point compact finite-difference stencil [44]. For example, for a variable ψ, its partial derivatives can be calculated by

$$\frac{\partial \psi(\vec{x})}{\partial x_a} = \frac{1}{c_s^2} \sum_i w_i \psi(\vec{x} + \vec{e}_i \delta_t) e_{ia} \quad . \tag{24}$$

Although the forcing term generates an interfacial tension, it does not guarantee the immiscibility of both fluids. To promote phase segregation and maintain a reasonable interface, the segregation (recolouring) algorithm of Latva-Kokko and Rothman is used [45]. It can overcome the lattice pinning problem and creates a symmetric distribution of particles around the interface so that unphysical spurious currents can be effectively reduced. Following Latva-Kokko and Rothman, the post-segregation (recoloured) distribution functions of the red and blue fluids are

$$\begin{aligned} f_i^R(\vec{x}, t) &= \frac{\rho_R}{\rho} f_i^{\dagger}(\vec{x}, t) + \beta \frac{\rho_R \rho_B}{\rho} w_i \cos(\varphi_i) |\vec{e}_i| \\ f_i^B(\vec{x}, t) &= \frac{\rho_B}{\rho} f_i^{\dagger}(\vec{x}, t) - \beta \frac{\rho_R \rho_B}{\rho} w_i \cos(\varphi_i) |\vec{e}_i| \quad , \end{aligned} \tag{25}$$

where β is the segregation parameter and is set to be 0.7 for numerical stability and model accuracy [46]; φ_i is the angle between the colour gradient and the lattice vector $\vec{e}_i$, which is defined by

$$\cos(\varphi_i) = \frac{\vec{e}_i \cdot \nabla \rho^N}{|\vec{e}_i||\nabla \rho^N|} \quad . \tag{26}$$

After the recolouring step, the red and blue distribution functions propagate to the neighbouring lattice nodes, known as propagation or streaming step:

$$f_i^k(\vec{x} + \vec{e}_i \delta_t, t + \delta_t) = f_i^k(\vec{x}, t), \quad k = R \text{ or } B \quad , \tag{27}$$

and the resulting distribution functions are then used to calculate the densities of both fluids i.e. $\rho^k = \sum_i f_i^k$

3. RESULTS AND DISCUSSION

In all the cases discussed below, a spherical droplet with the radius $R = 20$ lattices is initially placed in the centre of the computational domain, as illustrated in Fig. 1. The computational domain has a length of 71 lattices in the y-direction. The length in the x-direction is chosen as 101 in the study of droplet deformation, whereas in droplet break-up this length could be increased up to 401 in some cases in order to avoid the occurrence that the droplet touches itself via the periodic boundary conditions before the breakup occurs. In addition, the height of the computational domain is varied in the range of $51 \leq H \leq 223$ to obtain different confinement ratios $2R/H$. Periodic boundary conditions are applied in the x- and y-directions, while the half-way bounce-back scheme is used on the top and bottom walls [47]:

$$f_i(\vec{x}, t+\delta t) = f^{\dagger}_{ops(i)}(\vec{x}, t) + 6w_i\rho\frac{\vec{e}_i \cdot \vec{u}_W}{c}, \tag{28}$$

where $\vec{u}_W$ is the moving velocity of the solid wall with $\vec{u}_W$=(U,0,0) for the top wall and $\vec{u}_W$=(−U,0,0) for the bottom wall, and $ops(i)$ stands for the distribution function which has opposite direction to $\vec{e}_i$. For the sake of simplicity, both fluids are assumed to have equal densities. The Reynolds number is fixed at 0.1, which is small enough to characterise the flow as Stokes flow.

3.1. Droplet deformation

The numerical method is first tested by simulating the droplet deformation in a weakly confined shear flow. Simulations are conducted at $\lambda = 1$ and the confinement ratio $2R/H = 0.18$ for the capillary numbers ranging from 0.1 to 0.3. Fig. 2(a) plots the droplet deformation parameter as a function of the capillary number, in which our LBM results are compared with the previous experimental data [6,19], the numerical results obtained by the boundary integral method (BIM) [4,48] and the volume of fluid (VOF) method [20], as well as the predictions from Taylor theory and MMSH model. Note that the degree of confinement is not explicitly available for some of the data sources, in which the flow is considered to be unbounded. It can be seen from the comparison, the LBM results agree well with these experimental data, numerical results and model predictions; and good agreement among them suggests that the walls have a negligible influence on the droplet behaviour for the present confinement ratio, i.e. $2R/H = 0.18$, consistent with the previous numerical and experimental observations [4,8].

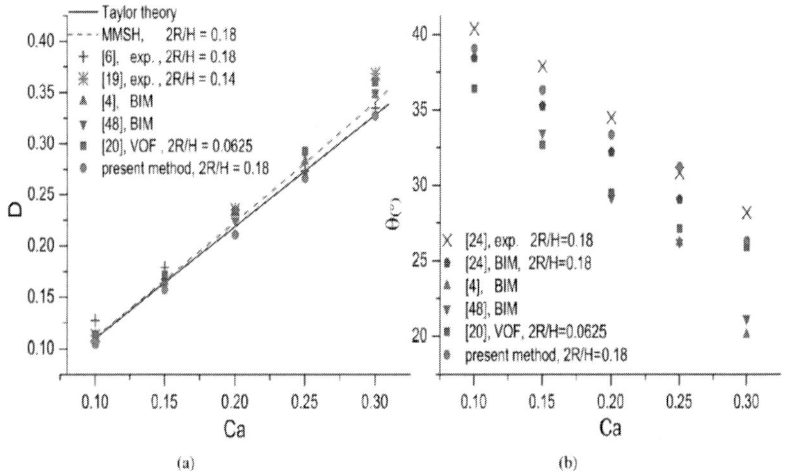

Figure. 2. The deformation parameter (a), and the orientation angle (b) as a function of the capillary number for $\lambda = 1$.

In addition to the deformation parameter, we also calculate the orientation angles of the deformed droplet (θ), defined as the angle between the longest axis of the droplet and the x-axis, at various capillary numbers, and compare them to those reported in Refs. [4,20,24,48](see Fig. 2(b)). Clearly, there is fairly good agreement especially between our results and those obtained by the experiment and the BIM *at an exactly matched confinement ratio, i.e.* $2R/H = 0.18$.

Next, the simulations were conducted for a high confinement of 0.69, which is achieved by adjusting the height of computational domain whilst keeping other parameters the same as in Fig. 2. Fig. 3 shows the comparison between the simulated deformation parameter and the Taylor theory, the MMSH model and the experimental results of Vananroye et al. [6]. Consistent with the previous findings, the MMSH model can provide experimently-matched predictions of the droplet deformation, but the predictions from the Taylor theory are no longer valid for high confinement ratios. Also, our simulation results again match quite well with the predictions from the MMSH model and the experimental data, verifying that the present LBM can successfully predict deformation of a highly confined droplet. Fig. 4 shows the final shape of the deformed droplet in the middle x–z plane for $Ca = 0.2$, 0.25 and 0.3, in which the contours with $\rho^N = -0.5$, 0, and 0.5 are displayed. It is observed that the droplet deviates from the ellipsoidal shape to some extent at high capillary numbers, e.g. $Ca = 0.25$ and 0.3. Therefore, the confinement ratio and the capillary number are restricted to $2R/H \leq 0.7$ and $Ca \leq 0.3$, and both of them will not take large values at the same time in the rest of the simulations in order to ensure that the deformed droplet maintains an ellipsoidal shape.

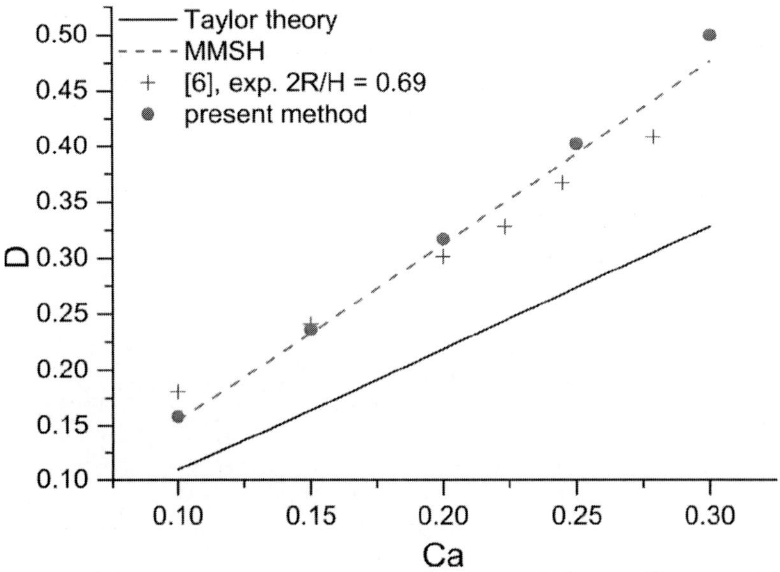

Figure. 3. Deformation parameter as a function of the capillary number for $\lambda = 1$ and $2R/H = 0.69$.

Then, the effect of viscosity ratio on droplet deformation is studied at two moderate confinement ratios, 0.4 and 0.5. For each confinement ratio, three different viscosity ratios are considered, i.e. $\lambda = 0.3$, 1, and 3. The variation of deformation parameter with the capillary number is presented in Fig. 5. For the viscosity ratio of 0.3, our LBM results agree well with the MMSH model at low Ca, but gradually deviate from the MMSH model with increasing Ca. Specifically, our LBM results lie between the Taylor theory and the MMSH model. The reason for the deviation remains an open question and needs further study. Note that similar observations were reported experimentally by Vananroye et al. [6]. For the viscosity ratio of

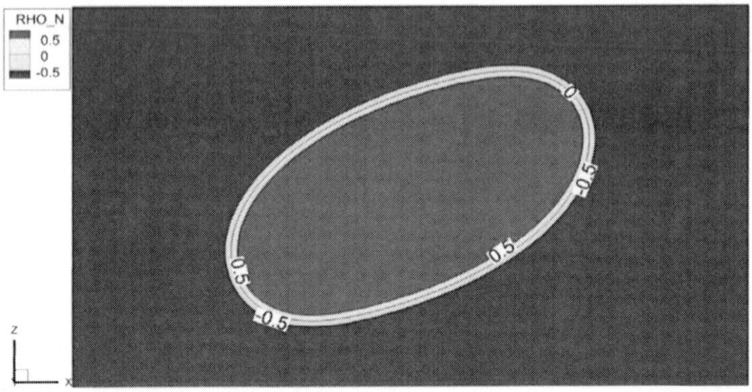

(a) $Ca = 0.2$

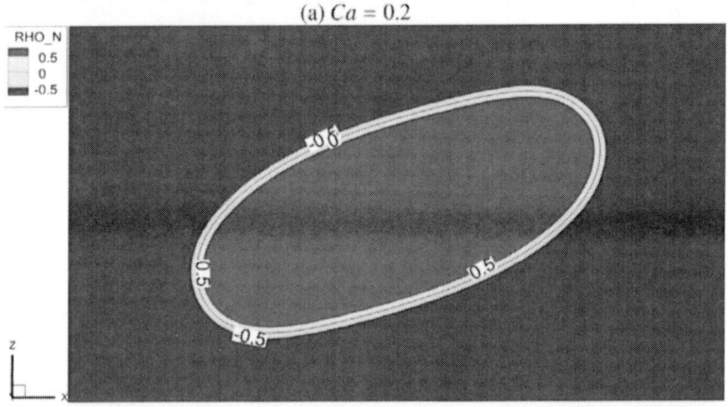

(b) $Ca = 0.25$

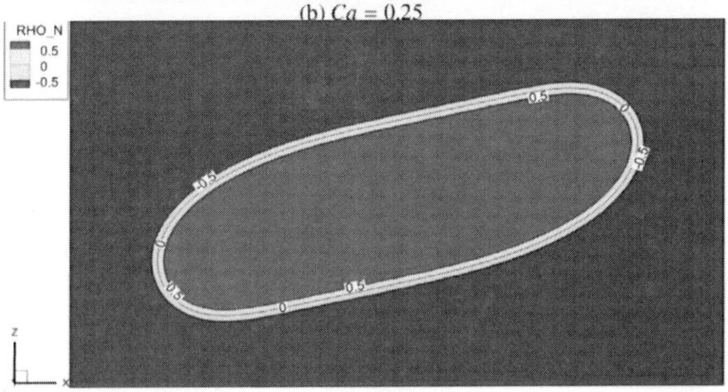

(c) $Ca = 0.3$

Figure. 4. Final shape of the deformed droplet in the middle x–z plane for $\lambda = 1$ and $2R/H = 0.69$ at different capillary numbers: (a) $Ca = 0.2$, (b) $Ca = 0.25$, and (c) $Ca = 0.3$.

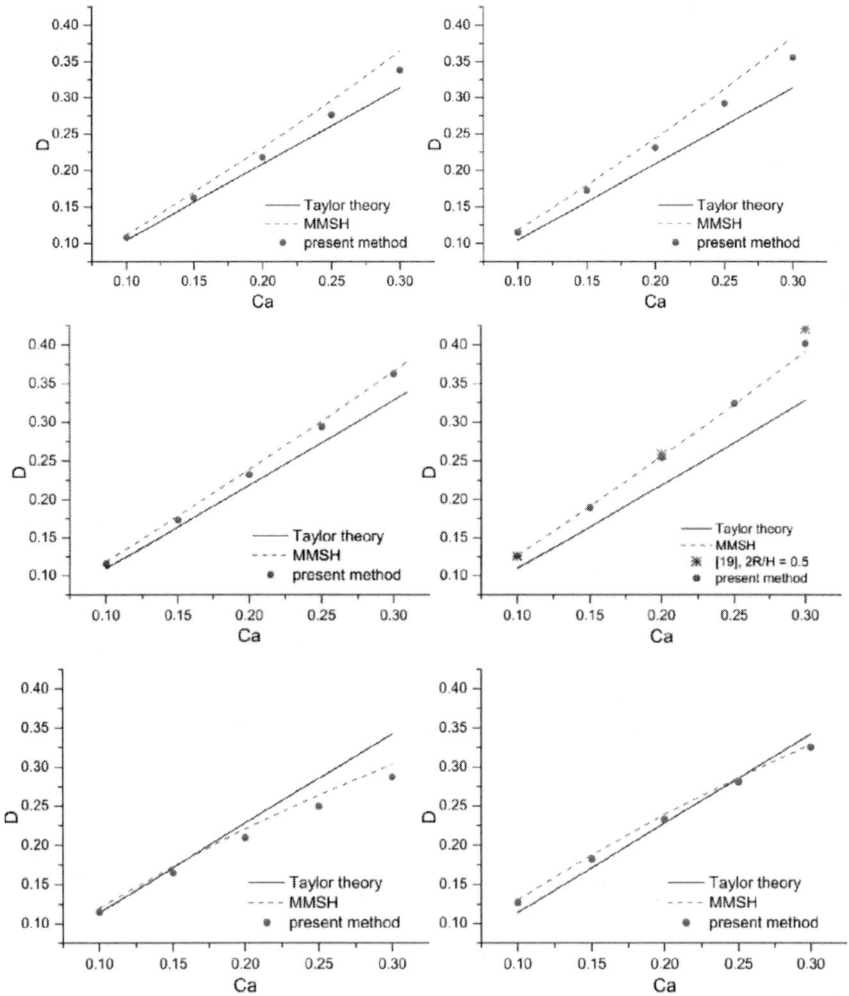

Figure. 5. Deformation parameter as a function of the capillary number for various parameters, each column has the same confinement ratio (left: $2R/H = 0.4$, right: $2R/H = 0.5$), and each row has the same viscosity ratio (top: $\lambda = 0.3$, middle: $\lambda = 1$, bottom: $\lambda = 3$).

unity, our LBM results agree quite well with the MMSH model, and the experimental data [19] where the data were only for the confinement ratio of 0.5. For the viscosity ratio of 3, the obtained results of droplet deformation are a little lower than the predictions from the MMSH model for the confinement ratio of 0.4, whereas for the confinement ratio of 0.5 one can clearly observe excellent agreement between the simulations and the MMSH model. Moreover, our results deviate significantly from the Taylor theory for both $2R/H = 0.4$ and $2R/H = 0.5$. Thus, it can be concluded that the Taylor theory cannot describe droplet deformation in confinement with the confinement ratio as small as 0.4, which is somehow different from the previous finding i.e. the wall effects can be neglected when $2R/H \leq 0.4$ [4].

The effect of the confinement and viscosity ratios is further investigated at a fixed capillary number of 0.2. The confinement ratio ranges from 0.18 to 0.7. Fig. 6(a) shows the variation of the deformation parameter for $\lambda = 0.3$, 0.5, 1, 3, and 5. It is observed that the deformation increases with increasing viscosity ratio up to unity. Upon a further increase of the viscosity ratio, the deformation decreases. In addition, we observe that increasing confinement ratio leads to bigger deformation of the droplet for all the viscosity ratios. Interestingly, the droplet deformation does not present a linear correlation with the proximity of the walls. It increases almost linearly, with different rates, before and after $2R/H = 0.4$. Up to this value the deformation exhibits a slow increase, which becomes more rapid upon higher values, especially for the high viscous droplets. This can be explained together with the corresponding orientation angles. The variation of the corresponding orientation angle (θ) with the confinement ratio is shown in Fig. 6(b). Increasing viscosity ratio causes the droplet to align more with the shear flow, thus resulting in a smaller orientation angle. For $\lambda \geq 1$, both the deformation parameter (see Fig. 6(a)) and the orientation angle (Fig. 6(b)) reveal more significant variations with the confinement ratio $2R/H \geq 0.4$. This suggests that the wall confinement has a stronger effect on droplet deformation for more viscous droplets. For $\lambda = 3$, 5, the orientation angles do not reduce further once $2R/H \geq 0.3$. Therefore, the carrier fluid has a narrower flow passage to pass the droplet which leads to built-up shear stress around the droplet. Consequently, the deformation rate increases more rapidly for $\lambda = 3$, 5 with increasing confinement.

In order to illustrate the effect of the viscosity ratio and the confinement ratio on the flowfield inside and around the droplet, Fig. 7 depicts the velocity contours and streamlines in the middle x–z plane for $2R/H = \{0.4, 0.7\}$ and $\lambda = \{0.3, 1, 5\}$ at $Ca = 0.2$. Note that we here limit our investigation to $2R/H = 0.4$ and 0.7 since the impact of wall becomes more evident with the confinement ratio larger than 0.4. In this figure, the thick solid lines represent the droplet interface where $\rho^N = 0$, while the black solid lines represent the streamlines, with arrows indicating the direction of the local flow. It is confirmed that the effect of the viscosity ratio and the confinement ratio are not only on the droplet shape and orientation angle, but also on the flowfield patterns inside and around the droplet. Increasing viscosity ratio leads to the streamline pattern inside the droplet varying from a double vortex to a single vortex structure. Interestingly, an analogous variation in the streamline pattern was observed in the previous numerical studies [20,21,34,35] when the Reynolds number is decreased from a relatively high value. In addition, as the viscosity ratio increases, the inclination angle of the internal vortices (streamlines) and the orientation angle of the droplet both decrease, and their difference also reduces. For two different confinement ratios, the streamline patterns are qualitatively similar but quantitatively different, e.g. the inclination of streamlines inside the droplet is smaller at the confinement ratio of 0.7, which corresponds to a larger deformation and smaller orientation angle of the droplet.

3.2. Droplet breakup

In this section, we investigate the critical capillary number of the droplet breakup at various viscosity and confinement ratios. The simulations are conducted for the confinement ratios ranging from 0.37 to 0.8 at three typical viscosity ratios, i.e. $\lambda = 0.3$, 1 and 5. Although the LBM has been used extensively to simulate the droplet breakup, e.g., Refs. [32–35], to the best of our knowledge, no studies to date have considered such a broad range of parameters. This is partially because of the unsolved issues of the existing models themselves, such as artificially enlarged interface thickness and case-dependent mobility in the phase-field-based model, high spurious currents, numerical instability in the interparticle-potential model.

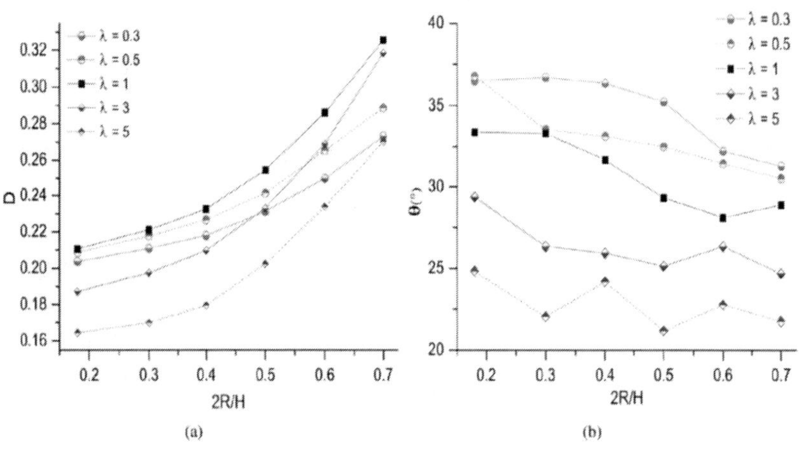

Figure. 6. Deformation parameter (a), and orientation angle (b) versus the confinement ratio for various viscosity ratios.

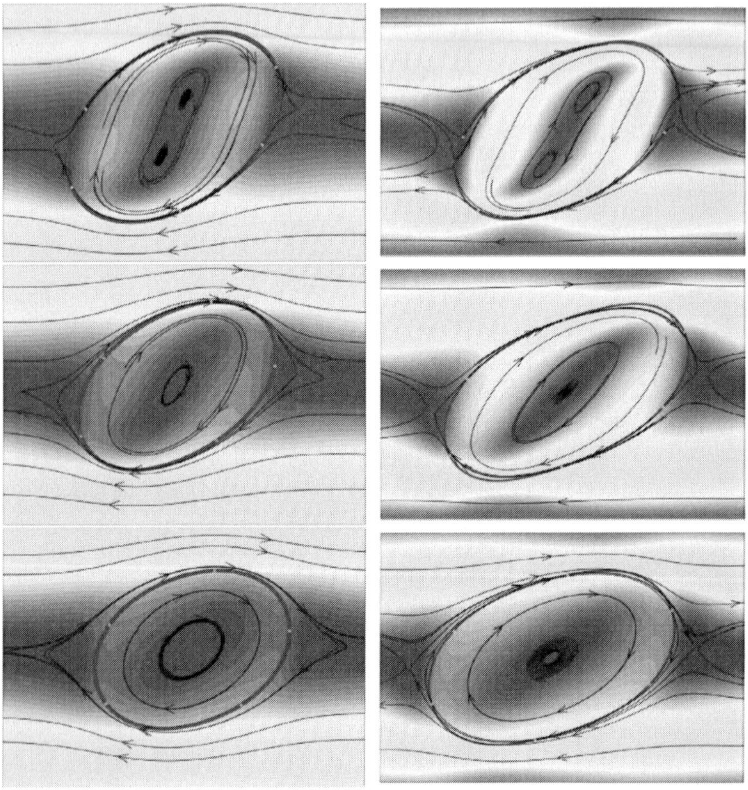

Figure. 7. Velocity vectors and streamlines around the droplet for *Ca* = 0.2: the columns are for the same confinement ratio (left: 0.4, right: 0.7); and the rows are for the same viscosity ratio (top: $\lambda = 0.3$, middle: $\lambda = 1$, bottom: $\lambda = 5$).

Fig. 8 shows the effect of the confinement ratio on the droplet breakup at the viscosity ratios of 0.3, 1, and 5, which are represented by circles, squares, and diamonds, respectively. Empty symbols are used in the cases where no droplet breakup is observed. To distinguish different modes of breakup, we use half-filled symbols to denote the binary breakup, and full-filled symbols to denote the multiple breakup, where the droplet splits into three or more daughter droplets. In addition, the breakup is depicted at the moments before and after it occurs, for a number of cases. Following the aforementioned definition, the critical capillary number, Ca_{cr}, lies between the empty and filled symbols. The confinement ratio has different effects on the critical capillary number, depending on the viscosity ratio. For the low viscosity ratio, i.e. $\lambda = 0.3$, the critical capillary number is varied from

$0.4 \leq Ca_{cr} \leq 0.5$ at $2R/H = 0.37$, to $0.5 \leq Ca_{cr} \leq 0.6$ at $2R/H = 0.5$ and finally to $0.55 \leq Ca_{cr} \leq 0.625$ at $2R/H = 0.625$, suggesting that Ca_{cr} increases with increasing confinement. Contrary to the low viscosity ratio case, Ca_{cr} decreases with increasing confinement for the high viscosity ratio, i.e. $\lambda = 5$. Specifically, the critical capillary number decreases from $0.5 \leq Ca_{cr} \leq 0.6$ to $0.3 \leq Ca_{cr} \leq 0.35$ when one increases the confinement ratio from 0.55 to 0.8. Fig. 9 shows the snapshots of the droplet breakup process for $Ca = 0.35$, $\lambda = 5$ and $2R/H = 0.8$. The droplet deforms continuously and eventually breaks up into two equal-sized daughter droplets, as opposed to the unbounded case where the high viscosity ratio droplet ($\lambda > 4$) does not break regardless of the capillary number. Also, it is interestingly observed in Fig. 8 that the confinement ratio has small effect on the critical capillary number for $\lambda = 1$, which is around 0.4. These findings agree well with the previous experimental and numerical studies [3,24]. However, in these reported works [3,24] through a large number of experiments and/or simulations, it was found that for $\lambda = 1$ that the critical capillary number first decreases and then increases with the confinement ratio. The minimum critical capillary number occurs at $2R/H = 0.5$. To capture this subtle variation of the critical capillary number with the confinement ratio, a large number of simulations are required which will be studied separately by exploiting massively parallel computing.

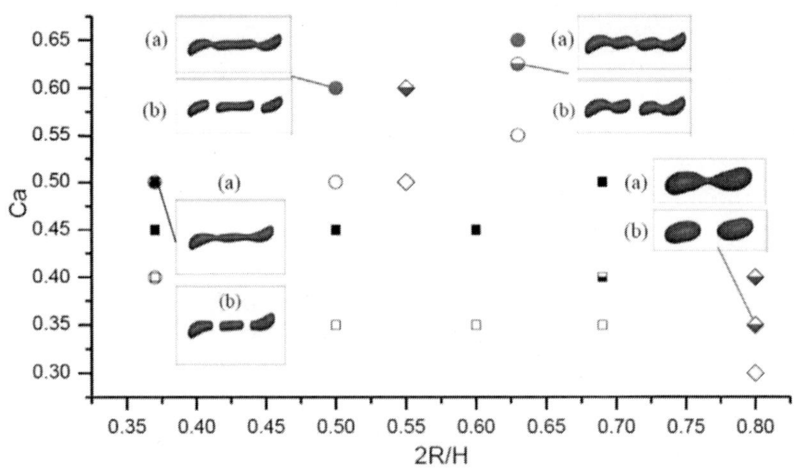

Figure. 8. Capillary numbers at which the droplet obtains a steady shape or experience break-up, for different viscosity ratios and confinements. The symbols are explained in the text. Inset: droplet shape before (a), and after (b) the breakup.

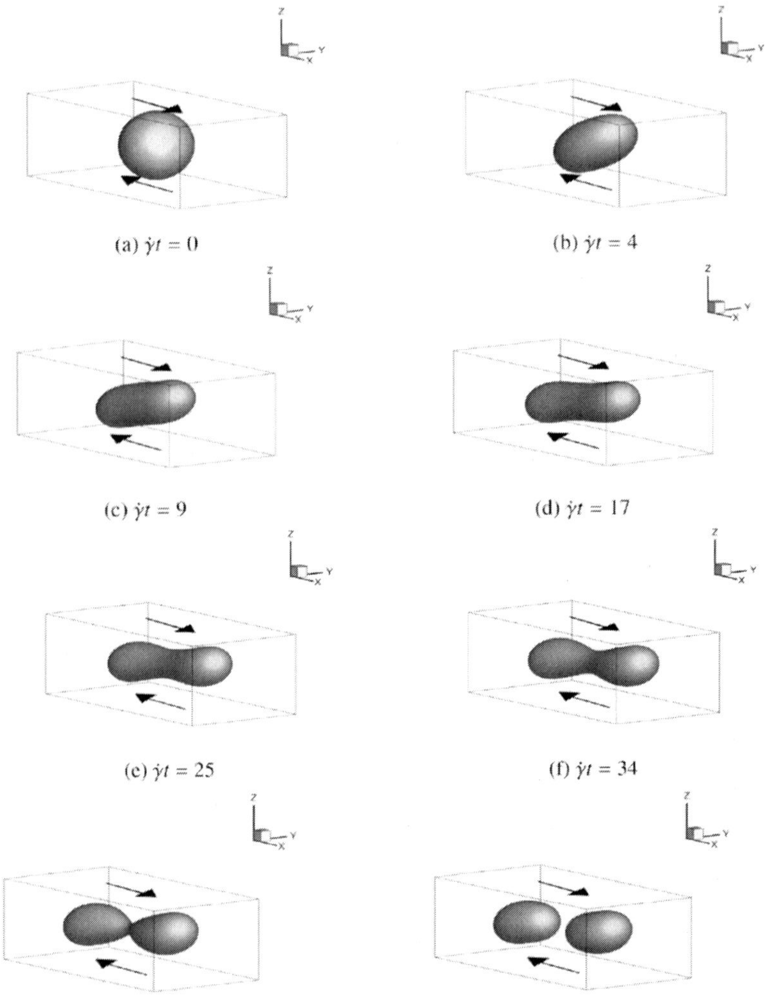

(a) $\dot{\gamma}t = 0$ (b) $\dot{\gamma}t = 4$ (c) $\dot{\gamma}t = 9$ (d) $\dot{\gamma}t = 17$ (e) $\dot{\gamma}t = 25$ (f) $\dot{\gamma}t = 34$ (g) $\dot{\gamma}t = 38$ (h) $\dot{\gamma}t = 39$

Figure. 9. Transient droplet shape for $Ca = 0.35$, $\lambda = 5$ and $2R/H = 0.8$.

Even though the overall agreement is quite satisfactory between the present results and those obtained by Janssen et al. [3], some quantitative differences are also noticed. For example, Janssen et al. [3] experimentally found that binary breakup is expected to occur for a droplet with $\lambda = 1.0$ at $Ca = 0.45$ and $0.37 \leq 2R/H \leq 0.45$, whereas in our simulations ternary breakup is observed. Fig. 10 shows the snapshots of the droplet breakup

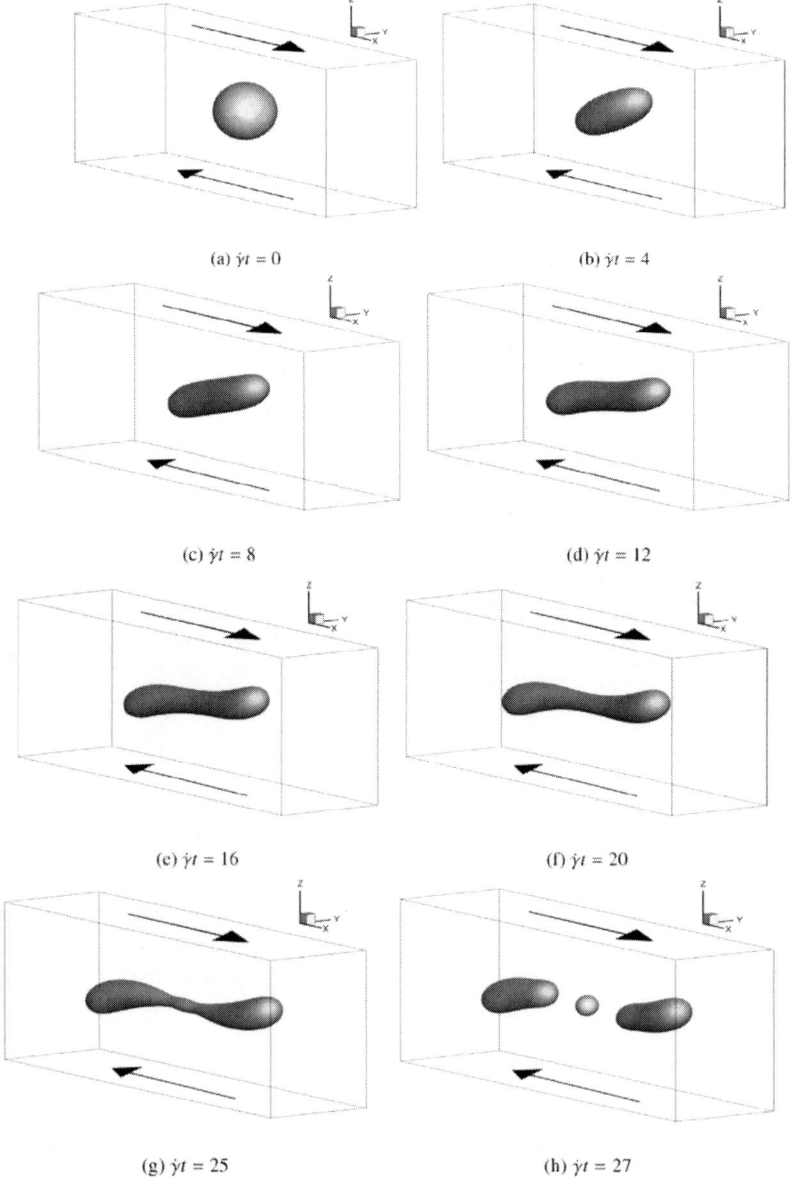

Figure. 10. Transient droplet shape for Ca = 0.45, λ = 1 and $2R/H$ = 0.4.

process for $Ca = 0.45$, $\lambda = 1$ and $2R/H = 0.45$. As the droplet is stretched by the external shear flow, it first deforms to an ellipsoidal shape ($\dot{\gamma}$t=4). The major axis lengthens while the minor axis shrinks, leading to the formation of a waist near the centre of droplet at $\dot{\gamma}$t=16. The droplet is then elongated to a 'dumbbell' shape, and at the same time the bulbs at the end of droplet achieve a stable shape ($\dot{\gamma}$t=20). At $\dot{\gamma}$t=25, two visible necks are formed between the central portion of the droplet and bulbs. The droplet continues to thin and eventually breaks up into three parts (i.e., ternary breakup) with two daughter droplets having much larger size ($\dot{\gamma}$t=27).

In addition, the previous experimental study indicated that the lowest Ca_{cr} for all of viscosity ratios ($0.3 \le \lambda \le 5$) and confinement ratios ($0.1 \le 2R/H \le 0.9$) is around 0.4 [3], but in our simulations the lowest Ca_{cr} is found between 0.3 and 0.35, which can be clearly seen in Fig. 8 for the case of $\lambda = 5$ and $2R/H = 0.8$. The reason for these differences remains unclear, and is likely to be the result of a combination of factors, including the experimental errors and model accuracy in describing the interfacial dynamics.

4. CONCLUSIONS

The influence of capillary number, viscosity ratio and confinement ratio have been systematically studied for droplet deformation and breakup in a simple shear flow. Increasing confinement leads to an increase in droplet deformation for a wide range of viscosity ratios, and the droplet with the same viscosity as the carrier fluid presents the largest deformation at all confinement ratios. The orientation angle of the deformed droplet decreases with increasing viscosity ratio. It is observed that the presence of walls has a more significant impact on both droplet deformation and orientation angle when the confinement ratio increases to 0.4 or higher. The flowfield inside and around the droplet shows that confinement and viscosity ratios can lead to a perplex pattern which may affect droplet behaviour. Finally, the critical capillary number, above which the droplet breakup occurs, is investigated for a broad range of flow conditions. Depending on the value of viscosity ratio, the confinement ratio has different effects on the critical capillary number: as the confinement increases, the critical capillary number increases for $\lambda < 1$ but decreases for $\lambda > 1$; and for $\lambda = 1$ the critical capillary number is kept at a value of around 0.4, regardless of the confinement ratio.

ACKNOWLEDGEMENTS

The results were obtained using the UK EPSRC funded ARCHIE-WeSt High Performance Computer (http://www.archie-west.ac.uk) under grant no. EP/K000586/1. The research is also supported by EPSRC under grant no. EP/M021475/1 and EP/L00030X/1. H.L. gratefully acknowledges the financial supports from the National Natural Science Foundation of China(Grant No. 51506168) and the Thousand Talents Program for Distinguished Young Scholars.

REFERENCES

1. G.I. TaylorTHE FORMATION OF EMULSIONS IN DEFINABLE FIELDS OF FLOWJ. Colloid Interface Sci., 38 (1934), pp. 501-523
2. F.A. MorrisonUNDERSTANDING RHEOLOGYOxford University Press (2001)
3. P.J.A. Janssen, A. Vananroye, P.V. Puyvelde, P. Moldenaers, P.D. AndersonGENERALIZED BEHAVIOR OF THE BREAKUP OF VISCOUS DROPS IN CONFINEMENTSJ. Rheol., 54 (5) (2010), pp. 1047-1060URL: http://scitation.aip.org/content/sor/journal/jor2/54/5/10.1122/1.3473924
4. M.R. Kennedy, C. Pozrikidis, R. SkalakMOTION AND DEFORMATION OF LIQUID DROPS, AND THE RHEOLOGY OF DILUTE EMULSIONS IN SIMPLE SHEAR FLOWComput. Fluids, 23 (2) (1994), pp. 251-278URL: http://www.sciencedirect.com/science/article/pii/004579309490040X
5. S. Guido, M. VilloneTHREE-DIMENSIONAL SHAPE OF A DROP UNDER SIMPLE SHEAR FLOWJ. Rheol., 42 (1998), pp. 395415URL: http://scitation.aip.org/content/sor/journal/jor2/42/2/10.1122/1.550942
6. Vananroye, P. Van Puyvelde, P. MoldenaersEFFECT OF CONFINEMENT ON THE STEADY-STATE BEHAVIOR OF SINGLE DROPLETS DURING SHEAR FLOWJ. Rheol., 51 (1) (2007), pp. 139153, 10.1122/1.2399089URL: http://scitation.aip.org/content/sor/journal/jor2/51/1/10.1122/1.2399089
7. V. Cristini, Y.-C. TanTHEORY AND NUMERICAL SIMULATION OF DROPLET DYNAMICS IN COMPLEX FLOWS – A REVIEWLab Chip, 4 (2004), pp. 257264 URL: http://pubs.rsc.org/en/content/articlehtml/2004/lc/b403226h
8. S. Guido, F. GrecoDYNAMICS OF A LIQUID DROP IN A FLOWING IMMISCIBLE LIQUIDRheol. Rev. (2004), pp. 99-142URL: http://www.bsr.org.uk/rheology-reviews/ RheologyReviews/liquid-drop-immiscible-liquid-GuidoGreco.pdf
9. P.V. Puyvelde, A. Vananroye, R. Cardinaels, P. MoldenaersREVIEW ON MORPHOLOGY DEVELOPMENT OF IMMISCIBLE BLENDS IN CONFINED SHEAR FLOWPolymer, 49 (25) (2008), pp. 5363-5372, 10.1016/j.polymer.2008.08.055URL: http://www.sciencedirect.com/science/article/pii/S0032386108007544
10. M. MinaleMODELS FOR THE DEFORMATION OF A SINGLE ELLIPSOIDAL DROP: A REVIEW Rheol. Acta, 49 (8) (2010), pp. 789-806, 10.1007/s00397-010-0442-0
11. C.E. Chaffey, H. BrennerA SECOND-ORDER THEORY FOR SHEAR DEFORMATION OF DROPS J. Colloid Interface Sci., 24 (2) (1967), pp. 258-269, 10.1016/0021-

9797(67)90229-9 URL: http://www.sciencedirect.com/sci ence/article/pii/0021979767902299

12. J.M. RallisonNOTE ON THE TIME-DEPENDENT DEFORMATION OF A VISCOUS DROP WHICH IS ALMOST SPHERICAL J. Fluid Mech., 98 (1980), pp. 625-633, 10.1017/S0022112080000316 URL: http://journals.cambridge.org/article_S0022112080000316
13. J.M. RallisonA NUMERICAL STUDY OF THE DEFORMATION AND BURST OF A VISCOUS DROP IN GENERAL SHEAR FLOWS J. Fluid Mech., 109 (1981), pp. 465-482 URL: http://journals.cambridge.org/download.php?file=%2FFLM%2FFLM109%2FS002211208100116Xa.pdf&code=afc88fbdcd8cfaf45e74d9e8bd98cb77
14. F. GrecoSECOND-ORDER THEORY FOR THE DEFORMATION OF A NEWTONIAN DROP IN A STATIONARY FLOW FIELD Phys. Fluids, 14 (3) (2002), pp. 946-954, 10.1063/1.1445182 URL: http://scitation.aip.org/content/aip/journal/pof2/14/3/10.1063/1.1445182
15. S. GuidoSHEAR-INDUCED DROPLET DEFORMATION: EFFECTS OF CONFINED GEOMETRY AND VISCOELASTICITY Curr. Opin. Colloid Interface Sci., 16 (1) (2011), pp. 61-70, 10.1016/j.cocis.2010.12.001 URL: http://www.sciencedirect.com/science/article/pii/S1359029410001226
16. P. Maffettone, M. MinaleEQUATION OF CHANGE FOR ELLIPSOIDAL DROPS IN VISCOUS FLOW J. Non-Newtonian Fluid Mech., 78 (1998), pp. 227-241 URL: http://ac.els-cdn.com/S0377025798000652/1-s2.0-S0377025798000652-main.pdf?_tid=dff6b528-f0e2-11e4-9bfa-00000aab0f 6c&acdnat=1430581973_58d00b70d8ae847f6b0278389a56cf08
17. M. MinaleA PHENOMENOLOGICAL MODEL FOR WALL EFFECTS ON THE DEFORMATION OF AN ELLIPSOIDAL DROP IN VISCOUS FLOW Rheol. Acta, 47 (2008), pp. 667-675
18. URL: http://download-v2.springer.com/static/pdf/111/art%253A10.1007%252Fs00397-007-0237-0.pdf?token2=exp=1430583106~acl=%2Fstatic%2Fpdf%2F111%2Fart%25253A10.1007%25252Fs00397-007-0237- 0.pdf*~hmac=ff39c5831a50644cd93dd91a50ff042eefced0064b60370d315695fcd2cb594c
19. M. Shapira, S. HaberLOW REYNOLDS NUMBER MOTION OF A DROPLET IN SHEAR FLOW INCLUDING WALL EFFECTS Int. J. Multiphase Flow, 16 (2) (1990), pp. 305-321 URL: http://www.sciencedirect.com/science/article/pii/030193229090061M
20. V. Sibillo, G. Pasquariello, M. Simeone, V. Cristini, S. GuidoDROP DEFORMATION IN MICROCONFINED SHEAR FLOW Phys. Rev. Lett., 97 (5) (2006), pp. 1-4, 10.1103/PhysRevLett.97.054502 054502
21. J. Li, Y.Y. Renardy, M. RenardyNUMERICAL SIMULATION OF BREAKUP OF A VISCOUS DROP IN SIMPLE SHEAR FLOW THROUGH A VOLUME-OF-FLUID METHOD Phys. Fluids, 12 (2) (2000), pp. 269-282 URL: https://www.math. vt.edu/ people /renardyy/Research/Publications/feb2000.pdf
22. Y.Y. Renardy, V. CristiniEFFECT OF INERTIA ON DROP BREAKUP UNDER SHEAR Phys. Fluids, 13 (1) (2001), pp. 7-13, 10.1063/1.1331321
23. Y. RenardyTHE EFFECTS OF CONFINEMENT AND INERTIA ON THE PRODUCTION OF DROPLETS Rheol. Acta, 46 (4) (2007), pp. 521-529, 10.1007/s00397-006-0150-y
24. P.J.A. Janssen, P.D. AndersonBOUNDARY-INTEGRAL METHOD FOR DROP DEFORMATION BETWEEN PARALLEL PLATES Phys. Fluids, 19 (2007), pp. 1-11 043602, URL: http://www.mate.tue.nl/mate/pdfs/7660.pdf
25. Vananroye, P.J.A. Janssen, P.D. Anderson, P. Van Puyvelde, P. MoldenaersMICROCONFINED EQUIVISCOUS DROPLET DEFORMATION: COMPARISON OF EXPERIMENTAL AND NUMERICAL RESULTS Phys. Fluids, 20 (2008), pp. 1-10, 10.1063/1.2835312 013101, URL: http://scitation. aip.org/content/aip/journal/pof2/20/1/10.1063/1.2835312
26. V. Cristini, J. Blawzdziewics, M. LoewenbergAN ADAPTIVE MESH ALGORITHM FOR EVOLVING SURFACES: SIMULATIONS IN DROP BREAKUP AND COALESCENCE J.

Comput. Phys., 168 (2001), pp. 445-463 URL: http://www.sciencedirect .com /science/article/pii/S0021999101967130
27. O. Shardt, J. Derksen, S.K. MitraSIMULATIONS OF DROPLET COALESCENCE IN SIMPLE SHEAR FLOW Langmuir, 29 (21) (2013), pp. 6201-6212, 10.1021 /la304919p
28. A.K. Gunstensen, D.H. Rothman, S. Zaleski, G. ZanettiLATTICE BOLTZMANN MODEL OF IMMISCIBLE FLUIDS Phys. Rev. A, 43 (1991), pp. 4320-4327, 10. 1103/ PhysRevA.43.4320
29. M.R. Swift, W.R. Osborn, J.M. YeomansLATTICE BOLTZMANN SIMULATION OF NONIDEAL FLUIDS Phys. Rev. Lett., 75 (1995), pp. 830-833, 10. 1103/ Phys Rev Lett.75.830
30. X. Shan, H. ChenLATTICE BOLTZMANN MODEL FOR SIMULATING FLOWS WITH MULTIPLE PHASES AND COMPONENTS Phys. Rev. E, 47 (1993), pp. 1815-1819, 10.1103/PhysRevE.47.1815
31. X. He, S. Chen, R. ZhangA LATTICE BOLTZMANN SCHEME FOR INCOMPRESSIBLE MULTIPHASE FLOW AND ITS APPLICATION IN SIMULATION OF RAYLEIGH–TAYLOR INSTABILITY J. Comput. Phys., 152 (2) (1999), pp. 642-663, 10. 1006/jcph.1999.6257
URL: http://www.sciencedirect.com/science/article/pii/S0021999199962575
32. H. Xi, C. DuncanLATTICE BOLTZMANN SIMULATIONS OF THREE-DIMENSIONAL SINGLE DROPLET DEFORMATION AND BREAKUP UNDER SIMPLE SHEAR FLOW Phys. Rev. E, 59 (1999), pp. 3022-3026, 10.1103/PhysRevE.59.3022
33. R. van der Sman, S. van der GraafEMULSION DROPLET DEFORMATION AND BREAKUP WITH LATTICE BOLTZMANN MODEL Comput. Phys. Commun., 178 (7) (2008), pp. 492-504, 10.1016/j.cpc.2007.11.009 URL: http:// www. sciencedirect.com/science/article/pii/S0010465507004675
34. H. Liu, A.J. ValocchiTHREE-DIMENSIONAL LATTICE BOLTZMANN MODEL FOR IMMISCIBLE TWO-PHASE FLOW SIMULATIONS Phys. Rev. E, 85 (2012), pp. 1-14, 10.1103/PhysRevE.85.046309 046309
35. S. Farokhirad, T. Lee, J.F. MorrisEFFECTS OF INERTIA AND VISCOSITY ON SINGLE DROPLET DEFORMATION IN CONFINED SHEAR FLOW Commun. Comput. Phys., 13 (3) (2013), pp. 706-724 URL: http://www.academia.edu/ 8043985/ Effects_of_Inertia_and_Viscosity_on_Single_Droplet_Deformation_in_Confined_Shear_Flow
36. Komrakova, O. Shardt, D. Eskin, J.J. DerksenLATTICE BOLTZMANN SIMULATIONS OF DROP DEFORMATION AND BREAKUP IN SHEAR FLOW Int. J. Multiphase Flow, 59 (2014), pp. 24-43URL: http://www.sciencedirect.com/science /article/ pii/S0301932213001547
37. Gupta, M. SbragagliaDEFORMATION AND BREAKUP OF VISCOELASTIC DROPLETS IN CONFINED SHEAR FLOW Phys. Rev. E, 90 (2014), pp. 1-14, 10. 1103 /PhysRevE.90.023305 023305
38. H. Liu, Y. Ju, G. Xi, Y. ZhangLATTICE BOLTZMANN MODELING OF CONTACT ANGLE AND ITS HYSTERESIS IN TWO-PHASE FLOW WITH LARGE VISCOSITY DIFFERENCE Phys. Rev. E, 92 (2015)
39. H. Liu, Y. Zhang, A.J. ValocchiLATTICE BOLTZMANN SIMULATION OF IMMISCIBLE FLUID DISPLACEMENT IN POROUS MEDIA: HOMOGENEOUS VERSUS HETEROGENEOUS PORE NETWORK Phys. Fluids, 27 (2015), 10.1063/1.4921611
40. J. Brackbill, D. Kothe, C. ZemachA CONTINUUM METHOD FOR MODELING SURFACE TENSION J. Comput. Phys., 100 (1992), pp. 335-354
41. URL: http://ac.els-cdn.com/002199919290240Y/1-s2.0-002199919290240Y main.pdf?_tid=c67f7a46-f0e0-11e4-9008-00000aab0f27&acdnat=1430581072_91f0c5465fa22f45755df6ed0232b289
42. Halliday, A.P. Hollis, C.M. CareLATTICE BOLTZMANN ALGORITHM FOR CONTINUUM MULTICOMPONENT FLOW Phys. Rev. E, 76 (2007), pp. 1-13, 10.1103 /Phys RevE.76.026708 026708

43. H. Liu, Y. Zhang, A.J. ValocchiMODELING AND SIMULATION OF THERMOCAPILLARY FLOWS USING LATTICE BOLTZMANN METHOD J. Comput. Phys. (2012), pp. 4433-4453 URL: http://www.sciencedirect.com/science/article/pii/ S00219991 12000940
44. Z. Guo, C. Zheng, B. ShiDISCRETE LATTICE EFFECTS ON THE FORCING TERM IN THE LATTICE BOLTZMANN METHOD Phys. Rev. E, 65 (2001), pp. 1-6, 10.1103/PhysRevE.65.046308 046308
45. Y.Q. Zu, S. HePHASE-FIELD-BASED LATTICE BOLTZMANN MODEL FOR INCOMPRESSIBLE BINARY FLUID SYSTEMS WITH DENSITY AND VISCOSITY CONTRASTS Phys. Rev. E, 87 (2013), pp. 1-23, 10.1103/PhysRevE.87.043301 043301
46. H. Liu, Y. ZhangPHASE-FIELD MODELING DROPLET DYNAMICS WITH SOLUBLE SURFACTANTS J. Comput. Phys., 229 (24) (2010), pp. 9166-9187, 10.1016 /j.jcp.2010.08.031
URL: http://www.sciencedirect.com/science/article/pii/S002199911000478X
47. M. Latva-Kokko, D.H. RothmanDIFFUSION PROPERTIES OF GRADIENT-BASED LATTICE BOLTZMANN MODELS OF IMMISCIBLE FLUIDS Phys. Rev. E, 71 (2005), pp. 1-8, 10.1103/PhysRevE.71.056702 056702
48. H. Liu, Y. ZhangDROPLET FORMATION IN MICROFLUIDIC CROSS-JUNCTIONS Phys. Fluids, 23 (2011), pp. 1-12, 10.1063/1.3615643 082101
49. T. Zhang, B. Shi, Z. Guo, Z. Chai, J. LuGENERAL BOUNCE-BACK SCHEME FOR CONCENTRATION BOUNDARY CONDITION IN THE LATTICE-BOLTZMANN METHOD Phys. Rev. E, 85 (2012), pp. 1-14, 10.1103/PhysRevE.85.016701 016701
50. S. Kwak, C. PozrikidisADAPTIVE TRIANGULATION OF EVOLVING, CLOSED, OR OPEN SURFACES BY THE ADVANCING-FRONT METHOD J. Comput. Phys., 145 (1) (1998), pp. 61-88, 10.1006/jcph.1998.6030 URL: http://www.s ciencedirect .com/ science/article/pii/S0021999198960302

CHAPTER 2

Performance Evaluation of Explicit Finite Difference Algorithms with Varying Amounts of Computational and Memory Intensity

Satya P. Jammy*, Christian T. Jacobs, Neil D. Sandham

Aerodynamics and Flight Mechanics Group, Faculty of Engineering and the Environment, University of Southampton, University Road, Southampton SO171BJ, United Kingdom

ABSTRACT

Future architectures designed to deliver exascale performance motivate the need for novel algorithmic changes in order to fully exploit their capabilities. In this paper, the performance of several numerical algorithms, characterised by varying degrees of memory and computational intensity, are evaluated in the context of finite difference methods for fluid dynamics problems. It is shown that, by storing some of the evaluated derivatives as single thread- or process-local variables in memory, or recomputing the derivatives on-the-fly, a speed-up of ~2 can be obtained compared to traditional algorithms that store all derivatives in global arrays.

Keywords: Finite Computational fluid dynamics difference methods Algorithms Exascale Parallel computing Performance

1. INTRODUCTION

Explicit finite difference methods are an important class of numerical methods for the solution of partial or ordinary differential equations. For example, they are used for numerically solving the governing equations in computational fluid dynamics (CFD), astrophysics, seismic wave simulations, financial simulations, etc.

In CFD they are used by many researchers for the direct numerical simulation (DNS) or large eddy simulation (LES) of compressible flows. DNS is often performed to study boundary layers, aerofoils (involving both hydrodynamics and noise computations) [1], mixing analysis [2], shock-wave boundary layer interactions [3] or benchmark test cases such as the Taylor–Green vortex [4], decaying homogeneous isotropic turbulence, etc. Even with the advances in computing hardware during the past decade, the current capabilities of DNS are limited to moderate Reynolds number flows [5].

It is expected that computing architectures will be capable of exaFLOPs (10^{18}Floating Point Operations) by 2018 and 30 exaFLOPs by 2030 [6]. Exascale architectures have the capability to perform DNS of the aforementioned examples (amongst others) at higher Reynolds numbers, or potentially wall-modelled LES of the full model of an aircraft at operating Reynolds numbers. However, while there is a consensus [6] that future architectures would not look much like the present IBM Blue Gene, Cray, or IBM Productive, it is hard to predict the architectural design of such exascale systems. For example, they are expected to comprise less memory per core than the existing architectures. Exploiting the full potential of the exascale architectures poses many challenges to researchers, such as the sustainability of the solver's implementation with the uncertainty of architectures, the need for new revolutionary algorithms/numerical methods, increasing computation to communication ratio and the likelihood of I/O bottlenecks.

To address the problem of sustainability, taking into account the uncertainty in future architectures, one solution adopted by the CFD community involves decoupling the work of a domain scientist and a computational/computer scientist [7]. In this approach, Domain Specific Languages (DSL) are developed by the computational/computer scientists, and the specifics of the problem and the numerical solution method are specified in the DSL by the domain scientist. Using source-to-source translation the numerical solver is targetted towards different parallel hardware backends (e.g. MPI, CUDA, OpenMP, OpenCL, and OpenACC) [8,9]. This ensures that, for new architectures, only the backend that interfaces with the new architecture needs to be written and supported by the translator. The underlying implementation of the solver remains the same, thereby introducing a separation of concerns.

On the algorithms front, a lot of effort has gone into rewriting CFD solvers to exploit the available FLOPS of existing architectures. While the architectures have changed drastically in the last decade, algorithms have not advanced at a similar pace [6]. Some algorithmical changes have been attempted by [10,11] to reduce the data transfer on Graphics Processing Units (GPUs), but a complete and detailed study on the performance of such algorithms on the existing CPU-based architectures is currently lacking. A first step towards exascale computing would be to evaluate the performance of algorithms characterised by varying intensities of memory usage and computational cost on current CPU-based architectures for a relevant hydrodynamic test case, solved using a finite difference scheme.

To facilitate these investigations, the capabilities of the recently developed OpenSBLI framework [12] are extended to easily generate algorithms with varying amounts of computational and memory intensity. OpenSBLI is a framework for the automated derivation and parallel execution of finite difference-based models. It is written in Python and uses SymPy to generate a symbolic representation of the governing equations and discretisation. The framework generates OPS-compliant C code that is targetted towards MPI via the OPS active library [9]. A similar approach can also be applied to any set of compute-intensive equations solved using finite difference methods.

The aims of this paper are to: (a) study the performance of various algorithms on current multi-core CPU-based architectures, (b) identify the best possible algorithm for the solution of explicit finite difference methods on current multi-core CPU-based architectures, and (c) demonstrate the ease at which algorithmic manipulations can be achieved with OpenSBLI framework to overcome the challenges exascale architectures can pose.

The rest of the paper is organised as follows. The various algorithms are described in Section 2. The validation of the algorithms is presented in Section 3. The performance and scaling results are presented in Section 4. Some conclusions are drawn in Section 5.

2. ALGORITHMS

All the algorithms presented in this paper solve the three-dimensional unsteady compressible Navier–Stokes equations, with constant viscosity, given by

$$\frac{\partial \rho}{\partial t} = -\frac{\partial}{\partial x_j}\left[\rho u_j\right], \quad (1)$$

$$\frac{\partial \rho u_i}{\partial t} = -\frac{\partial}{\partial x_j}\left[\rho u_i u_j + p\delta_{ij} - \tau_{ij}\right], \quad (2)$$

and

$$\frac{\partial \rho E}{\partial t} = -\frac{\partial}{\partial x_j}\left[\rho E u_j + u_j p - q_j - u_i \tau_{ij}\right], \quad (3)$$

for the conservation of mass, momentum and energy, respectively. The quantity ρ is the fluid density, u_i is the velocity vector, p is pressure and E is the total energy. The stress tensor τ_{ij} is defined as,

$$\tau_{ij} = \frac{1}{Re}\left(\frac{\partial u_i}{\partial x_j} + \frac{\partial u_j}{\partial x_i} - \frac{2}{3}\delta_{ij}\frac{\partial u_k}{\partial x_k}\right), \quad (4)$$

where δ_{ij} is the Kronecker Delta and Re is the Reynolds number. The heat flux term $q_j$is given by,

$$q_j = \frac{1}{(\gamma - 1)\ M^2\ \ Pr\ \ Re}\frac{\partial T}{\partial x_j}, \quad (5)$$

where T is temperature, M is the Mach number of the flow, Pr is Prandtl number and $\gamma$is the ratio of specific heats. The pressure and temperature are given by,

$$p = (\gamma - 1)\left(\rho E - \frac{1}{2}\rho u_j^2\right), \quad (6)$$

and

$$T = \frac{\gamma M^2 p}{\rho}, \quad (7)$$

respectively. The variables that are advanced in time $(\rho, \rho u_i, \rho E)$ are referred to as the conservative variables, and the right-hand sides (RHS) in the mass, momentum and energy equations are referred to as the residuals of the equations.

The mass, momentum and energy equations are discretised in space using a fourth-order central finite-difference scheme and a low storage Runge–Kutta (RK) scheme with three stages of temporal discretisation. For improved stability, the convective terms in the governing equations are rewritten using the formulation of [13],

$$\frac{\partial}{\partial x_j}\rho\phi u_j = \frac{1}{2}\left(\frac{\partial}{\partial x_j}\rho\phi u_j + u_j\frac{\partial}{\partial x_j}\rho\phi + \rho\phi\frac{\partial}{\partial x_j}u_j\right), \quad (8)$$

where ϕ is 1, u_i or internal energy (e) for the mass, momentum and energy equations, respectively. To improve the stability of the present scheme, the viscous terms in the momentum and energy equations are expanded to second derivatives as used by [2,10,14].

A generic pseudo-code of the solution algorithm is shown in Fig. 1. The time loop is the most computationally expensive part of the algorithm. It consists of evaluating the primitive variables (p, u_i, T), spatial derivatives, the residual for each equation and advancing the solution in time. This is achieved by iterating over the solution points of the grid, referred to as the grid loop in the rest of the paper. Various algorithms used for the evaluation of the residual of the equations are presented herein. Starting with a memory-intensive algorithm representing a typical handwritten CFD solver, the amount of memory used and the computational intensity are varied, either by re-evaluating the derivatives on-the-fly or evaluating the derivatives using process-local variables. In all the algorithms presented, the primitive variables are evaluated and stored in memory.

```
set-the-initial-condition
for each-iteration do
  save-state
  for each-rk-substep do
    evaluate-u_i,p,T
    evaluate-the-derivatives
    evaluate-the-residual-of-the-equations
    boundary-conditions
    advance-solution-in-time
  end for // end of rk sub loop
end for // end of iteration loop
```

Figure. 1. Pseudo-code for the solution of the compressible Navier–Stokes equations.

Baseline algorithm (BL). This algorithm incorporates features similar to a typical handwritten static algorithm (i.e. the derivatives in the residual of each equation are evaluated and stored in memory as arrays of grid point values; these are referred to as work arrays in the rest of the paper) on CPUs, to run as a sequential or parallel using MPI or OpenMP. For multi-threaded parallel programs, this requires the algorithm to be thread-safe in order to avoid race conditions; these occur when a variable is updated in the grid loop and the updated variable is used to update another variable in the same loop. For example, in the evaluation of the primitive variables from the conservative variables, the equation for pressure (6) is dependent on the evaluated velocity components, and the equation for temperature (7) is dependant on the evaluated pressure. When running on threaded architectures, this potentially results in race conditions. This means that temperature could be evaluated before evaluating the pressure, and pressure could be evaluated before the velocity components are evaluated. Similar candidates for race conditions exist in the update equations (which advance the conservative variables forward in time) of the RK scheme.

To remove the race conditions, the code is generated such that no variable is updated and used in the same loop. This is achieved by separating the evaluations into multiple loops over grid points. For example, in the evaluation of primitive variables, the velocity components (u_0, u_1, u_2) are grouped into a single loop as the evaluations are independent, but the pressure and temperature are evaluated in different loops.

When generating the code that implements the BL algorithm, the first and second derivatives in the equations are evaluated and stored in work arrays in order to compute the RHS residual. The evaluation of the derivative of a combination of variables e.g. $\partial(\rho u_0 u_0/\partial x_0)$ is achieved in two stages. In the first stage the function $\rho u_0 u_0$ is evaluated and stored in a work array. In the second stage the derivative of the work array is evaluated using the central finite difference formula, and this result is stored in a new work array. The work array used in the first stage is not freed in memory, but is overwritten/reused when evaluating other quantities.

The baseline algorithm is optimised such that computationally-expensive divisions are minimised. Rational numbers (e.g. finite difference stencil weights) and all the negative powers of constants in the equations are evaluated and stored at the start of the simulation. Typically, these are γ^{-1}, Pr^{-1}, Re^{-1}, and so on.

A sample setup file used to generate an implementation of this algorithm in OpenSBLI is shown in Listing 1. All the algorithms presented next are also optimised to reduce computationally expensive divisions. The setup file for other algorithms is similar to the BL algorithm with extra attributes to control the combinations of memory used and computational intensity.

```
ndim=3 # Problem dimension
# Define the compressible Navier-Stokes in Einstein notation.
mass="Eq(Der(rho,t),- Skew(rho*u_j,x_j))"
momentum="Eq(Der(rhou_i,t),-Skew(rhou_i*u_j,x_j)-Der(p,x_i)+Der(
    tau_i_j,x_j))"
energy="Eq(Der(rhoE,t),-Skew(rhoE*u_j,x_j)-Conservative(p*u_j,
    x_j)+Der(q_j,x_j)+Der(u_i*tau_i_j,x_j))"
equations=[mass,momentum,energy]
# Substitutions
stress_tensor="Eq(tau_i_j,(1/Re)*(Der(u_i,x_j)+ Der(u_j,x_i)
    -(2/3)* KD(_i,_j)* Der(u_k,x_k)))"
heat_flux="Eq(q_j,(1/((gama-1)*Minf*Minf*Pr*Re))*Der(T,x_j))"
substitutions=[stress_tensor,heat_flux]
# Define all the constants in the equations
constants=["Re","Pr","gama","Minf"]
# Define coordinate direction symbol (x) this will be x_i,x_j,
    x_k
coordinate_symbol="x"
# Formulas for the variables used in the equations
velocity="Eq(u_i,rhou_i/rho)"
pressure="Eq(p,(gama-1)*(rhoE - (1/2)*rho*(u_j*u_j)))"
temperature="Eq(T,gama*Minf*Minf/rho)"
formulas=[velocity,pressure,temperature]
# Create the problem and expand the equations.
problem=Problem(equations,substitutions,ndim,constants,
    coordinate_symbol,metrics,formulas)
ex_eq=problem.get_expanded(problem.equations)
ex_form=problem.get_expanded(problem.formulas)
ss=Central(4) # Fourth-order central differencing
ts=RungeKutta(3) # Third-order RK scheme
# Create a numerical grid of solution points
np=[64]*ndim; deltas=[2.0*pi/np[i] for i in range(len(length))]
grid=Grid(ndim,{'delta':deltas,'number_of_points':np})
# Perform the discretisation
sd=SpatialDiscretisation(ex_eq,ex_form,grid,ss)
td=TemporalDiscretisation(ts,grid,constant_dt=True,sd)
# Boundary conditions
bc=PeriodicBoundaryCondition(grid)
for dim in range(ndim):
    bc.apply(td.prognostic_variables,dim)
# Constant initial conditions
ics=["Eq(grid.work_array(rho),1.0)","Eq(grid.work_array(rhou0)
    ,1.0)","Eq(grid.work_array(rhou1),0.0)","Eq(grid.work_array(
    rhou2),0.0)","Eq(grid.work_array(rhoE),1.0)"]
ics=GridBasedInitialisation(grid,initial_conditions)
```

```
io=FileIO(td.prognostic_variables)# I/O
# Simulation parameters
var=['niter','Re','Pr','gama','Minf','precision','name','deltat'
    ]
values=[1000,1600,0.71,1.4,0.1,"double","test",3.385*10**-3]
sp=dict(zip(vars,values))# dictionary
code=OPSC(grid,sd,td,bc,ics,io,sp)#code generation
```

Listing 1. Key lines of the setup file for obtaining the BL algorithm.

Recompute All algorithm (RA). In contrast to the BL algorithm, the evaluation of pressure and temperature are first rewritten using the conservative variables,

$$p = (\gamma - 1)\left(\rho E - \frac{1}{2}\rho\left(\frac{\rho u_j}{\rho}\right)^2\right), \tag{9}$$

and

$$T = \frac{\gamma M^2 p}{\rho} = \frac{\gamma(\gamma - 1)M^2\left(\rho E - \frac{1}{2}\rho\left(\frac{\rho u_j}{\rho}\right)^2\right)}{\rho}, \tag{10}$$

within the code to avoid race condition errors while fusing loops for the evaluation of the primitive variables. Then, to evaluate the residual of the equation, all the continuous spatial derivatives in the residual are replaced by their respective finite difference formulas in the code generation stage. This differs from the BL algorithm in that, instead of evaluating the derivatives to work arrays and using them to compute the RHS residual, the code generation process directly replaces the derivatives by their respective finite difference formulas such that they are recomputed every time.

This algorithm results in a code in which no work arrays are used for storing the derivatives. The memory required for this algorithm is therefore the least of all algorithms and the computational intensity is the highest of all. The control parameters to generate code for this algorithm are shown in Listing 2.

```
grid = Grid(ndim,{"delta":deltas,"number_of_points":np})
grid.store_derivatives = False # Do not store derivatives
```

Listing 2. Control parameters to generate the code for RA algorithm.

Store None algorithm (SN). This algorithm is similar to the RA algorithm. The difference is that, in the loop over grid points where the residuals are evaluated, each derivative in the RHS is evaluated to a single thread- or process-local variable. These variables are then used to update the residuals on a point-by-point basis, rather than storing all evaluations in a global-scope, grid-sized work array. To generate the code that implements this algorithm in OpenSBLI, the grid attribute local_variables should be set to True along with the control parameters given in Listing 2. The pseudo-code for the residual evaluation as described here is provided in Fig. 2.

```
for each-solution-point do
  double t1 = central difference formula for ∂u0/∂x0
  double t2 = central difference formula for ∂u1/∂x1
  ⋮
  residual = t1 + t2 ...
end for
```

Figure. 2. Pseudo-code for residual evaluation using SN algorithm.

The memory footprint of this algorithm is similar to that of the RA algorithm, but is slightly less computationally-intensive. This is because, for example, if a derivative is evaluated to a process-local variable then it can be reused if that derivative appears in any of the equations more than once.

Recompute Some algorithm (RS). In this algorithm, some of the derivatives (in this case, the first derivatives of the velocity components) are stored in work arrays and the remaining derivatives are replaced by their respective finite difference formulas in the residual. The evaluation of primitive variables follows the same procedure as the RA algorithm. Listing 3 shows the control parameters used to generate code for the RS algorithm. The memory usage for this algorithm is moderate, i.e. it is more than the RA algorithm but less than the BL algorithm.

```
grid = Grid(ndim,{"delta":deltas,"number_of_points":np})
grid.store_derivatives = False # Do not store derivatives
grid.derivatives_to_store = set(problem.
    get_expanded_term_in_equations("Der(u_i,x_j)"))
```

Listing 3. Control parameters to generate the code for the RS algorithm.

Store Some algorithm (SS). This algorithm is a fusion of the RS and SN algorithms, such that the derivatives that are not stored in the RS algorithm are evaluated and stored in thread- or process-local variables as per the SN algorithm. Listing 4 shows the control parameters used to generate code for this algorithm. Compared to the SN algorithm, an additional nine work arrays would be required for the SS algorithm for the 3D cases, and an additional four work arrays for 2D cases, since the first derivatives of the velocity components are now stored.

```
grid = Grid(ndim,{"delta":deltas,"number_of_points":np})
grid.store_derivatives = False
grid.derivatives_to_store = set(problem.
    get_expanded_term_in_equations("Der(u_i,x_j)"))
grid.local_variables = True
```

Listing 4. Control parameters in setup file to generate the code for the SS algorithm.

3. VALIDATION

The baseline (BL) algorithm is validated for a 3D compressible Taylor–Green vortex problem, to check the correctness of the solver. The initial conditions and the post-processing procedure are described in [4]. The simulations are performed in a cube of non-dimensional length 2π, with periodic boundary conditions in all three directions for grids containing 64^3, 128^3, 256^3 and 512^3 solution points. The Mach number, Prandtl number and Reynolds number of the flow are taken as 0.1, 0.71 and 1600, respectively. The non-dimensional time-step for the 64^3 grid size was set to 3.385×10^{-3}, and was halved for each increase in the grid size by a factor of 2^3. Double-precision is used throughout all simulations presented in this paper.

Fig. 3 shows the evolution of kinetic energy and enstrophy compared with the reference data [15]. The results from the BL algorithm agree very well with the reference data for the 512^3 case. For computational expedience, the other algorithms are validated on the 128^3 grid. For each one, the results relative to the BL algorithm are found to be the same up to machine precision.

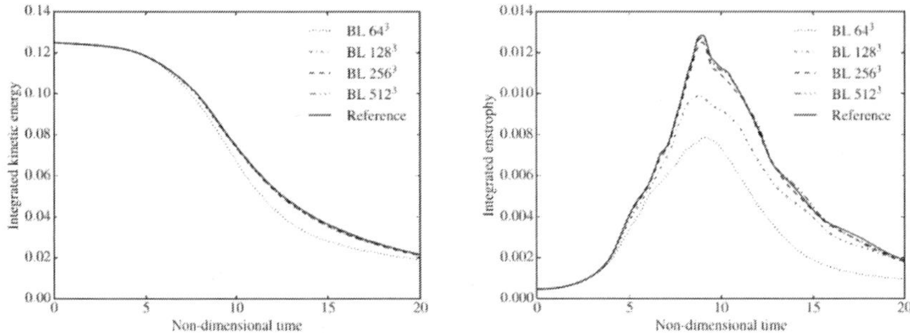

Figure. 3. Left: Evolution of the integral of kinetic energy. Right: Evolution of the integral of enstrophy.

4. PERFORMANCE EVALUATION

After checking that the results from the various algorithms match, the performance of the different algorithms were evaluated using the same Taylor–Green vortex test case described in Section 3. All simulations are performed on ARCHER (the UK National Supercomputing Service) and the code that implements the various algorithms is compiled using the Cray C compiler (version 2.4.2) with the -O3 optimisation flag. Each ARCHER node comprises 24 cores, with each MPI process being mapped to its own individual core. All simulations for performance evaluation purposes are run in parallel using 24 MPI processes/cores (one ARCHER node). The run-time of the time iteration loop was recorded for 500 iterations and is summarised in Table 1 for a range of grid sizes.

Table 1. Total run-time in seconds for different grid sizes for all algorithms on ARCHER using 24 MPI processes.

Nx	**Ny**	**Nz**	**BL**	**RA**	**RS**	**SN**	**SS**
64	64	64	16.21	9.29	10.76	8.44	9.78
128	128	128	182.55	98.18	97.36	90.72	88.95
256	256	256	1561.52	765.42	802.76	693.66	685.25

The data in Table 1 is plotted in Fig. 4; from this figure it can be inferred that when the amount of memory access is reduced, the current CPU-based architectures perform better, even though the computational intensity of such algorithms is higher. The baseline algorithm is a factor of ~2 slower than all the other algorithms presented in this paper. For larger grid sizes the benefit of the SS algorithm becomes more pronounced.

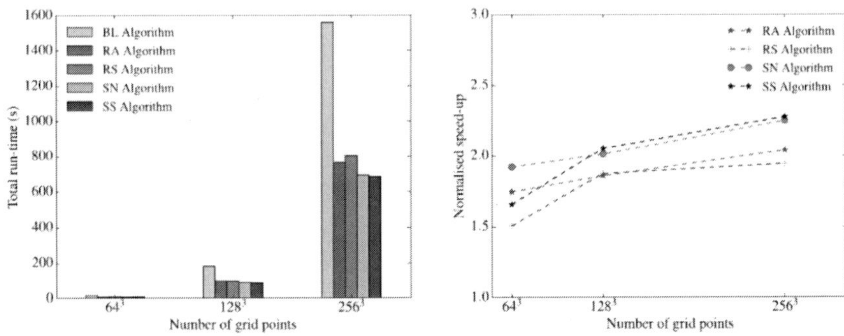

Figure. 4. Left: Figure showing data in Table 1. Right: Speed-up of algorithms normalised with the BL algorithm.

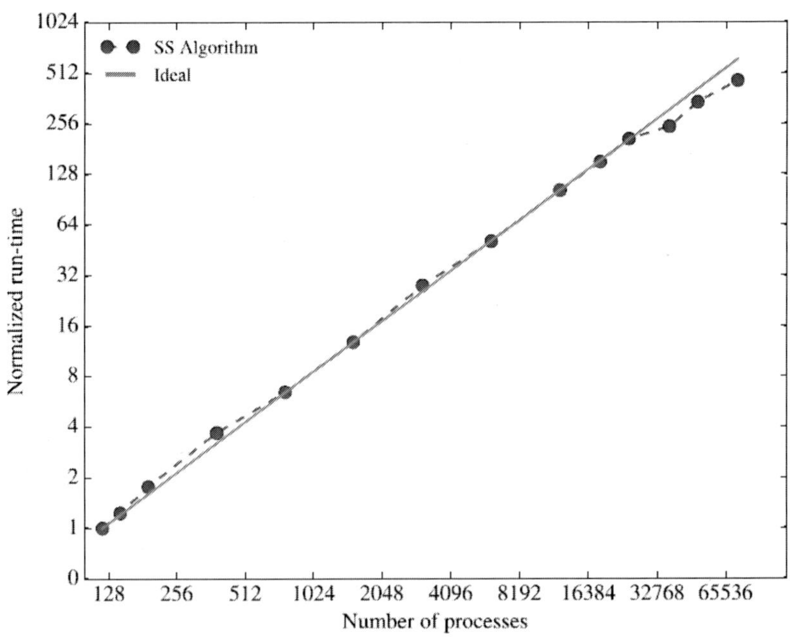

Figure. 5. Strong scaling of the SS algorithm on ARCHER up to 73,728 cores using 1.07 × 10^9 grid points. The run-time has been normalised by that of the 120-process case.

4.1. Scaling

Strong scaling tests were performed for the best performing algorithm (i.e. the SS algorithm) on ARCHER for the test problem with a total of 1.07×10^9 grid points and the runtime of the time iteration loop was recorded for 10 iterations. Fig. 5 shows the strong scaling results on ARCHER up to 73,728 MPI processes/cores (i.e. 3072 ARCHER nodes). The minimum number of processes required for running the problem is 120. The algorithm shows a near-linear scaling (speed-up of 2) until 36,864 MPI processes (i.e. 1536 ARCHER nodes) and thereafter the speed-up is ~1.5 as the process count doubles.

Weak scaling tests were also performed for the SS algorithm. Here, the number of MPI processes was varied from 192 to 65,856 (i.e. 8 to 2744 ARCHER nodes), while the number of grid points per MPI process was kept fixed at 64^3 and the runtime of the time iteration loop was recorded for 10 iterations. The largest grid size considered comprises ~17 billion solution points. Fig. 6 demonstrates that the normalised run-time is near-ideal.

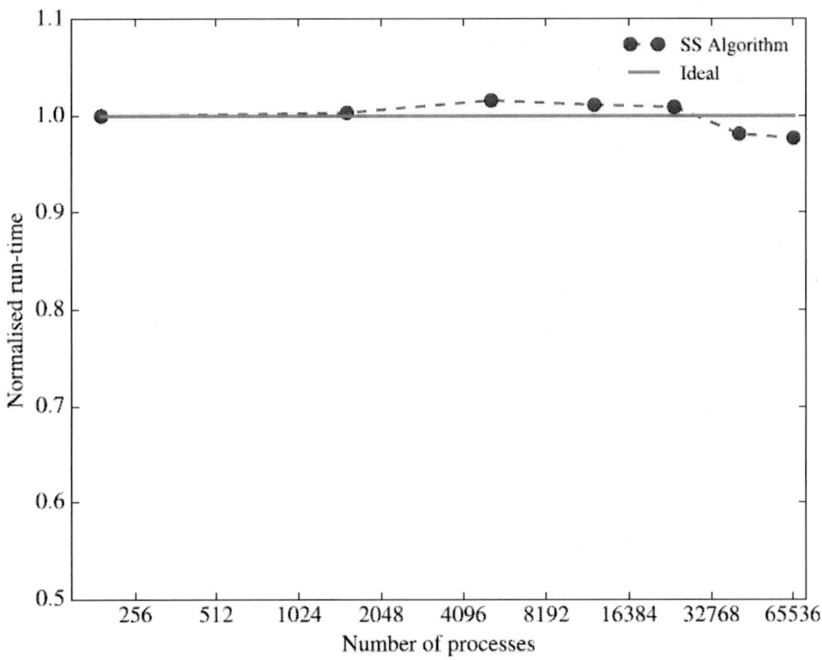

Figure. 6. Weak scaling of the SS algorithm on ARCHER with 64^3 grid points per MPI process, up to 65,856 processes. The results have been normalised by the run-time from the 192-process case.

5. CONCLUSION

In this paper the automated code generation capabilities of the OpenSBLI framework have been extended to easily modify the memory usage and computational intensity of the solution algorithm. It was found that the baseline (BL) algorithm featured in traditional CFD codes, in which all derivatives are evaluated and stored in work arrays, is not the best algorithm in terms of performance on current multi-core CPU-based architectures. Recomputing all or some of the derivatives performs better than the baseline algorithm. The best algorithm found here for the solution of the compressible Navier–Stokes equations is to store only the first derivatives of velocity components in work arrays, and compute the remaining spatial derivatives and store them in thread- or process-local variables. The run-time of such an algorithm has been shown to be ~2 times smaller than the BL algorithm. Through the use of modern code generation techniques in the OpenSBLI framework, it has been demonstrated that by changing just a few attributes (three in this case) in the problem setup file, different algorithms with varying degrees of memory and computational intensity can be readily generated automatically. The methodology presented in this paper can also be used to find the best possible algorithm for other existing architectures such as GPUs or Intel Xeon Phi coprocessors. Moreover, existing numerical models that use finite difference methods for the solution of any governing equations can be optimised for the current CPU-based architectures. When exascale systems become available, depending on their architecture and amount of available memory, users can readily tune the memory and computational intensity in the OpenSBLI framework to determine the best performing algorithm on such systems.

ACKNOWLEDGEMENTS

SPJ was supported by an EPSRC grant entitled “Future-proof massively-parallel execution of multi-block applications” (EP/K038567/1). CTJ was supported by a European Commission Horizon 2020 project grant entitled “ExaFLOW: Enabling Exascale Fluid Dynamics Simulations” (grant reference 671571). The data behind the results presented in this paper are available from the University of Southampton's institutional repository (http://dx.doi.org/10.5258/SOTON/401892). The authors acknowledge the use of the UK National Supercomputing Service (ARCHER), with computing time provided by the UK Turbulence Consortium (EPSRC grant EP/L000261/1). The authors would also like to thank the anonymous reviewers whose constructive comments helped improve the quality of this paper.

REFERENCES

1. L.E. Jones, R.D. Sandberg**Acoustic and hydrodynamic analysis of the flow around an aerofoil with trailing-edge serrations** J. Fluid Mech., 706 (2012), pp. 295-322, 10.1017/jfm.2012.254
2. S. Pirozzoli, M. Bernardini, S. Marié, F. Grasso**Early evolution of the compressible mixing layer issued from two turbulent streams** J. Fluid Mech., 777 (2015), pp. 196-218, 10.1017/jfm.2015.363
3. B. Wang, N.D. Sandham, Z. Hu, W. Liu**Numerical study of oblique shock-wave/boundary-layer interaction considering sidewall effects** J. Fluid Mech., 767 (2015), pp. 526-561, 10.1017/jfm.2015.58
4. J. DeBonis**Solutions of the Taylor–Green vortex problem using high-resolution explicit finite difference methods** 51st AIAA Aerospace Sciences Meeting including the New Horizons Forum and Aerospace Exposition (February) (2013), pp. 1-9, 10.2514/6.2013-382
5. W.C. Reynolds**Whither turbulence?** Turbulence at the Crossroads: Proceedings of a Workshop Held at Cornell University, Ithaca, NY, March 22–24, 1989, Springer, Berlin, Heidelberg (1990), pp. 313-343, 10.1007/3-540-52535-1_52 Ch. The potential and limitations of direct and large eddy simulations
6. J. Slotnick, A. Khodadoust, J. Alonso, D. Darmofal, W. Gropp, E. Lurie, D. Mavriplis**CFD Vision 2030 Study: A Path to Revolutionary Computational Aerosciences** Nasa Cr-2014-21878 (2014), pp. 1-73
7. C.T. Jacobs, M.D. Piggott**Firedrake-Fluids v0.1: numerical modelling of shallow water flows using an automated solution framework** Geosci. Model Dev., 8 (3) (2015), pp. 533-547, 10.5194/gmd-8-533-2015
8. F. Rathgeber, G.R. Markall, L. Mitchell, N. Loriant, D.A. Ham, C. Bertolli, P.H. Kelly**PyO P2: a high-level framework for performance-portable simulations on unstructured meshes** High Performance Computing, Networking Storage and Analysis, SC Companion, IEEE Computer Society (2012), pp. 1116-1123
9. I.Z. Reguly, G.R. Mudalige, M.B. Giles, D. Curran, S. McIntosh-Smith**The OPS domain specific abstraction for multi-block structured grid computations** Proceedings of the 2014 Fourth International Workshop on Domain-Specific Languages and High-Level Frameworks for High Performance Computing, IEEE Computer Society (2014), pp. 58-67, 10.1109/WOLFHPC.2014.7
10. F. Salvadore, M. Bernardini, M. Botti**GPU accelerated flow solver for direct numerical simulation of turbulent flows** J. Comput. Phys., 235 (2013), pp. 129-142, 10.1016/j.jcp.2012.10.012
11. J.C. Thibault**Implementation of a Cartesian grid incompressible Navier–Stokes solver on multi-Gpu desktop platforms using CUDA** New Horizons (2009), pp. 1-15
12. C.T. Jacobs, S.P. Jammy, N.D. Sandham**OpenSBLI: a framework for the automated derivation and parallel execution of finite difference solvers on a range of computer architectures** J. Comput. Sci. (2016), 10.1016/j.jocs.2016.11.001
13. G.A. Blaisdell, E.T. Spyropoulos, J.H. Qin**The effect of the formulation of nonlinear terms on aliasing errors in spectral methods** Appl. Numer. Math., 21 (3) (1996), pp. 207-219, 10.1016/0168-9274(96)00005-0
14. N. Sandham, Q. Li, H. Yee**Entropy splitting for high-order numerical simulation of compressible turbulence** J. Comput. Phys., 178 (2) (2002), pp. 307-322, 10.1006/jcph.2002.7022

15. Z. Wang, K. Fidkowski, R. Abgrall, F. Bassi, D. Caraeni, A. Cary, H. Deconinck, R. Hart mann, K.Hillewaert, H. Huynh, N. Kroll, G. May, P.-O. Persson, B. van Leer, M. Visbal**High-order CFD methods: current status and perspective** Int. J. Numer. Methods Fluids, 72 (8) (2013), pp. 811-845, 10.1002/fld.3767

CHAPTER 3

Green Synthesis of Tetrazines and their Role as Human Cytomegalovirus (HCMV) Protease Inhibitor

Singh P[1*] and Kumari K[2]

[1]ARSD College, University of Delhi, New Delhi, India

[2]DDU College, University of Delhi, New Delhi, India

ABSTRACT

Herein, we report a one-pot, efficient, and environmentally benign synthesis of biological and pharmacological potent tetrazines and tetrazines were synthesized via reaction of benzyl cyanide and hydrazine in the presence of sodium nitrite and hydrochrloic acid at room temperature in bronsted acidic ionic liquid as a green solvent and efficient catalytic media. Further, computational studies of the synthesized tetrazines have been performed against Human cytomegalovirus (HCMV) protease and found to inhibitors for HCMV. Therefore, tetrazines have been proposed as potent antiviral agents.

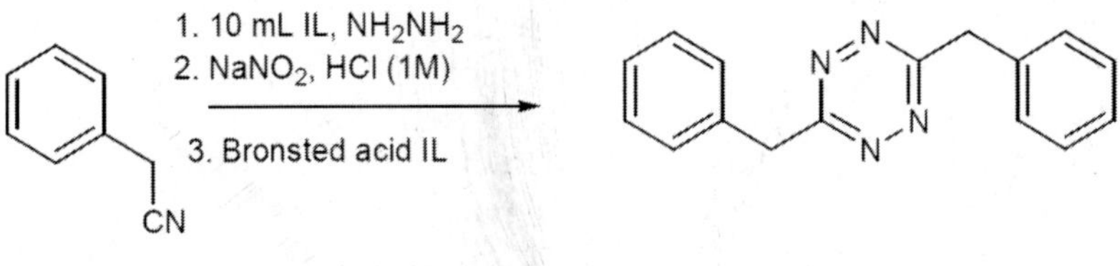

Keywords: Ionic liquid; Tetrazines; One-pot synthesis

INTRODUCTION

Designing of one-pot reactions is a major challenge for the organic and pharmaceutical chemist in academic and industrial world. Such reactions are used for carrying out two or more than two steps reactions in one pot. A huge library of heterocyclic compounds and complicated structures from simple raw materials can be achieved by these methodologies. These processes are very near to an "ideal organic or inorganic synthesis", and via these methodologies, duration of reaction time can be decreased, excellent yields and selectivities can be obtained [1]. Herein no typical equipment/ set-up, expensive protection/ deprotection reagents or steps, purification procedures are required. These methodologies are known as environmentally benign and atom economical processes [2-4].

Tetrazines are heterocyclic compounds and they have attracted the researchers due to their interesting biological activities like intracellular small molecule imaging, genetically targeted protein tagging, post-synthetic DNA labeling, and in-vivo imaging etc. [5,6]. The easy and efficient route for the synthesis of 1,2,4,5-tetrazines can be done by the addition of hydrazine to aromatic cyanide or aromatic nitriles followed by the oxidation of the 1,2-dihydrotetrazine product [7,8]. Although, this methodology has some demerits like it does not work for producing tetrazines from unactivated nitriles means with alkyl nitriles [9-12]. Therefore, developing of easy, efficient and highly selective one-pot processes is urgency in the world of organic synthetic chemistry. Ionic liquids (ILs) have attracted the extensive interest as benign reaction media as well catalyst for carrying the organic reactions due to their unique properties of non-volatility, non-flammability, recyclability, and ability to dissolve a wide range of materials [13-15]. Due to their green credentials and potential to enhance yield, rates of reaction and selectivity, they have been found increasing applications to organic synthesis [16,17].

Human cytomegalovirus (HCMV), a betaherpesvirus, is an opportunistic pathogen that commonly afflicts immuno compromised (HIV) or immunosuppressed (organ/bone marrow transplant) persons. Unfortunately for these patients, current treatments (ganciclovir, foscarnet sodium, cidofovir, and fomivirsen sodium) suffer from cross-resistance, dose-limiting toxic side effects and are only moderately effectual. As a result, in an effort to identify inhibitors with novel mechansims of action, a number of laboratories have sought inhibitors of the essential protease encoded by the HCMV UL80 gene as potentially effective antiviral chemotherapeutic agents. Thus far, several mechanism and peptide-based inhibitors of this enzyme have been disclosed [18-20].

We speculated that the use of bronsted acid ILs might promote this reaction by binding to the nitrile to hydrazine. This is the first report on the bronsted acid ILs, can activate nitriles for nucleophilic addition and promote the formation of 1,2,4,5-tetrazines. Further, they have been used against Human cytomegalovirus (HCMV) protease and proposed them as potent antiviral agents.

EXPERIMENTAL

Synthesis of Bronsted acid ionic liquid {N-Methyl-2-oxo- N-(3-sulfo-propyl)-pyrrolidinium triflate, MOSPPOTf}

A recyclable bronsted acid ionic liquid was synthesized as in **Scheme 1**. Equimolar amount of N-methyl-pyrrolidone (20 mmol) and 1,3-propanesultone (20 mmol) were allowed to react in the absence of solvent at 35°C to produce zwitter ionic pyrrolidinium salt in quantitative yield. Further, obtained zwitter ionic pyrrolidoium salt was reacted with an equimolar amount of trifluoromethanesulfonic acid at 135°C with continuous stirring, which yield bronsted acid ionic liquid in quantitative yield.

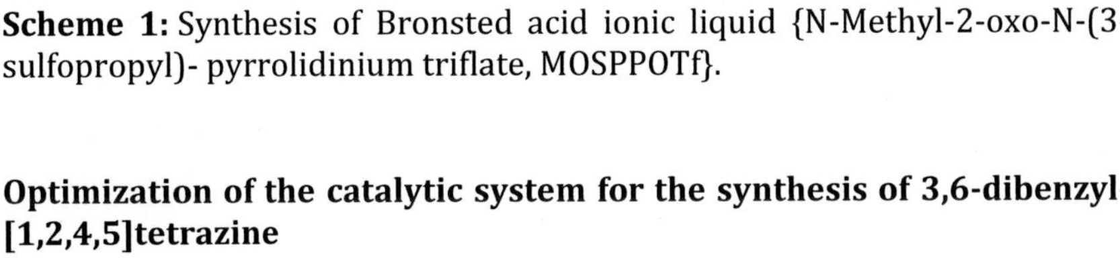

Scheme 1: Synthesis of Bronsted acid ionic liquid {N-Methyl-2-oxo-N-(3-sulfopropyl)- pyrrolidinium triflate, MOSPPOTf}.

Optimization of the catalytic system for the synthesis of 3,6-dibenzyl-[1,2,4,5]tetrazine

Different catalytic system has been used to catalyze the synthesis of tetrazines i.e., 3,6-dibenzyl-[1,2,4,5]-tetrazine via the reaction between aromatic/ aliphatic nitriles with hydrazines and the reaction was carried out for appropriate time in the presence of sodium nitrite and hydrochloric acid. Conversion of reactants into product was determined by thin layer chromatography and after the completion of reaction; the compound was extracted with diethyl ether followed by hexane from the reaction mixture. The organic layer was evaporated under vaccume, and yields the corresponding product as in **Scheme 2**. Structural assignments of the products are based on their ^{1}H-NMR, ^{13}C-NMR, FTIR and mass spectroscopic techniques. Other ionic liquids carrying other counter anions, such as N-Methyl-2-oxo-N-(3-sulfo-propyl)-pyrrolidinium tetrafluoroborate ($MOSPPBF_4$) and N-Methyl-2-oxo-N-(3-sulfo-propyl)- pyrrolidinium hexafluorophosphite ($MOSPPPF_6$), organic solvents like dioxane, acetonitrile, DMF, tolouene and other catalyst were used for the synthesis of 3,6-dibenzyl-[1,2,4,5]tetrazine as in **Table 1**. We observed maximum yield was obtained in case MOSPPOTf in short span of time at a temperature of 35°C. Similarly, other products i.e., symmetrical and asymmetrical tetrazines were synthesized and characterized as in **Table 2**.

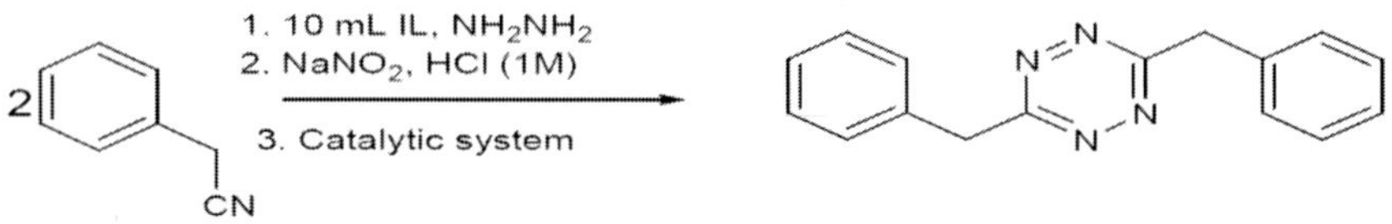

Scheme 2: Synthesis of 3,6-dibenzyl-[1,2,4,5] tetrazine in the catalytic system (Entry 3, Table 2).

Table 1: Optimization of catalytic condition for the synthesis of 3,6-dibenzyl-[1,2,4,5] tetrazine (Entry 3, Table 2).

S No	Solvent/catalyst	t (h)	Temp. (°C)	Yield (%)
1	Dioxane	36	35	75
2	CH_3CN	30	35	78
3	DMF	32	35	52
4	Toluene	38	35	70
5	Et_3N	25	35	45
6	None	36	35	Complex
7	None	48	75	Complex
8	DMSO	36	35	50
9	THF	36	45	60

10	MOSPPOTf	8	35	94
11	CH_3CN	48	60	80
12	$Ni(OTf)_2$	24	60	90
13	NiI_2	24	60	85

Table 2: Synthesis of tetrazines at 35°C using bronsted acid ionic liquid.

C. No.	Analytical data of selected compounds synthesized as in Table 2	
1	^1H NMR ($CDCl_3$) 1.21 (t, 6H, CH_3), 1.75 (m, 4H, CH_2), 2.76 (q, 4H, CH_2)	^{13}C NMR ($CDCl_3$) Aliphatic carbons (15.2, 27.3, 68.4), aromatic carbon (175.8)
2	^1H NMR ($CDCl_3$) 1.08 (t, 6H, CH_3), 1.45 (m, 4H, CH_2), 1.80 (m, 4H, CH_2), 2.67 (q, 4H, CH_2)	^{13}C NMR ($CDCl_3$) Aliphatic carbons (10.5, 29.5, 37.7, 40.5), aromatic carbon (172.7)
3	^1H NMR ($CDCl_3$) 3.92 (s, 4 H, CH_2), 7.08 (m, 10 H, Aromatic)	^{13}C NMR ($CDCl_3$) Aliphatic carbons (45.5), aromatic carbons (122.2, 125.6, 130.4, 139.6, 168.6)
4	^1H NMR ($CDCl_3$) 3.89 (s, 4 H, CH_2), 4.75 (s, 4 H, CH_2), 7.12 (m, 8H, Aromatic)	^{13}C NMR ($CDCl_3$) Aliphatic carbons (40.2, 52.6), aromatic carbons (122.2, 125.6, 130.4, 139.6, 168.6)
5	^1H NMR ($CDCl_3$) 3.95 (s, 4 H, CH_2), 6.85 (m, 8H, Aromatic)	^{13}C NMR ($CDCl_3$) Aliphatic carbons (43.4), aromatic carbons (121.5, 135.1, 139.5, 156.2, 165.2)
9	^1H NMR ($CDCl_3$) 3.91 (s, 4 H, CH_2), 7.18 (m, 8H, Aromatic)	^{13}C NMR ($CDCl_3$) Aliphatic carbons (43.4), aromatic carbons (105.6, 119.1, 132.5, 141.8, 166.8)
7	^1H NMR ($CDCl_3$) 1.12 (t, 6H, CH_3), 1.45 (m, 8H, CH_2), 1.92 (m, 4H, CH_2), 2.50 (q, 4H, CH_2)	^{13}C NMR ($CDCl_3$) Aliphatic carbons (11.5, 25.3, 35.4, 37.2, 40.1), aromatic carbon (169.1)
8	^1H NMR ($CDCl_3$) 2.46 (s, 6H, CH_3)	^{13}C NMR ($CDCl_3$) Aliphatic carbon (21.2), aromatic carbon (164.5)
9	^1H NMR ($CDCl_3$) 1.35 (t, 6H, CH_3), 2.69 (t, 4H, CH_2)	^{13}C NMR ($CDCl_3$) Aliphatic carbons (18.2, 28.5), aromatic carbon (168.7)
10	^1H NMR ($CDCl_3$) 3.95 (d of triplet, 4H, CH_2), 2.82 (t, 4H, CH_2), 1.96 (t, 1H, OH)	^{13}C NMR ($CDCl_3$) Aliphatic carbons (41.5, 62.8), aromatic carbon (169.2)
11	^1H NMR ($CDCl_3$) 4.65 (d, 4H, CH_2), 7.35 (m, 8H, aromatic), 1.98 (t, 1H, OH)	^{13}C NMR ($CDCl_3$) Aliphatic carbons (69.8), aromatic carbon (125.5, 128.5, 136.8, 141.6, 165.7)

Reactant 1	Reactant2	Product	C. No.	Yield (%)
		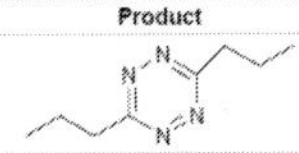	1	92
			2	90
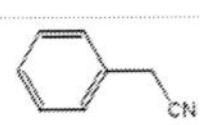	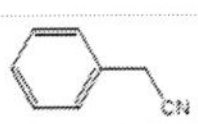	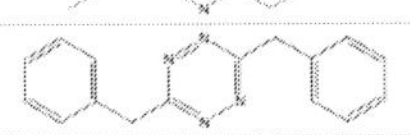	3	94
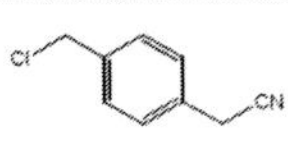	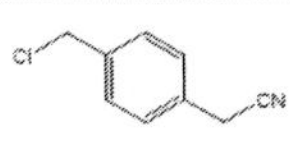	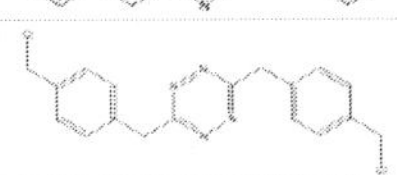	4	95
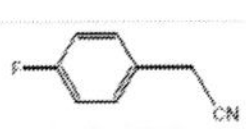	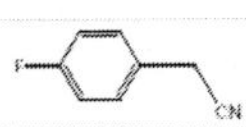	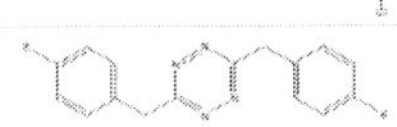	5	92
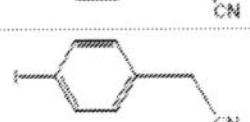	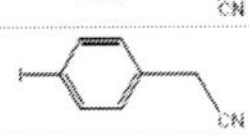	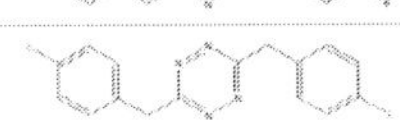	6	92

			7	90
CH_3CN	CH_3CN	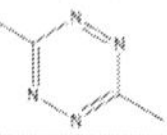	8	90
		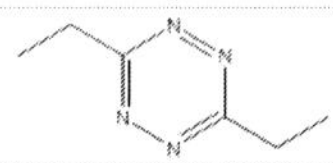	9	92
		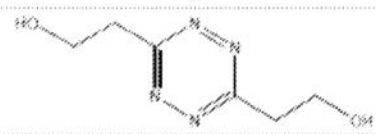	10	96
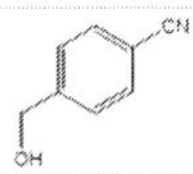	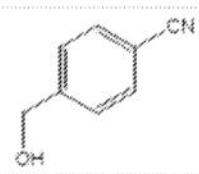	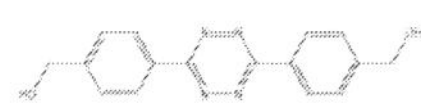	11	96
		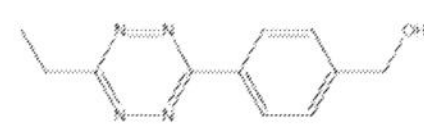	12	96
	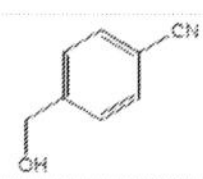	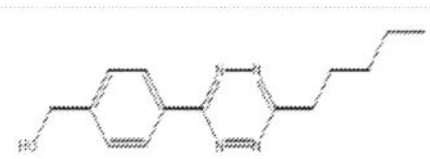	13	97
		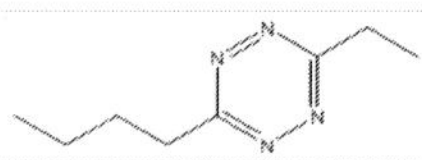	14	95
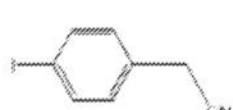	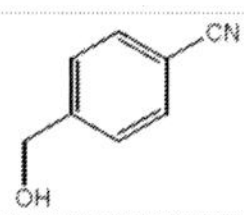	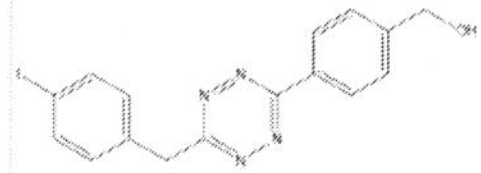	15	95
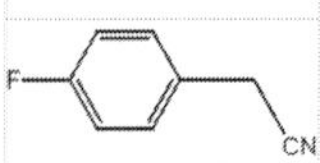		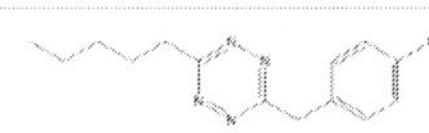	16	96
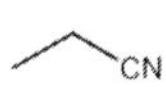		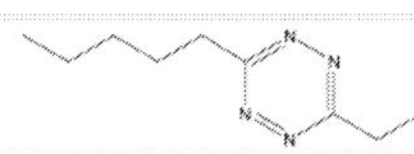	17	94
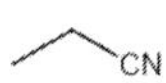		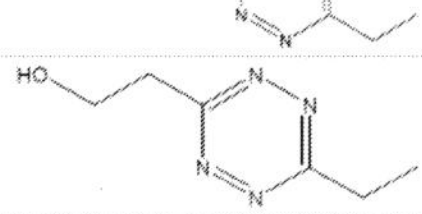	18	94

Computational study of tetrazines as Human cytomegalovirus (HCMV) protease inhibitors

Optimization of tetrazines: All the tetrazines (18 compounds) were drawn on Chem Draw and were imported to Chem Draw 3D followed by their optimization using MOPAC (a computational tool) suing AM1 calculation and closed shell restricted with RMS gradient 0.1.

Protein preparation: The crystal structure of Human cytomegalovirus (HCMV) protease having PDB ID: 1CMV was taken (http://www.rcsb.org/). The protein was prepared using Molegro Molecular Viewer 2.5 and the same was used for the virtual drug screening.

Virtual screening: Herein, iGEMDock was used for the drug screening as in **Table 3**. iGEMDock is a molecular docking tool and generates diversity of tetrazines conformations from different seeds with high temperature molecular dynamics. Then, it orients the tetrazines conformations within the defined protein active site by translating the center of the surfactant. Each orientation is subjected to simulated annealing molecular dynamics and sorted according to the interaction energy. iGEMDOCK energy function consists of electrostatic, steric, and hydrogen- bonding potentials. Steric and hydrogen bonding potentials use a linear model. There are four main steps which are used here. Parameters used for drug screening in iGEMDOCK were as followed: initial step sizes (σ=0.8 and ψ=0.2), family competition length (L=2), population size (N=200), and recombination probability (pc=0.3). Optimization is set to generate 70 iterations for which it generated 1200 solutions in one generation process and if exceeded then it terminated after 84,000 solutions. (Yen, Fu 2007) Parameters in iGEMDOCK were set for the successful screening of drug like molecules are as follow: initial step sizes (r=0.8 and w=0.2), family competition length (L=2), population size (N=300), and recombination probability (pc=0.3). Optimization is set to generate 70 iterations for which it generated 1200 solutions in one generation process and if exceeded then it terminated after 84,000 solutions.

Table 3: Docking studies of the synthesized tetrazines as in Table 2.

C. No.	Energy	VDW	H-Bond	E (pharma)	H-M-ARG-22	H-M-GLN-25	H-M-LEU-32	V-M-ARG-22	V-M-TYR-23	V-M-GLN-25	V-S-PRO-27	V-M-LEU-34	V-S-ARG-166
1	-85.6	-67.27	-18.27	-124.4	-6.92	-4.53	-6.81	-6.09	-4.82	-5.86	-6.26	1.62	-6.36
2	-80.2	-58.73	-21.44	-80.2	0	0	0	0	0	0	0	0	0
3	-84.3	-74.99	-9.30	-84.3	0	0	0	0	0	0	0	0	0
4	-98.9	-73.99	-24.94	-98.9	0	0	0	0	0	0	0	0	0
5	-87.7	-79.05	-8.60	-87.7	0	0	0	0	0	0	0	0	0
6	-92.3	-74.86	-17.4	-112.6	-5.00	-2.03	0	-4.32	0.68	1.46	-6.02	-5.74	-5.49
7	-88	-72.08	-15.89	-108.2	-5.63	0	0	-4.89	-4.89	-1.11	-5.32	-2.19	-4.30
8	-70.8	-49.77	-21	-111.4	-7	-7	-7	-4.50	-2.30	-4.55	-5.26	-3.40	-4.79
9	-79.4	-59.78	-19.60	-118.2	-5.60	-7	-7	-5.31	-2.36	-4.99	-4.08	-6.26	-4.87
10	-89.5	-62.47	-27.00	-128.1	-5.56	-6.72	-7	-5.27	-4.39	-4.98	-6.25	-0.93	-5.63
11	-93.2	-80.43	-12.75	-100.7	0	0	0	1.46	-6.24	-5.36	-2.98	-9.59	-0.11
12	-98.6	-78.07	-20.54	-140	-7	-7	-6.54	-4.73	-5.82	-5.04	-4.58	-8.14	-1.97
13	-99.3	-80.78	-18.49	-99.3	0	0	0	0	0	0	0	0	0
14	-89.3	-68.35	-20.92	-130.9	-6.92	-7	-7	-4.91	-5.90	-4.69	-4.61	-4.66	-4.17
15	-111.7	-91.39	-20.28	-157.5	-7	-7	-6.28	-5.04	-5.24	-10.35	-6.21	-8.47	-5.75
16	-110.3	-91.79	-18.52	-138.2	-7	0	0	-8.41	-2.05	-6.48	-4.02	-4.48	-8.11
17	-86.8	-68.87	-17.95	-122.8	-4.01	-6.93	-7	-5.20	-2.81	-5.36	-5.97	-0.09	-7.72
18	-83.8	-57.46	-26.36	-122.6	-9.64	-6.98	0	-6.32	-4.76	-4.56	-1.92	-6.23	-3.44

RESULTS AND DISCUSSION

Herein, a bronsted acidic ionic liquid was synthesized and was used a green and efficient solvent for the synthesis of biological potent symmetrical and

Figure 1: Computational studies of the synthesized tetrazines.

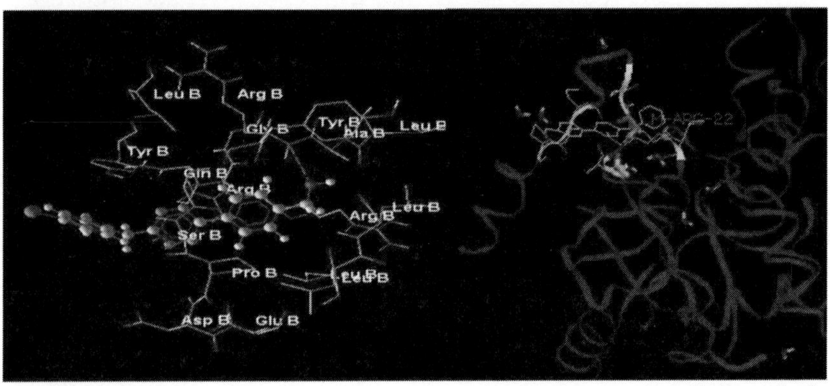

Figure 2: Picture of the interaction of synthesized tetrazine (15) with the Human cytomegalovirus (HCMV) protease having PDB ID:1CMV.

asymmetrical tetrazines. A mixture of phenylacetonitrile (nitrile) was reacted in its own in the presence of hydrazine underwent one-pot using MOSPPOTf, a solvent producing compound as in Entry 3, **Table 1**. The activity of synthesized IL, MOSPPOTf, was compared with other solvent as mentioned in **Table 1** in terms of yield and duration of reaction. We observed that the IL, MOSPPOTf showed the best catalytic activity amongst all mentioned in **Table 1**. Similarly, different tetrazines have been prepared as in **Table 2**. The advantage of this method is that compounds as in **Table 2** were obtained in quantitative yields directly without formation of any byproduct (**Scheme 2**). Further, docking studies of tetrazines have been carried out using computational tools, iGEMDock. Docking studies are used extensively in drug discovery such as in the prediction of ligandreceptor complex structures and also to rank the ligand molecules based upon the binding energies of the corresponding ligand-enzyme complex. The objective of our docking study is to elucidate the potential interaction mode of the tetrazines with HCMV-protease. The binding energies of the best-docked inhibitor-enzyme complexes are shown in **Table 3**. A general conclusion derived from these docking results is that the side chain of the Leu 34 forms a hydrogen bond with the inhibitors. The nature of these hydrogen bonds involves an interaction between the amidic nitrogen of HCMV-protease and the nitrogen of tetrazines. Interestingly, this interaction is almost conserved with all studied inhibitors. The docking calculations provided us with a general static picture of the most energetically favourable binding orientation of inhibitors to the enzyme. To obtain further insight into the dynamic changes of the docked inhibitors within the enzyme active site pocket over time, the lowest energy docked complex of the most active inhibitor, 15, was subjected to unconstrained MD simulations and shown in **Figures 1 and 2**. The post dynamic monitoring of the overall hydrogen bond networks and the electrostatic interaction between the 15 inhibitor and nearby residue clearly revealed that these inhibitors settle comfortably inside the HCMV-protease active site.

CONCLUSION

In conclusion we have discovered that bronsted acid ionic liquid can be used to catalyze the one-pot synthesis of 1,2,4,5-tetrazines from activated and unactivated nitriles means aliphatic and aromatic nitrile respectively. These tetrazines were not practically accessible and cannot be obtained in high yield even after multistep syntheses, also takes plenty of time. This potential of the process to get a variety of symmetric and asymmetric tetrazines from nitriles can be benefited for the academicians and industrialists. This process will surely be benefited in coordination chemistry, explosives research, materials science, and total synthesis. This process will allow the syntheses of the typical compounds, which were not so easy or able to synthesis in satisfactory yield. Further, computational studies of the synthesized compounds have been performed against Human cytomegalovirus (HCMV) protease and proposed them as potent antiviral agents. Among the all tetrazines, compound no. 15 has shown potent atypical anti-protease activity probably by inhibition of Human cytomegalovirus (HCMV) protease receptors.

ACKNOWLEDGEMENTS

Author is highly thankful to Department of Science and Technology (DST), India for their financial support.

REFERENCES

1. Reichert WM, Trulove PC, De Long HC (2010) 3-(1-Methyl-3-imidazolio) propanesulfonate: a precursor to a Bronsted acid ionic liquid. Acta Crystallogr Sect E Struct Rep Online 66: 0591.
2. Zhanga L, Xiana M, Hea Y, Lia L, Yanga J (2009) A Bronsted acidic ionic liquid as an efficient and environmentally benign catalyst for biodiesel synthesis from free fatty acids and alcohols. Bioresour Technol 100: 4368-4373.
3. Hasaninejad A, Zare A, Shekouhy M, Ameri Rad J (2010) Catalyst-free one-pot four component synthesis of polysubstituted imidazoles in neutral ionic liquid 1-butyl-3-methylimidazolium bromide. J Comb Chem 12: 844-849.
4. Chaskar A, Patil L, Phatangare K, Padalkar V, Takale S (2011) Bismuth (III) Salts Promoted and Ionic Liquid Assisted an Efficient and Environmentally Benign One-Pot Synthesis of 1,5-Benzodiazepine Derivatives. ISRN Org Chem 2011: 604348.
5. Sammelson RE, Olmstead MM, Haddadin MJ, Kurth MJ (2000) 1,2,4,5-Tetrazines as oxidant and reactant with DBU: An unexpected formation of a novel fused tetraheterocyclic azepine J Org Chem 65: 9265-9267.
6. Yang J, Karver MR, Li W, Sahu S, Devaraj NK (2012) Metal-catalyzed one-pot synthesis of tetrazines directly from aliphatic nitriles and hydrazine. Angew Chem Int Ed Engl 51: 5222-5225.
7. Akbari J, Heydari A, Kalhor RH, Kohan SA (2009) Sulfonic acid functionalized ionic liquid in combinatorial approach, a recyclable and water tolerant-acidic catalyst for one-pot Friedlander quinoline synthesis. J Comb Chem 121: 137-140.
8. Karver MR, Weissleder R, Hilderbrand SA (2011) Synthesis and evaluation of a series of 1,2,4,5-tetrazines for bioorthogonal conjugation. Bioconjug Chem 22: 2263-2270.
9. Novák Z, Kotschy A (2003) First cross-coupling reactions on tetrazines. Org Lett 5: 3495-3497.
10. Pellegatti L, Vedrenne E, Leger JM, Jarry C, Routier S (2010) Efficient pallado-catalyzed C6-(het)arylation of Imidazo[1,2-b][1,2,4,5]tetrazines under microwave irradiations. J Comb Chem 12: 604-608.
11. Chavez DE, Hanson SK, Veauthier JM, Parrish DA (2013) Electroactive explosives: nitrate ester-functionalized 1,2,4,5-tetrazines. Angew Chem Int Ed Engl 52: 6876-6879.
12. Hutke V, Wankhade G, Waghmare PJ, Harinath BC (2013). Elisa protocol for rapid screening of potential anti-tubercular drugs based on antigenic reactivity of mycobacterial ES-31 serine protease - a drug target supported by axenic culture of Mycobacterium tuberculosis H37 Ra strain in the presence of inhibitor. Indian J Tuberc 60: 138-141.
13. Jiang X, Ye W, Song X, Ma W, Lao X, et al. (2011) Novel ionic liquid with both Lewis and Brønsted acid sites for Michael addition. Int J Mol Sci 12: 7438-7444.
14. Wang XR, Xing J, Yan CX, Cheng Y (2011) The reaction of beta-lactam carbenes with 3,6-dipyridyltetrazines: switch of reaction pathways by 2-pyridyl and 4-pyridyl substituents of tetrazines. Org Biomol Chem 10: 970-977.

15. Nia RH, Mamaghani M, Tabatabaeian K, Shirini F, Rassa M (2013) A rapid one-pot synthesis of pyrido[2,3-d]pyrimidine Derivatives using bronsted-acidic ionic liquid as catalyst. Acta Chim Slov 60: 889-895.
16. Lang SA Jr, Cohen JE, Sloboda AE, Greenblatt E (1976) Aryl-s-tetrazines with antiinflammatory activity. J Med Chem 19: 1404-1409.
17. Nhu D, Duffy S, Avery VM, Hughes A, Baell JB (2010) Antimalarial 3-arylamino-6-benzylamino-1,2,4,5-tetrazines. Bioorg Med Chem Lett 20: 4496-4498.
18. Smith DG, Gribble AD, Haigh D, Ife RJ, Lavery P, et al. (1999) The inhibition of human cytomegalovirus (hCMV) protease by hydroxylamine derivatives. Bioorg Med Chem Lett 9: 3137-3142.
19. Vražič D, Jereb M, Laali KK, Stavber S (2012) Brønsted acidic ionic liquid accelerated halogenation of organic compounds with N-Halosuccinimides (NXS). Molecules 18: 74-96.
20. Yen-Fu C, Yu-Ju C, Jinn-Moon Y (2007) National Innovation Award, "GEMDOCK: An Integrated Environment for Computer-aided Drug Design and Its Applications".

CHAPTER 4

Fuzzy Varying Coefficient Bilinear Regression of Yield Series

Ting He, Qiujun Lu

College of Science, University of Shanghai for Science and Technology, Shanghai, China

ABSTRACT

We construct a fuzzy varying coefficient bilinear regression model to deal with the interval financial data and then adopt the least-squares method based on symmetric fuzzy number space. Firstly, we propose a varying coefficient model on the basis of the fuzzy bilinear regression model. Secondly, we develop the least-squares method according to the complete distance between fuzzy numbers to estimate the coefficients and test the adaptability of the proposed model by means of generalized likelihood ratio test with SSE composite index. Finally, mean square errors and mean absolutely errors are employed to evaluate and compare the fitting of fuzzy auto regression, fuzzy bilinear regression and fuzzy varying coefficient bilinear regression models, and also the fore- casting of three models. Empirical analysis turns out that the proposed model has good fitting and forecasting accuracy with regard to other regression models for the capital market.

Keywords: Fuzzy Varying Coefficient Bilinear Regression Model, Fuzzy Financial Assets Yield, Least-Squares Method, Generalized Likelihood Ratio Test, Forecast

1. INTRODUCTION

Researchers usually expect a more reliable estimate of dynamic indicators of the market data through rational design and adjustment with more flexible and applicable models when they study the financial assets price changes. The actual financial data which are often given in the form of interval are not only random but also contain fuzziness. Therefore, taking the innate fuzziness of actual financial data into account, starting with interval financial observed data and combining the features of interval financial series and finally establishing an analytical mode are big problems in analyzing assets price changes.

Since Zadeh [1] put forward the fuzzy set in 1965, the fuzzy set theory has been widely used in social science research, especially in the economic construction, financial investment, capital market operation and management etc. For interval financial observed data, Li Zhuyu et al. [2] defined the stationary of fuzzy financial time series as well as fuzzy financial assets yield series and then constructed a pth-order fuzzy auto regression (FAR) model, and also estimated the unknown coefficients by using fuzzy linear program (FLP) with satisfying the minimum fuzzy index and real meaning of financial yields. The result of estimation in [2] shows that yields convergence and fluctuation have the same trends; in addition, the model only can apply to centralized fuzzy data. This leads to deviations with the reality. In order to reflect the dynamic changes in financial yields in a period of time, Li Zhuyu et al. [4] learned the idea from D'Urso and T. Gastaldi [3] that fluctuations depended on the centers to some extent in a dynamic process, then built the fuzzy bilinear regression (FBR) model with fuzzy financial yields centers and fluctuations respectively, and estimated the unknown coefficients by fuzzy least- squares (FLS) method. Wang Donghua [5] suggested that the method of fuzzy linear program (FLP) was relatively simple so as to get a wide range result and was worse for our application.

Due to the influence of various kinds of social factors, financial asset price changes with non-linear dynamic characteristics, so the traditional method of estimation in the description of the non-linear problems tends to have larger error when modelling. Financial assets yield prediction model in literature [4] still needs to combine the explanatory ability on model, the relationship between the dynamic data, the fitting precision and prediction precision more ideally. Varying coefficient models (also called functional coefficient model) can effectively avoid the problem of the curse of dimensionality because of its obvious flexible model structure, and have a distinct advantage in exploring non-linear dynamic characteristics, reducing model specification errors, describing the features of data as well as forecasting. This article generalizes the fuzzy model in literature [4] to varying coefficient model, which is called fuzzy varying coefficient bilinear regression (FVCBR) model, by the correspondence between financial yields and the symmetric numbers used to depict its fuzziness and use fuzzy least-squares (FLS) method to deduce the estimator. Additionally, we test the adaptability of the proposed model by means of generalized likelihood ratio test. Mean square errors and mean absolutely errors are employed to evaluate and compare the fitting of fuzzy auto regression (FAR), fuzzy bilinear regression (FBR) and fuzzy varying coefficient bilinear regression (FVCBR) models, and also the forecasting of three models in and out of the sample period.

2. THE DISTANCE OF FUZZY NUMBER SPACE

In order to introduce the distance of fuzzy number space, we first need to give the introduction to the symmetric numbers and its operational properties.

Definition 1 [6]. A fuzzy number $\tilde{A}$ is a fuzzy set of the real line R with satisfying the following conditions:

1) there exists an $x_0 \in R$ such that $A(x_0) = 1$;

2) for any $\alpha \in (0, 1]$, the α-level set $A\alpha = \{x \mid A(x) \geq \alpha\}$ is an interval number.

The set of all the fuzzy numbers is denoted by $\tilde{R}$.

Definition 2 [6]. A fuzzy number $\tilde{A}$ with the following membership function:

$$A(x) = L\left(\frac{x-a}{\alpha}\right) \tag{1}$$

is called a symmetric fuzzy number and usually denoted by $(a,\alpha)_L$, where a $(a \in R)$ and α $(\alpha > 0)$ are the center and spread of $\tilde{A}$ respectively. Besides, $L(\cdot)$ is a strictly decreasing function on $[0,+\infty)$ with $L(0) = 1$ and $L(x) = L(-x)$. When $\alpha = 0$, symmetric fuzzy number $\tilde{A}$ becomes an ordinary real number a and is denoted by $(a,0)L = a$.

The set of all the symmetric fuzzy numbers is denoted by $\tilde{R}_L$. Symmetric fuzzy number $(a,\alpha)_L$ is called symmetric triangle fuzzy number if $L(x) = \max(0,1-|x|)$.

Proposition 1 [6]. Suppose (a, α)L, $(b, \beta)_L \in \tilde{R}_L$ and $k \in R$, therefore:

1) $k\cdot(a,\alpha) = (k\alpha, |k|\alpha)_L$;

2) $(a,\alpha)_L + (b, \beta)_L = (a+b,\alpha+\beta)$.

Let $\tilde{A}$ and $\tilde{B}$ be two fuzzy numbers. Xu [7] proposed a formula that defines complete distance between fuzzy numbers generalized by distance between interval numbers:

$$d\left(\tilde{A},\tilde{B}\right)=\left(\int_0^1 f(\lambda)\overline{d}^2(A_\lambda,B_\lambda)\mathrm{d}\lambda\right)^{\frac{1}{2}} \tag{2}$$

where $f(\lambda)$ is an increasing function on [0,1] satisfying $f(0) = 0$ and $\int_0^1 f(\lambda)\mathrm{d}\lambda = 1/2$, $A\lambda = [a_1(\lambda), a_2(\lambda)]$

and $B\lambda = [b_1(\lambda), b_2(\lambda)]$ are λ-level sets of $\tilde{A}$ and $\tilde{B}$ respectively and

$$\overline{d}^2(A_\lambda,B_\lambda)=\left[a_1(\lambda)-b_1(\lambda)\right]^2+\left[a_2(\lambda)-b_2(\lambda)\right]^2.$$

As pointed out in [7], the monotonically increasing function $f(\lambda)$ emphasizes the contribution of higher values of λ to the distance between $\tilde{A}$ and $\tilde{B}$. Furthermore, $f(0) = 0$ and $\int_0^1 f(\lambda)\mathrm{d}\lambda = 1/2$ ensure

$d(\tilde{A}, \tilde{B})$ is generalized by ordinary distance. In fact, we usually let $f(\lambda) = \lambda$.

When $\tilde{A} = (a, s_A)_L$ and $\tilde{B} = (b, s_B)_L$ are two symmetric fuzzy numbers, the λ-level sets of $\tilde{A}$ and $\tilde{B}$ are respectively $A_\lambda=\left[a-L^{-1}(\lambda)s_A,a+L^{-1}(\lambda)s_A\right]$ and $B_\lambda=\left[b-L^{-1}(\lambda)s_B,b+L^{-1}(\lambda)s_B\right]$, therefore

$\overline{d}^2(A_\lambda,B_\lambda)=2(a-b)^2+2\left(L^{-1}(\lambda)\right)^2(s_A-s_B)^2$ and the distance (2) becomes

$$d\left(\tilde{A},\tilde{B}\right)=\left((a-b)^2+\Delta(L,f)(s_A-s_B)^2\right)^{\frac{1}{2}} \tag{3}$$

where $\Delta(L,f)=2\int_0^1 f(\lambda)\left(L^{-1}(\lambda)\right)^2\mathrm{d}\lambda$.

Proposition 2[1]. If $\tilde{A}, \tilde{B}, \tilde{C} \in \tilde{R}$ and $k \in R$, therefore:

1) $d(k\tilde{A}, k\tilde{B}) = |k|\, d(\tilde{A}, \tilde{B})$;

2) $d(\tilde{A}, \tilde{B}) = d(\tilde{A} + \tilde{C}, \tilde{B} + \tilde{C})$.

Proposition 3[2]. $(\tilde{R}_L, d)$ is separable distance space.

3. THE ESTIMATION AND TEST OF MODEL

Suppose that $\left\{\left[x_t^l, x_t^u\right], t=1,2,\cdots,n\right\}$ is a given interval series, in which x_t^l and x_t^u are respectively the lowest

and highest prices of financial products. Fuzzy financial time series $\left\{\tilde{x}_t=\left(x_t, s_t\right)_L, t=1,2,\cdots,n\right\}$ is the fuzzy

depiction of interval series $\left\{\left[x_t^l, x_t^u\right], t=1,2,\cdots,n\right\}$, where $x_t=0.5\left(x_t^l+x_t^u\right)$ represents the average price of fi-

nancial product on t-th day and the series of centers $\{x_t\}$ depicts the trend of convergence of financial product

price; $s_t=0.5\left(x_t^u-x_t^l\right)$ is the radius of fluctuations in price and the series of spreads $\{S_t\}$ depicts the uncer-

tainty of $\{\tilde{x}_t\}$ on t-th day, namely the magnitude of nonrandomfluctuations. Series $\{\tilde{r}_t\}$ is called fuzzy finan- cial yields series and denoted by $\tilde{r}_t=\nabla \ln \tilde{x}_t$. Furthermore, $\{\tilde{r}_t\}$ can also be put in another form of sym-

metric fuzzy numbers series $\{\tilde{r}_t\}=\left\{\left(c_t, u_t\right)_L, t=2,\cdots,n\right\}$, then the series of centers reflects the convergence be-

ing the following formula:

$$\left\{c_t\right\}=\left\{\frac{1}{2}\left[\ln \frac{x_t^u}{x_{t-1}^l}+\ln \frac{x_t^l}{x_{t-1}^u}\right], t=2,\cdots,n\right\} \tag{4}$$

the series of spreads reflects the magnitude of nonrandomfluctuations and is expressed as following:

$$\left\{u_t\right\}=\left\{\frac{1}{2}\left[\ln \frac{x_t^u}{x_{t-1}^l}-\ln \frac{x_t^l}{x_{t-1}^u}\right], t=2,\cdots,n\right\} \tag{5}$$

3.1. Fuzzy Auto Regression (FAR) Model

If $\{\tilde{r}_t\} = \{(c_t, u_t)_L, t = 2, \cdots, n\}$ is a conditional stationary series, a p-th order fuzzy auto regression model [2] can be established through $\{\tilde{r}_t\}$ and $\tilde{r}_{t-1}, \cdots, \tilde{r}_{t-p}$:

$$\tilde{r}_t = A_1\tilde{r}_{t-1} + A_2\tilde{r}_{t-2} + \cdots + A_p\tilde{r}_{t-p} + \tilde{\varepsilon}_t, t = p+1, \cdots, n \tag{6}$$

where $A_1, A_2, \ldots, A_p \in R$ are auto regressive parameters and $\tilde{\varepsilon}_t$ are the fuzzy errors on t-th day.

The p-th order fuzzy auto regression model has two obvious limitations, one of those is that the fuzzy auto regression as (6) will finally leads to the same trends of centers and spreads, another is that formula (6) can be equivalent to the combination of auto regression models built by two ordinary time series with centers and spreads respectively. However, only centralized data can omit the constant terms in auto regression models with traditional time series while in the common case of data, the constant terms usually cannot be ignored.

3.2. Fuzzy Bilinear Regression (FBR) Model

The fuzzy bilinear regression (FBR) model in [4] set up by the centers and spreads of fuzzy financial yields series respectively solves those two problems fundamentally.

$$\begin{cases} c_t = \alpha_0 + \alpha_1 c_{t-1} + \cdots + \alpha_p c_{t-p} + \varepsilon_t \\ u_t = \beta_0 + \beta_1 u_{t-1} + \cdots + \beta_q u_{t-q} + \gamma c_t + e_t \end{cases}, \quad t = k+1, \cdots, n \tag{7}$$

Where $k = \max(p,q)$, $\alpha = (\alpha_0, \alpha_1, \ldots, \alpha_p)^T$ and $\beta = (\beta_0, \beta_1, \ldots \beta_q, \gamma)^T$ are unknown coefficient vectors and ε_t and e_t are error terms of the series of centers and spreads respectively on t-th day.

The model (7) respectively describes the auto regression relationship between the convergence of fuzzy financial yields on t-th day and the p-th order lagging value of yields, as well as the auto regression relationship between fluctuations and its p-th order lagging value. Meanwhile, the model also expresses the interdependent relationship between the fluctuations and the convergence of fuzzy financial yields. When $p = q$, $\alpha_0 = \beta_0 = 0$, $\alpha_i = \beta_i$

$(1 \le i \le p)$ and $\gamma = 0$, the model (7) will turn into the model (6) in form. Therefore, on one hand, the model (7) is formally generalized by the model (6), and on other hand, the model (7) can improve the explanatory ability to fuzzy financial yields.

3.3. Fuzzy Varying Coefficient Bilinear Regression (FVCBR) Model

In the FAR model and FBR model, the regression coefficients being constants suggests that explanatory variables impact on explained variables constantly during the sample period. While in the study of financial assets yields prediction, the yields is a time series and the level of influence of various factors in different intervals will change. Consequently, linear models with constant coefficients are unfit for prediction and the varying coefficient models should put to use. The form of FVCBR model shows below:

$$\begin{cases} c_t = \alpha_0(t) + \alpha_1(t)c_{t-1} + \cdots + \alpha_p(t)c_{t-p} + \varepsilon_t \\ u_t = \beta_0(t) + \beta_1(t)u_{t-1} + \cdots + \beta_q(t)u_{t-q} + \gamma(t)c_t + e_t \end{cases}, \quad t = k+1, \cdots, n \tag{8}$$

In which the model coefficients are the function of t.

3.3.1. Coefficients Estimate of Fuzzy Varying Coefficient Bilinear Regression (FVCBR) Model

Considering the number of unknown coefficients of varying coefficient model will multiply as the sample size gets large, thus fitting models with traditional methods of estimation are not appropriate. Because of the above- mentioned disadvantages, the restricted weighted least-squares estimation is much used at present. Generally speaking, suppose that t_0 is a given point in the domain of variable t, the existing observations $\{(c_t, u_t)_L\}$ all

provide information to the model coefficients $\alpha_0(t_0), \alpha_1(t_0), \ldots, \alpha_p(t_0)$, $\beta_0(t_0)$, $\beta_1(t_0), \ldots, \beta_q(t_0)$, $\gamma(t_0)$ around t_0, but different observations work different effects. The importance is shown by assigning each observation a weight, and the t-th weight corresponds to the t-th observation $(c_t, u_t)_L$.

The kernel estimation is used to estimate the unknown coefficients $\alpha_0(t_0)$, $\alpha_1(t_0), \ldots, \alpha_p(t_0)$, $\beta_0(t)$, $\beta_1(t)$, $\beta 2(t), \ldots, \beta_q(t)$, $\gamma(t)$. On the basis of distance (3) and the principle of the kernel smoothing in statistics, we formulate the following restricted weighted least-square problem. That is, the objective function is

$$\Psi\left(\alpha_0(t_0),\alpha_1(t_0),\cdots,\alpha_p(t_0),\beta_0(t_0),\cdots,\beta_q(t_0),\gamma(t_0)\right)$$

$$=\sum_{t=k+1}^{n} d^2\left((c_t,u_t)_L,\left(\left(\alpha_0(t_0)+\sum_{i=1}^{p}\alpha_i(t_0)c_{t-i}\right),\left(\beta_0(t_0)+\sum_{j=1}^{q}\beta_j(t_0)u_{t-j}+\gamma(t_0)c_t\right)\right)_L\right)K_h(t-t_0)$$

$$=\sum_{t=k+1}^{n}\left(\left(c_t-\alpha_0(t_0)-\sum_{i=1}^{p}\alpha_i(t_0)c_{t-i}\right)^2+\Delta(L,f)\left(u_t-\beta_0(t_0)-\sum_{j=1}^{q}\beta_j(t_0)u_{t-j}-\gamma(t_0)c_t\right)^2\right)K_h(t-t_0)$$

$$=\sum_{t=k+1}^{n}\left(c_t-\alpha_0(t_0)-\sum_{i=1}^{p}\alpha_i(t_0)c_{t-i}\right)^2K_h(t-t_0)+\Delta(L,f)\sum_{t=k+1}^{n}\left(u_t-\beta_0(t_0)-\sum_{j=1}^{q}\beta_j(t_0)u_{t-j}-\gamma(t_0)c_t\right)^2K_h(t-t_0)$$

And it is minimized with respect to $\alpha_0(t_0)$, $\alpha_1(t_0)$,..., $\alpha_p(t_0)$, $\beta_0(t_0)$,..., $\beta_q(t_0)$, $\gamma(t_0)$ where $K_h(\cdot) = K(\cdot/h)/h$, with $K(\cdot)$ being a given kernel function and h being the smoothing parameter.

The restricted weighted least-squares problem is equivalent to minimizing the following equations:

$$\begin{cases}\sum_{t=k+1}^{n}\left(c_t-\alpha_0(t_0)-\sum_{i=1}^{p}\alpha_i(t_0)c_{t-i}\right)^2K_h(t-t_0)\\ \sum_{t=k+1}^{n}\left(u_t-\beta_0(t_0)-\sum_{j=1}^{q}\beta_j(t_0)u_{t-j}-\gamma(t_0)c_t\right)^2K_h(t-t_0)\end{cases} \tag{9}$$

Let

$\mathbf{1}=(1,1,\cdots,1)^{\mathrm{T}}$, $\boldsymbol{c}=(c_{k+1},c_{k+2},\cdots,c_n)^{\mathrm{T}}$, $\boldsymbol{u}=(u_{k+1},u_{k+2},\cdots,u_n)^{\mathrm{T}}$, $\boldsymbol{c}_{-i}=(c_{k+1-i},c_{k+2-i},\cdots,c_{n-i})^{\mathrm{T}}$ $(1\le i\le p)$
$\boldsymbol{u}_{-j}=(u_{k+1-j},u_{k+2-j},\cdots,u_{n-j})^{\mathrm{T}}$ $(1\le j\le q)$, $\boldsymbol{X}=(\mathbf{1},\boldsymbol{c}_{-1},\boldsymbol{c}_{-2},\cdots,\boldsymbol{c}_{-p})$, $\boldsymbol{Y}=(\mathbf{1},\boldsymbol{u}_{-1},\boldsymbol{u}_{-2},\cdots,\boldsymbol{u}_{-q},\boldsymbol{c})$
$\boldsymbol{\alpha}(t_0)=(\alpha_0(t_0),\alpha_1(t_0),\cdots,\alpha_p(t_0))^{\mathrm{T}}$, $\boldsymbol{\beta}(t_0)=(\beta_0(t_0),\beta_1(t_0),\cdots,\beta_q(t_0),\gamma(t_0))^{\mathrm{T}}$

$$\boldsymbol{W}(t_0)=\mathrm{diag}\left(K_h(k+1-t_0),K_h(k+2-t_0),\cdots,K_h(n-t_0)\right).$$

We here assume that the inverse matrix of $X^{\mathrm{T}}W(t_0)X$ and $Y^{\mathrm{T}}W(t_0)Y$ exist for

each .Then the solution of the weighted least-squares problem (9), that is, the estimation of the vectors $\alpha(t_0)$ and $\beta(t_0)$ of the fuzzy coefficients can be obtained using matrix notation as

$$\begin{cases}\hat{\boldsymbol{\alpha}}(t_0)=\left(\boldsymbol{X}^{\mathrm{T}}\boldsymbol{W}(t_0)\boldsymbol{X}\right)^{-1}\boldsymbol{X}^{\mathrm{T}}\boldsymbol{W}(t_0)\boldsymbol{c}\\ \hat{\boldsymbol{\beta}}(t_0)=\left(\boldsymbol{Y}^{\mathrm{T}}\boldsymbol{W}(t_0)\boldsymbol{Y}\right)^{-1}\boldsymbol{Y}^{\mathrm{T}}\boldsymbol{W}(t_0)\boldsymbol{u}\end{cases} \tag{10}$$

From (10) we can see the factor $\Delta(L, f)$ is independent of the unknown parameters.

If the observed values of explanatory variables are known, we can obtain the fitted values of explained variables at t_0. Furthermore, performing the above estimation procedure at $t_0 = k + 1, k + 2, \ldots$, n respectively, we can obtain the estimation of explained variables during the whole study period.

$$\begin{cases} \hat{c}_t = \hat{\alpha}_0(t) + \hat{\alpha}_1(t)c_{t-1} + \cdots + \hat{\alpha}_p(t)c_{t-p} \\ \hat{u}_t = \hat{\beta}_0(t) + \hat{\beta}_1(t)u_{t-1} + \cdots + \hat{\beta}_q(t)u_{t-q} + \hat{\gamma}(t)c_t \end{cases}, t = k+1, k+2, \cdots, n \tag{11}$$

As in statistical nonparametric regression, two kinds of kernel functions are commonly used and one of them is Gaussian kernel:

$$K(t) = \frac{1}{\sqrt{2\pi}} \exp\left(-\frac{1}{2}t^2\right) \tag{12}$$

and the other is Beta kernel [8] :

$$K_\gamma(t) = \frac{1}{\mathrm{Beta}(1/2, \gamma+1)}\left(1-t^2\right)^\gamma I(|t| \le 1) \tag{13}$$

Here, we use the distance (3) used to fuzzify the cross-validation procedure [8] in statistics for selecting the optimal value of the smoothing parameter h, that is, let

$$\begin{aligned} CV(h) &= \sum_{t=k+1}^{n} d^2\left((c_t, u_t)_L, \left(\hat{c}_{(-t)}(h), \hat{u}_{(-t)}(h)\right)_L\right) \\ &= \sum_{t=k+1}^{n}\left(c_t - \hat{\alpha}_0^{(-t)}(t,h) - \sum_{i=1}^{p}\hat{\alpha}_i^{(-t)}(t,h)c_{t-i}\right)^2 \\ &\quad + \Delta(L,f)\sum_{t=k+1}^{n}\left(u_t - \hat{\beta}_0^{(-t)}(t,h) - \sum_{j=1}^{q}\hat{\beta}_j^{(-t)}(t,h)u_{t-j} - \hat{\gamma}^{(-t)}(t,h)c_{(-t)}(h)\right)^2 \end{aligned} \tag{14}$$

where $\hat{\alpha}_i^{(-t)}(t,h)\,(0 \le i \le p)$, $\hat{\beta}_j^{(-t)}(t,h)\,(0 \le j \le q)$ and $\hat{\gamma}^{(-t)}(t,h)$ are the resulting estimates of the centers

and spreads of the fuzzy coefficients under h through deleting the t-th observation and computing the estimates according to the restricted weighted least-squares described above. Thus,

$$\begin{cases} \hat{c}_{(-t)}(h) = \hat{\alpha}_0^{(-t)}(t,h) + \sum_{i=1}^{p} \hat{\alpha}_i^{(-t)}(t,h) c_{t-i} \\ \hat{u}_{(-t)}(h) = \hat{\beta}_0^{(-t)}(t,h) + \sum_{j=1}^{q} \hat{\beta}_j^{(-t)}(t,h) u_{t-j} + \hat{\gamma}^{(-t)}(t,h) c_{(-t)}(h) \end{cases}, t = k+1, k+2, \cdots, n \tag{15}$$

Then, select h_0 as the optimal value of the smoothing parameter such that $CV(h_0) = \min_{h>0} CV(h)$.

3.3.2. The Test of the FVCBR Model

As point out in [4], FBR model can be used to analyze the fuzzy financial yields series, but whether the analysis model with constant coefficients are enough to embody the dynamic changes of yields is a question worth thinking about, that is, whether the effects of explanatory variables on explained variables will significantly change or nor as the time goes by? For that reason, we need to test the constant coefficients hypothesis, and

$$H_0 : \boldsymbol{\alpha}(t) = \hat{\boldsymbol{\alpha}}, \boldsymbol{\beta}(t) = \hat{\boldsymbol{\beta}}$$

where $\tilde{\alpha}$ and $\tilde{\beta}$ are estimates of coefficients of constant coefficients model.

Here we use the generalized likelihood ratio (GLR) test [9]. Let

$$\begin{aligned} RSS_0 &= \sum_{t=k+1}^{n} \left(c_t - \hat{\alpha}_0 - \sum_{i=1}^{p} \hat{\alpha}_i c_{t-i} \right)^2 + \sum_{t=k+1}^{n} \left(u_t - \hat{\beta}_0 - \sum_{j=1}^{q} \hat{\beta}_j u_{t-j} - \hat{\gamma} c_t \right)^2 \\ &= (\boldsymbol{c} - \boldsymbol{X}\hat{\boldsymbol{\alpha}})^{\mathrm{T}} (\boldsymbol{c} - \boldsymbol{X}\hat{\boldsymbol{\alpha}}) + (\boldsymbol{u} - \boldsymbol{Y}\hat{\boldsymbol{\beta}})^{\mathrm{T}} (\boldsymbol{u} - \boldsymbol{Y}\hat{\boldsymbol{\beta}}) \end{aligned} \tag{16}$$

$$\begin{aligned} RSS_1 &= RSS_1^{(1)} + RSS_1^{(2)} \\ &= \sum_{t=k+1}^{n} \left(c_t - \hat{\alpha}_0(t) - \sum_{i=1}^{p} \hat{\alpha}_i(t) c_{t-i} \right)^2 + \sum_{t=k+1}^{n} \left(u_t - \hat{\beta}_0(t) - \sum_{j=1}^{q} \hat{\beta}_j(t) u_{t-j} - \hat{\gamma}(t) c_t \right)^2 \\ &= (\boldsymbol{c} - \boldsymbol{X}\hat{\boldsymbol{\alpha}}(t))^{\mathrm{T}} (\boldsymbol{c} - \boldsymbol{X}\hat{\boldsymbol{\alpha}}(t)) + (\boldsymbol{u} - \boldsymbol{Y}\hat{\boldsymbol{\beta}}(t))^{\mathrm{T}} (\boldsymbol{u} - \boldsymbol{Y}\hat{\boldsymbol{\beta}}(t)) \end{aligned} \tag{17}$$

Here RSS_0 and RSS_1 are residual sum of squares of null hypothesis and the whole space respectively. The GLR statistic is

$$T = \frac{n-k}{2} \ln \frac{RSS_0}{RSS_1} \tag{18}$$

Then, the asymptotic distribution of statistic can be generalized by the method of Bootstrap. Specific steps are as follows:

Step 1. Let $\hat{\sigma}_1^2 = \frac{1}{n-k} RSS_1^{(1)}$, and a series of random numbers obeyed $N\left(0, \hat{\sigma}_1^2\right)$ will be generated, that is,

$\tilde{\varepsilon}_t \sim N\left(0, \hat{\sigma}_1^2\right)$ where $t = k+1, \cdots, n$, and let $\tilde{c} = X\hat{\alpha} + \tilde{\varepsilon}$

Let $\hat{\sigma}_2^2 = \frac{1}{n-k} RSS_1^{(2)}$, and a series of random numbers obeyed $N\left(0, \hat{\sigma}_2^2\right)$ will be generated, that is,

$\tilde{e}_t \sim N\left(0, \hat{\sigma}_2^2\right)$ where $t = k + 1, \ldots, n$, and let $\tilde{u} = Y\hat{\beta} + \tilde{e}$;

Step 2. Use the sample data to construct GLR statistics $\tilde{T}$;

Step 3. Repeat Step 1 and Step 2 m times and then get m GLR statistics $\tilde{T}_1, \tilde{T}_2, \cdots, \tilde{T}_m$;

Step 4. The asymptotic distribution of GLR statistic T of null hypothesis is expressed by the empirical distribution of $\tilde{T}_1, \tilde{T}_2, \cdots, \tilde{T}_m$, that is

$$\hat{F}_m(t) = \frac{1}{m} \sum_{i=1}^{m} I\left\{\tilde{T}_i \le t\right\} \tag{19}$$

4. EMPIRICAL ANALYSIS

Now, we will fit and forecast the fuzzy financial yields with the FVCBR model. The database consists of 119observations of the SSE Composite index from July 17, 2014 to January 9, 2015. Assume that $\left\{\left[x_t^l, x_t^u\right], t = 1, 2, \cdots, 119\right\}$ is a series of interval observations of the SSE Composite index, in which x_t^l and x_t^u represent the maximum and minimum of the SSE Composite index at time t respectively, as shown in Figure 1. In order to obtain the yields series $\left\{\tilde{r}_t\right\}, t = 2, 3, \cdots, 119$, as shown in Figure 2, we deal with the fuzzy series $\left\{\tilde{x}_t\right\}$ by logarithmic transformation and then first order difference. Further, we divide the data

$\{(c_t, u_t)_L, t = 2, \cdots, 119\}$ into two parts. One is the fitted samples $\{(c_t, u_t)_L, t = 2, \cdots, 104\}$ and the other is the forecast samples $\{(c_t, u_t)_L, t = 105, \cdots, 119\}$, and we will test the empirical results and prediction abilities of models in and out of the sample period.

4.1. The Varying Coefficient Bilinear Regression Model of Fuzzy Financial Yields Series

In Figure 1 we can see that the observations of the SSE Composite index $\{[x_t^l, x_t^u], t = 1, 2, \cdots, 119\}$ show an obvious increase with the time. Then, we adopt the method of run test to test the series of centers $\{x_t\}$ of $\{[x_t^l, x_t^u], t = 1, 2, \cdots, 119\}$ and the results shown in the column 2 of Table 1 rejecting the null hypothesis with sig-nificant level 5% suggests that the sequence of $\{x_t\}$ is non-stationary, in other words, the sequence of the SSE Composite index $\{x_t\}$ is non-stationary.

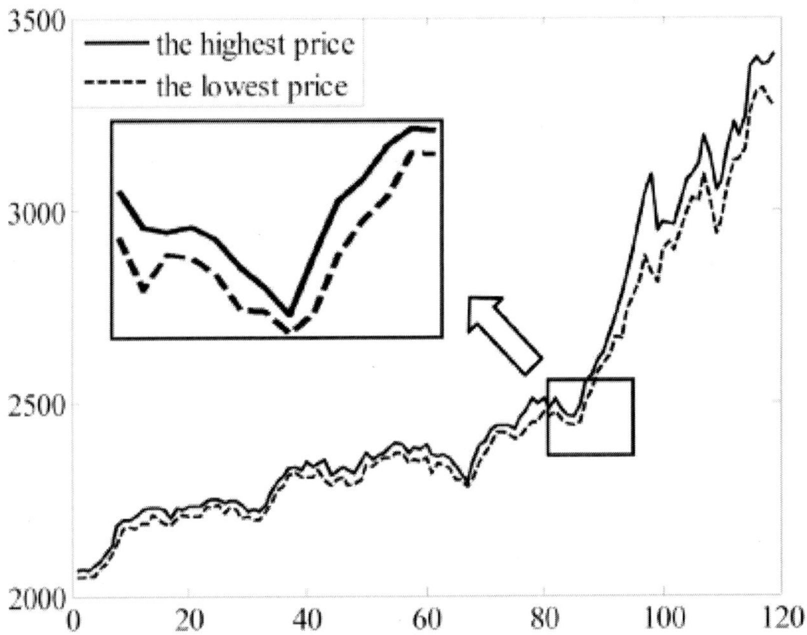

Figure 1. The series of interval observations of the SSE Composite index SSE Composite index (2014.7.17-2015.01.09) (2014.7.17-2015.01.09).

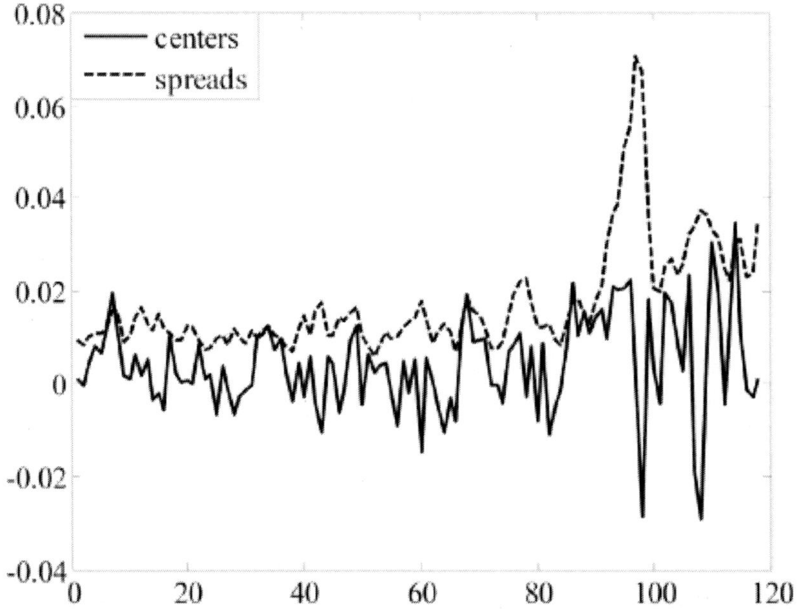

Figure 2. The fuzzy yields series of the SSE Composite index SSE Composite index (2014.7.17-2015.01.09) (2014.7.17- 2015.01.09).

Table 1. The fuzzy time series of the SSE Composite index/the run test of centers of fuzzy yields series.

	Centers of fuzzy time series	Centers of fuzzy yields series
Test values	2353.99300	0.00394
The number of samples that are less than test values	59	59
The number of samples that are equal or greater than test values	60	59
Simple size	119	118
Number of runs	4	51
Z	−10.403	−1.664
P value	0.000	0.096

Note: the test values are the averages of each sequence.

As can be seen in Figure 2, the sequence of yields $\{\tilde{r}_t\}$ has been detrended fluctuation basically and ranges about from −0.04 to 0.04. Then, we also employ the method of run test to the centers $\{c_t\}$ of $\{\tilde{r}_t\}$, and the results shown in the column 3 of Table 1 cannot reject the null hypothesis with significant level 5%, that is, the sequence $\{c_t\}$ is stationary so that we can infer $\{\tilde{r}_t\}$ is conditional stationary.

We can establish the FVCBR model for the conditional stationary financial yields $\{\tilde{r}_t\} = \{(c_t, u_t)_L\}$. Here we will not consider the order determination of the model and only modelling the fuzzy yields sequence of the SSE Composite index $\{\tilde{r}_t\} = \{(c_t, u_t)_L\}$ with FVCBR (1,1) and finally discuss the imitative effect.

$$\begin{cases} c_t = \alpha_0(t) + \alpha_1(t)c_{t-1} + \varepsilon_t \\ u_t = \beta_0(t) + \beta_1(t)u_{t-1} + \gamma(t)c_t + e_t \end{cases}, t = 2, 3, \cdots, 104 \tag{20}$$

We select Epanechnikov kernel, Briweight kernel, Triweight kernel and Gaussian kernel to analysis respectively, and in turn denote above kernel functions as K_1, K_2, K_3 and K_4. For each kernel function, we use the cross-validation procedure to find the right *h*.

Suppose that *h* is selected from 1 to 20 and steps by 0.2. Figure 3(a) illustrates the CV values computed by each of the kernel functions change with *h*. From Figure 3(a) we can see that the CV values of all kernel functions decrease first and then increase with the increase of *h*. Therefore, under the assumption of that *h* is chosen from 1 to 20, the optimum *h* for K_4 is 2 and are 4 for K_1, K_2, K_3 respectively. In order to find more better *H*, we further let *h* steps by 0.05 from 1 to 5, and the relationship between CV values and *h* is shown in Figure 3(b). So we can obtain the right *h* for every kernel functions are respectively 3.18, 3.82, 4.36 and 1.82.

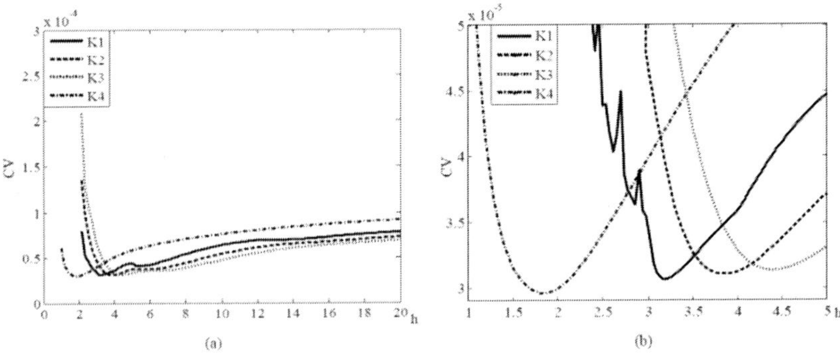

Figure 3. The changes of CV values of every kernel function with h increasing (a) h is from 1 to 20; (b) h is from 1 to 5.

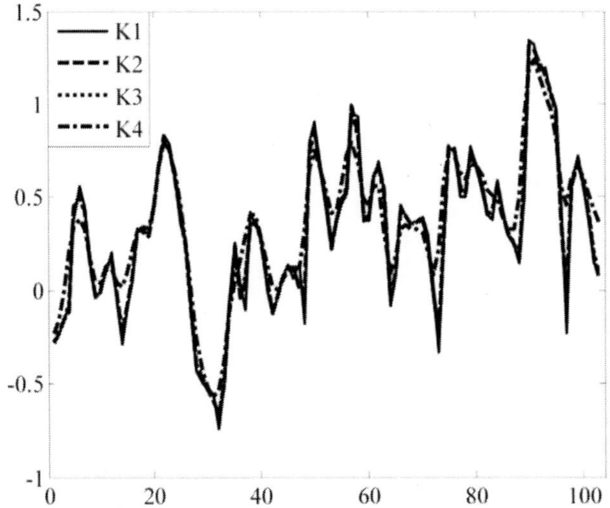

(a)

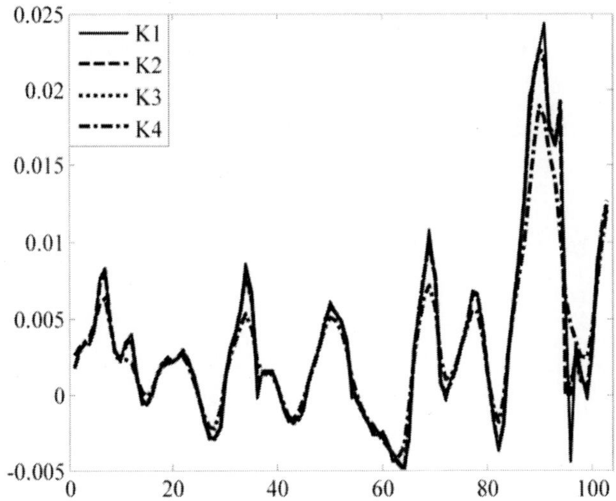

(b)

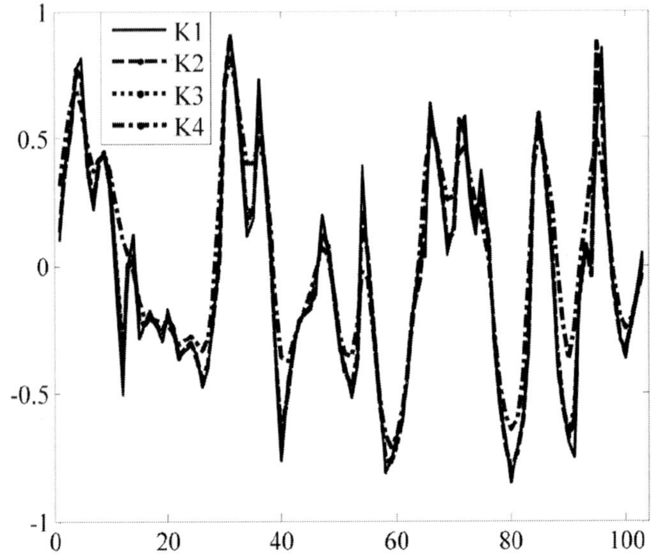
K1
K2
K3
K4
1
0.5
0
-0.5
-1
0
20
40
60
80
100

(c)

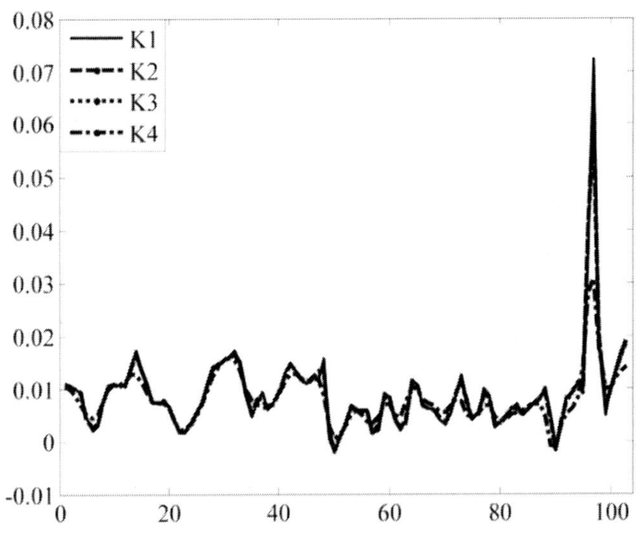
K1
K2
K3
K4
0.08
0.07
0.06
0.05
0.04
0.03
0.02
0.01
0
-0.01
0
20
40
60
80
100

(d)

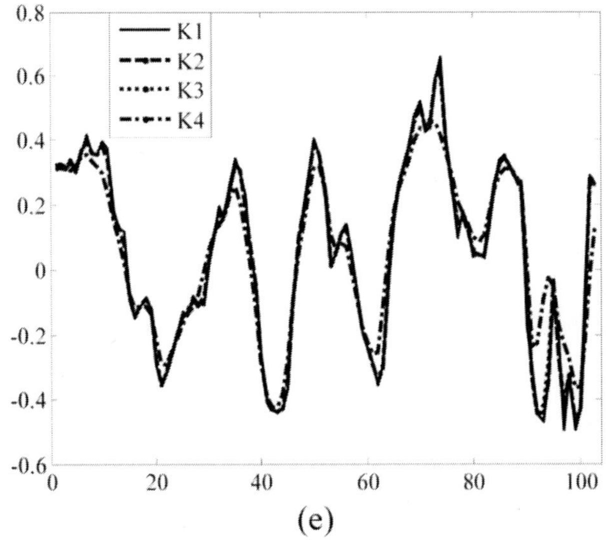

Figure 4. The estimates of regression coefficients of model (20). α_0; (*b*) α_1; (*c*) β_0;(*d*) β_1; (*e*) γ

For every kernel function, we select the optimal bandwidth and then obtain the estimates of regression coefficients of model (20) at any point in time. Figure 4 shows the estimate values of regression coefficients of different kernel functions and also indicates that the regression coefficients vary similarly with the time although the different kernel functions, that is, different kernel function has little effect on the estimates values of coefficients. Based on this result, we instead considering Gaussian kernel as the kernel function in this article. As we can see from the estimates of model (20) and Figure 4, the spreads of fuzzy yields have positive correlation to its first order lagging values while the relationship between the centers of fuzzy yields and its first order lagging values as well as the relationship between the spreads and current centers remain positive or negative in the dynamic changes.

Next, we will test the hypothesis of the regression coefficients are constants by using Gaussian kernel as the selected kernel function. Based on the data of SSE composite index and formula (18), we can get the generalized likelihood ratio (GLR) statistic T = 42.6781. When repeat the Step 1 and Step 2 of Bootstrap method for 100 times, that is, m = 100, the curve of asymptotic distribution of GLR statistic is as shown in Figure 5(a) and at this time the p value being lower than 0.01 suggests that the result reject the null hypothesis with significant level 0.01. Similarly, when m = 1000, the curve of asymptotic distribution of GLR statistic is as shown in Figure 5(b), and in this case the p value being lower than 0.001 illustrates the result reject the null hypothesis with significant level 0.001. In conclusion, we reject the null hypothesis and hold that the regression coefficients are change with time. As a result, the fuzzy varying coefficient regression model is a better choice.

4.2. Forecasting and Evaluation of Simulation

4.2.1. Forecasting

We forecast the real data from 105-th to 119-th with one-step-ahead prediction by using the model (20) and formula (10). The results are as shown in Table 2.

4.2.2. Evaluation of Simulation of the Model

In order to compare to the prediction of FAR model and FBR model, we respectively calculate the absolute error of prediction and show them with curves (as see in Figure 6) of FAR model, FBR model and FVCBR model.

As we can see from Figure 6(a), when predict the centers, the absolute errors of FVCBR model are a little greater than FAR model and FBR model only on the 1st, the 2nd, the 5th and the 9th periods; meanwhile, when predict the spreads, the predicted values are all much close to the observed values at every period except the 1st, the 12th and the 14th periods. So, we preliminary infer that the predictions of FVCBR model are more precise than FAR model and FBR model.

For the purpose of judging the predictions and evaluations of simulation of the three models more accurately, we use the mean square error (MSE) and mean absolute error (MAE) that are often used in regression analysis to evaluate the fitting effects and prediction accuracy. Specifically, we calculate the MSE and MAE of financial yields in and out of sample period respectively and the results are shown in Table 3.

From Table 3, we can see no matter in and out of the sample period, the MSE and MAE of FVCBR model are all lower than FAR model and FBR model. This result proves that the fitting effects and prediction accuracy of FVCBR model are superior to FAR model and FBR model.

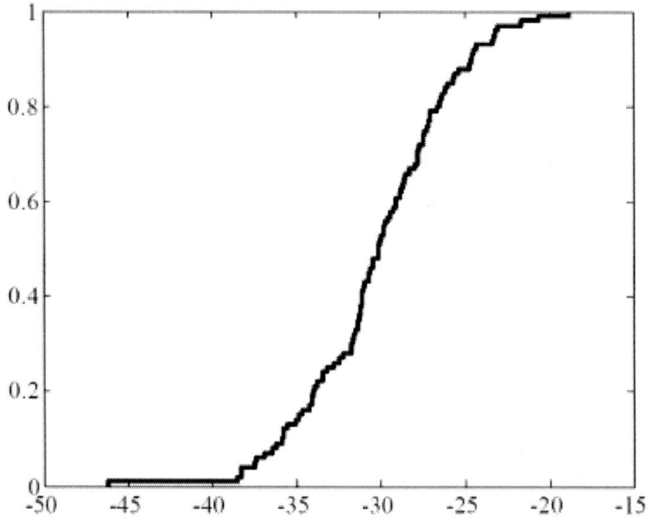

(a)

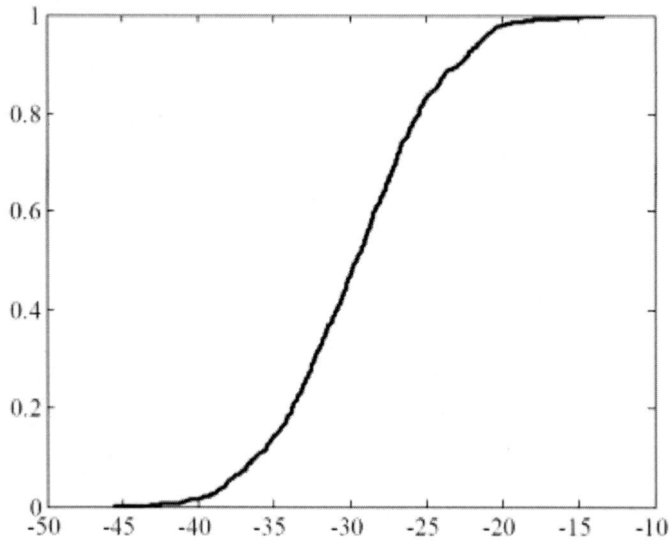

(b)

Figure 5. The asymptotic distribution of GLR statistic. (a) m = 100; (b) m = 1000.

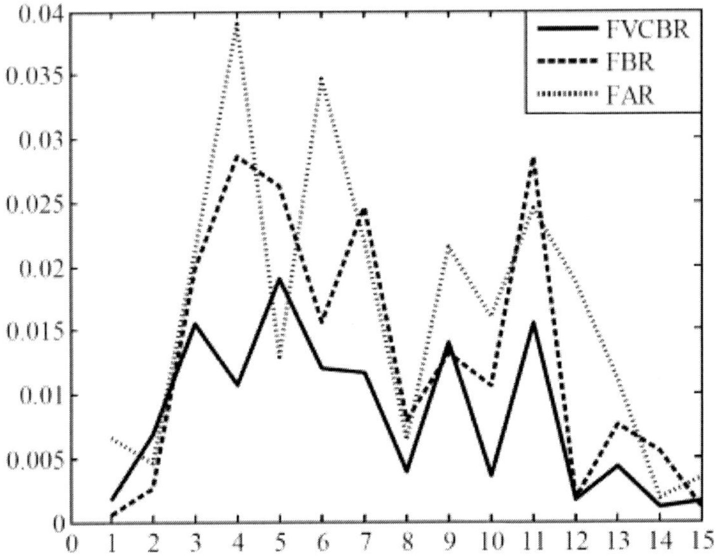

(a)

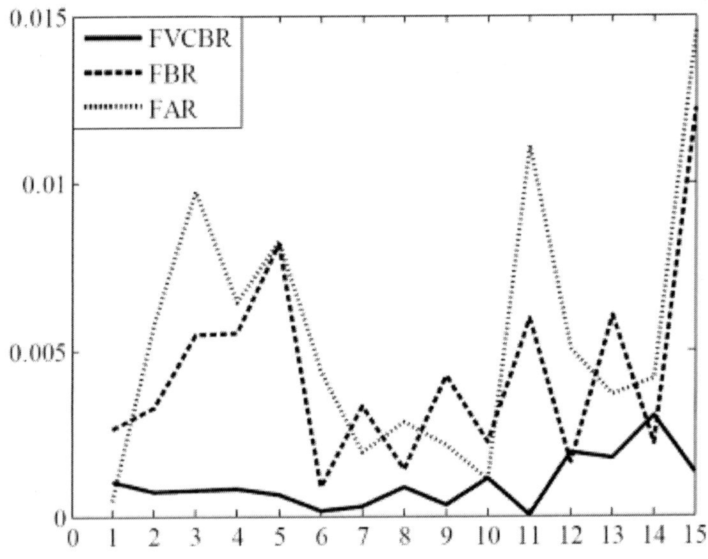

(b)

Figure 6. The comparisons of predictions of FAR model, FAR model and FVCBR model. (a) The sequence of centers; (b) The sequence of spreads.

Table 2. The results of one-step-ahead prediction of fuzzy varying coefficient bilinear regression model[3].

Periods	105	106	107	108	109
Observed values	(0.0083, 0.0234)	(0.0025, 0.0259)	(0.0233, 0.0320)	(−0.0190, 0.0339)	(−0.0292, 0.0374)
Predicted values	(0.0065, 0.0245)	(0.0093, 0.0252)	(0.0077, 0.0312)	(−0.0083, 0.0347)	(−0.0101, 0.0367)
Periods	110	111	112	113	114
Observed values	(0.0097, 0.0364)	(0.0302, 0.0332)	(0.0195, 0.0313)	(−0.0047, 0.0247)	(0.0119, 0.0223)
Predicted values	(−0.0024, 0.0363)	(0.0186, 0.0335)	(0.0156, 0.0304)	(0.0092, 0.0244)	(0.0155, 0.0235)
Periods	115	116	117	118	119
Observed values	(0.0347, 0.0303)	(0.0112, 0.0310)	(−0.0015, 0.0230)	(−0.0031, 0.0238)	(0.0007, 0.0350)
Predicted values	(0.0192, 0.0303)	(0.0130, 0.0291)	(0.0028, 0.0247)	(−0.0020, 0.0268)	(−0.0008, 0.0338)

Table 3. The errors measurements of fitting and forecasting of SSE composite index.

Model	In the sample period		Out of the sample period	
	MSE	MAE	MSE	MAE
FVCBR model	4.19E−05	0.0061	1.11E−04	0.0098
FBR model	9.46E−05	0.0099	3.15E−04	0.0185
FAR model	1.26E−04	0.0111	4.56E−04	0.0233

5. CONCLUSION

This article introduces the fuzzy financial yields series to deal with the interval observed samples in financial markets and constructs the fuzzy varying coefficient bilinear regression (FVCBR) model with satisfying the practical significance of financial yields. Besides, based on the complete distance between fuzzy numbers, we develop the fuzzy least squares method to obtain the estimates of the unknown coefficients. Empirical analysis shows that compared with constant coefficient regression model and fuzzy auto regression model, the varying coefficient regression has shown some improvements no matter in fitting effects or prediction. The fuzzy auto regression model has certain limitations in model fitting and forecasting because the auto regression of centers and spreads are considered independently rather than taking the effect that centers have on the spreads into account. Fuzzy varying coefficient bilinear regression explores the problem of financial fuzzy time series more flexibly and applicatively to make the fitted values and predictions be intervals and thus more consistent with the description of real financial market. Finally, this article only discusses the uncertainty of changes of financial assets price, that is, reflects the changes of yields only by fuzzy data but ignores the probability distribution of parameters. So, later research will focus on the coefficient estimates, statistical tests, model fitting and forecasting after introducing the random error, so as to give deciders more perspective to recognize and explain the changes of financial markets.

REFERENCES

1. Zadeh, L.A. (1965) Fuzzy Sets. Information and Control, 8, 338-353. http://dx. doi.org/10.1016/S0019-9958(65)90241-X
2. Li, Z.-Y., Zhang, C. and Wang, T.-J. (2010) Studying on the Interval Financial Times Series and Evaluating on the Forecast. Journal of Applied Statistics and Management, 29, 129-136.
3. D'Urso, P. and Gastaldi, T. (2000) A Least-Squares Approach to Fuzzy Linear Regression Analysis. Computational Statistics and Data Analysis, 34, 427-440. http://dx.doi.org/10.1016/S0167-9473(99)00109-7
4. Li, Z.-Y., Liu, W.-Y. and Wang, T.-J. (2009) Fuzzy Bilinear Regression of Yields Series. Statistical Research, 26, 68-73.
5. Wang, H.-D., Guo, S.-C. and Yue, L.-Z. (2014) An Approach to Fuzzy Multiple Linear Regression Mode Based on the Structured Element Theory. Systems Engineering—Theory & Practice, 34, 2628-2636.
6. Hu, B.-Q. (2010) Foundations of Fuzzy Theory. Wuhan University Press, Wuhan, 103-114.
7. Xu, R.N. (1991) A Linear Regression Model in Fuzzy Environment. Advances in Modelling Simulation, 27, 31-40.
8. Fan, J.Q. and Yao, Q.W. (2003) Nonlinear Time Series. Springer, Berlin, 243-245.
9. Cai, Z.W., Fan, J.Q. and Yao, Q.W. (2000) Functional-Coefficient Regression Models for Nonlinear Times Series. Journal of the American Statistical Association, 95, 941-956. http://dx.doi.org/10.1080/01621459.2000.10474284

NOTES

[1]It is proved by proposition 1.

[2]It is proved by the density of the set of rational number in the set of real numbers.

[3]The regression coefficients in one-step-ahead prediction are previous estimates.

CHAPTER 5

An Automated Multiscale Ensemble Simulation Approach for Vascular Blood Flow

Mohamed A. Itani[a,b], Ulf D. Schiller[b], Sebastian Schmieschek[b], James Hetherington[b], Miguel O. Bernabeu[b], Hoskote Chandrashekar[c], Fergus Robertson[c], Peter V. Coveney[b], Derek Groen[b,d,*]

[a]*Department of Electrical & Computer Engineering, American University of Beirut, P.O. Box 11-0236, Beirut, Lebanon*

[b]*Centre for Computational Science, University College London, 20 Gordon Street, WC1H 0AJ London, United Kingdom*

[c]*Lysholm Department of Neuroradiology, National Hospital for Neurology and Neurosurgery, University College London, London, United Kingdom*

[d]*CoMPLEX, University College London, Physics Building, Gower Street, London WC1E 6BT, United Kingdom*

ABSTRACT

Cerebrovascular diseases such as brain aneurysms are a primary cause of adult disability. The flow dynamics in brain arteries, both during periods of rest and increased activity, are known to be a major factor in the risk of aneurysm formation and rupture. The precise relation is however still an open field of investigation. We present an automated ensemble simulation method for modelling cerebrovascular blood flow under a range of flow regimes. By automatically constructing and performing an ensemble of multiscale simulations, where we unidirectionally couple a 1D solver with a 3D lattice-Boltzmann code, we are able to model the blood flow in a patient artery over a range of flow regimes. We apply the method to a model of a middle cerebral artery, and find that this approach helps us to fine-tune our modelling techniques, and opens up new ways to investigate cerebrovascular flow properties.

Keywords: Multiscale modelling, Blood flow, Ensemble simulation, Parallel programming, High-performance computing

1. INTRODUCTION

Stroke is a major cause of death and morbidity in the developed world. Subarachnoid haemorrhage (SAH) is a type of stroke characterised by bleeding into the fluid around the brain, for example due to the rupture of an intracranial aneurysm. An aneurysm is a congenital weakness in a blood vessel wall which gradually bulges out to form a balloon which can eventually burst. SAHs represent 5% of cases of stroke, but is relatively more important, as the mortality rate for these events is about 50%. Overall, approximately 5–10 people per 100,000 are affected by SAH due to bleeding in the intracranial arterial wall. [1]The mean age of the victims is 50 years and 10–15% fail to reach hospital. Unruptured aneurysms are much more prevalent, estimated to affect 1–5% of the population of the UK [2]. Indeed, unruptured/asymptomatic cerebral aneurysms are a relatively common finding when scanning the brain for other reasons [1]. Current methods of determining which aneurysms have a significant risk of subsequent rupture are based on crude measures such as aneurysm size and shape, and there is a clear need for a non-invasive tool to stratify risk more effectively in this large patient group.

Computational fluid dynamics (CFD) techniques may provide means to help quantification of the rupture risk, if they can incorporate the key conditions affecting brain aneurysms. Particularly high or low wall shear stress is believed to increase the risk of aneurysm rupture [3]. Researchers increasingly apply computational fluid dynamics to investigate these problems [4–6], and in particular Shojima et al. concluded that both a very high and a very low wall shear stress increases the chance of aneurysm growth and rupture in MCA aneurysms [7]. In alignment with these research efforts, we seek to establish computational diagnosis and prediction techniques, which may lead to major health benefits and reduce the costs of health care in the long term.

An essential driver for these CFD calculations is the flow solver, and over the last decade several sophisticated and scalable solvers have emerged. Within this work we rely on HemeLB (described in Section 2.1), which is highly optimized for modelling sparse geometries and has unique optimizations which allow it to achieve excellent load balance in the presence of complex boundary and in- and outflow conditions [8]. There are several other scalable flow solvers that are worth mentioning as well. These include the Nektar finite element package [9–11], the Palabos package [12–14], the Musubi environment [15,16], MuPhy [17] and WaLBerla [18]. Although the aforementioned works have provided valuable insight into the haemodynamic environment of brain aneurysms, little is known about how the intrinsic variablity of blood flow throughout the day affects aneurysm growth and rupture.

The purpose of this paper is to present a tool which automatically creates an ensemble of multiscale blood flow simulations based on a set of clinically measurable patient parameters, and runs these simulations using supercomputing resources. The tool allows us to automate the study of the blood flow in a vascular geometry under varying patient-specific conditions. In addition, an automated data processing component extracts velocity and wall shear stress (WSS) values, and generates plots and animations which allow us to visualize these properties in the vascular geometry. This paper builds on previous works where we simulated flow in arterial networks using a single flow configuration [19–21].

To showcase our approach, we construct and execute a range of multiscale simulations of a middle cerebral artery (MCA). We present the results of these simulations, and compare our approach to related efforts as an initial validation of our 1D-3D multiscale scheme. This work is organised as follows. In Section 2, we present the tools we developed to perform our multiscale ensemble simulations and how we integrate them in an automated workflow. We describe the setup of our simulation in Section 3, our results in Section 4 and provide a brief discussion in Section 5.

2. AUTOMATED MULTISCALE ENSEMBLE SIMULATIONS

Our automated workflow combines three existing components. These include the HemeLB and pyNS simulation environments, and the FabHemeLB automation environment. In this section we describe these three components, and how they interoperate in our automated multiscale ensemble simulation (MES) environment.

2.1. HemeLB

HemeLB is a 3 dimensional lattice-Boltzmann simulation environment developed to simulate fluid flow in complex systems. It is a MPI parallelised C++ code with world-class scalability for sparse geometries. It can efficiently model flows in sparse cerebral arteries using up to 32,768 cores [22,23] and utilises a weighted domain decomposition approach to minimize the overhead introduced by compute-intensive boundary and in-/outflow conditions [8]. HemeLB allows users to obtain key flow properties such as velocity, pressure and wall shear at predefined intervals of time, using a property-extraction framework.

HemeLB has previously been applied to simulate blood flow in healthy brain vasculature as well as in the presence of brain aneurysms [20,24]. Segmented angiographic data from patients is read in by the HemeLB Setup Tool, which allows the user to visually indicate the geometric domain to be simulated. The geometry is then discretized into a regular unstructured grid, which is used as the simulation domain for HemeLB. HemeLB supports predefined velocity profiles at the inlets of the simulation domain, which we generate using pyNS in this work.

2.2. pyNS: Python network solver

pyNS is a discontinuous Galerkin solver developed in Python, which simulates haemodynamic behaviour in vascular networks [25]. pyNS uses aortic blood flow input based on a set of patient-specific parameters, and combines one-dimensional wave propagation elements to model arterial vasculature with zero-dimensional resistance elements to model veins. The solver requires two XML files as input data, one with a definition of the vasculature and one containing the simulation parameters. Simulation parameters include mean blood pressure, cardiac output, blood dynamic viscosity and heart rate. pyNS has been used in several studies, e.g. to try to inform treatment decisions on haemodialysis patients [26] and as a large-scale model for distributed multiscale simulations of cerebral arteries [19].

2.3. FabHemeLB

FabHemeLB is a Python tool which helps automate the construction and management of ensemble simulation workflows. FabHemeLB is an extended version of FabSim [27] configured to handle HemeLB operations. Both FabSim and FabHemeLB help to automate application deployment, execution and data analysis on remote resources. FabHemeLB can be used to compile and build HemeLB on any remote resource, to reuse machine-specific configurations, and to organize and curate simulation data. It can also submit HemeLB jobs to a remote resource specifying the number of cores and the wall clock time limit for completing a simulation. The tool is also able to monitor the queue status on remote resources, fetch results of completed jobs, and can conveniently combine functionalities into single one-line commands. In general, the FabHemeLB commands have the following structure:

fab < target machine >< command >: < parameter > = < value >, ...

For example:

fab archer ensemble:config= /path/to/ config,cores= 1536, wall_time =05:00:00

In Table 1 we present a number of commands typically executed with FabHemeLB in the scope of this work. The commands are customised to run on local machines, continuous integration servers, regional, national or international supercomputing resources. The workflow is presented in the diagram of Fig. 1. Specifically, the *analyze* command processes the compressed output files to generate human readable files and visualisations. It also generates an image file showing the whole geometry, wall shear stress within the geometry, and velocity measurements on pre-selected planes inside the geometry over time as an animation.

Table 1. List of commands commonly used in FabHemeLB.

Command name	Brief description
cold	Copy HemeLB source to remote resource, compile and build everything.
run_pyNS	Execute instances of pyNS to generate a range of flow output files.
generate_LB	Convert pyNS output to HemeLB input.
submit_LB	Given a set of velocity profiles, submits the corresponding HemeLB jobs to the remote (supercomputer) resource.
fetch_results	Fetch all the simulation results from the remote resource and save them locally.
analyze	Performs data-analysis that allows for easy visualization of the results.
ensemble	Do all of the above, except cold.

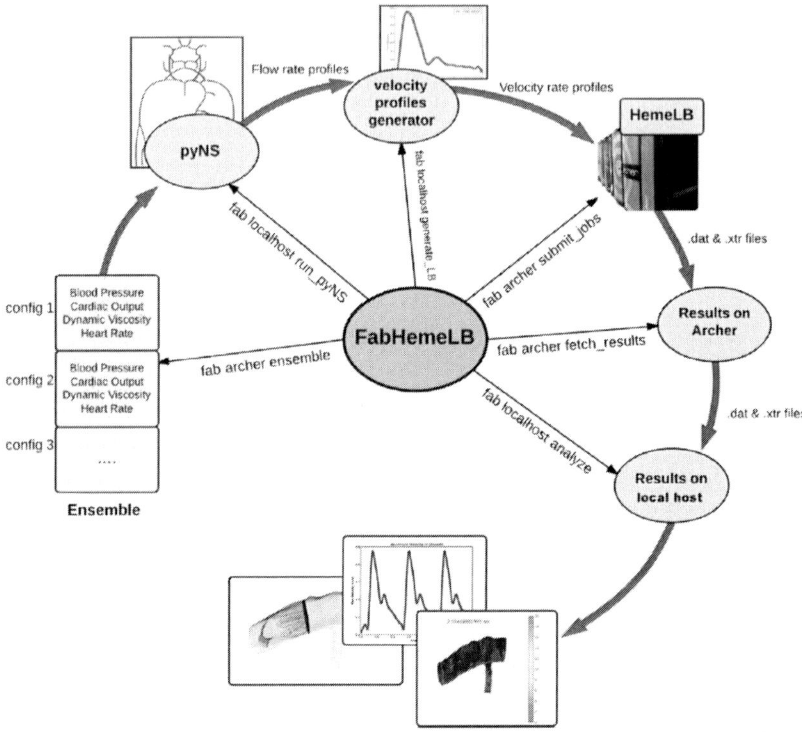

Figure. **1**. Workflow diagram showing the processes involved in the ensemble simulation method. The simulations were distributed, with PyNS simulations using a local workstation in London, and the HemeLB simulations using the ARCHER supercomputer.

3. SETUP

To apply our automated ensemble simulation tool, we employ patient-specific parameters measured by Sugawara et al. during a study to assess cardiac output during exercise [28]. They measured the blood pressure, cardiac output and heart rate of 16 young patients (9 male and 7 female) at different exercise intensities, being at rest or at 70%, 90%, 110% and 130% of the ventilatory threshold (VT). The VT is the point during exercise training at which pulmonary ventilation becomes disproportionately high with respect to oxygen consumption. This is believed to reflect onset of anaerobiosis and lactate accumulation. We add two more sets of values at 30% and 50% VT, linearly extrapolating the other parameters. The resulting seven sets of parameters are used in our automated workflow, resulting in seven HemeLB simulations being run. We present the seven sets of parameters in Table 2.

Table 2. Configuration data used as input for pyNS to run the ensemble of simulations. The values are based on the average of the measurements of 16 people at different exercise intensities measured by the percentage of the ventilatory threshold(VT) [28]. In the last column, we provide the mean flow velocity in the right MCA, as calculated using PyNS, for each configuration.

Configuration	Exercise intensity	Blood pressure mean (mmHg)	Cardiac output (L/min)	Heart rate (bpm)	Mean flow velocity (m s^{-1})
1	Rest	80	4.8	68	0.460
2	30% VT	87	6.2	79	0.451
3	50% VT	94	7.6	90	0.428
4	70% VT	100	9	101	0.393
5	90% VT	112	10.7	113	0.371
6	110% VT	116	11.9	120	0.351
7	130% VT	122	13.2	134	0.339

For our pyNS simulations, which we ran for 10 cardiac cycles, we set the blood density to 1050 kg/m^3, Poisson's ratio of transverse to axial strain to 0.5 and the time step to 5 ms. For our HemeLB runs we use a model derived from a patient-specific angiographic 3D geometry of a middle cerebral artery, supplied by the Lysholm Department of Neuroradiology, University College Hospital, London and segmented using the GIMIAS tool [29]. We use a voxel size of 18.9 μm, which results in a geometry containing 13,179,961 lattice sites. In Fig. 2 we show the setup tool interface with the MCA geometry. Here the inlet is given by the green plane and the outlets by the red planes. The location of interest for our WSS analysis is highlighted. For simulations of this particular voxelized simulation domain, we specify a time-step of 0.5014 μs and run each simulation for 7.9 million steps, or 4 s of simulated time.

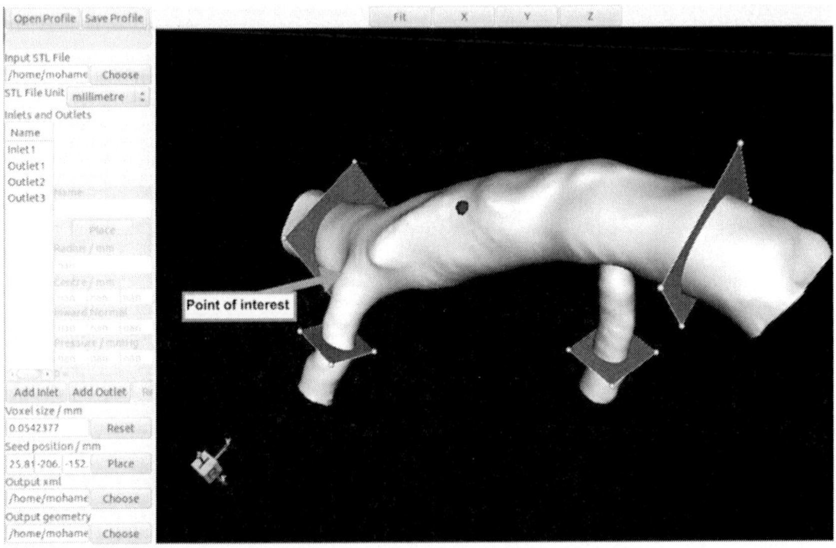

Figure. 2. HemeLB Setup Tool loaded with a middle cerebral artery geometry. The inlet is indicated by a green surface, and the three outlets with a red surface. The lattice site which we used for in-depth wall-shear stress analysis is highlighted with a light blue arrow. (For interpretation of the references to color in this figure legend, the reader is referred to the web version of the article.)

During our HemeLB runs, we store the WSS throughout the geometry for every 50,000 time steps. In addition, we define an output plane close to the outlet, at 49 mm from the ear, 2mm away from the outflow boundary (or *outlet*), to record velocity and pressure data every 50,000 steps. In all our runs we use interpolated Bouzidi wall boundary conditions [30] and zero-pressure outlet conditions (for details see [22]). Using the *ensemble* command, we run an ensemble of 7 simulations on the ARCHER supercomputer. In total, we used 10,752 (7 × 1536) cores for a duration of approximately 3.5 h.

4. RESULTS

We present the time series of the maximum velocity in the measurement plane in Figs. 3 and 4 for the parameters listed in Table 2. The curves demonstrate that the frequency of the 1D cardiac cycles generated by pyNS are accurately reproduced in the 3D high-resolution HemeLB simulations. The peak velocity in the measurement plane is higher than the input velocity because the measurement plane has a lower area but the total flux remains constant throughout the MCA due to the incompressible flow.

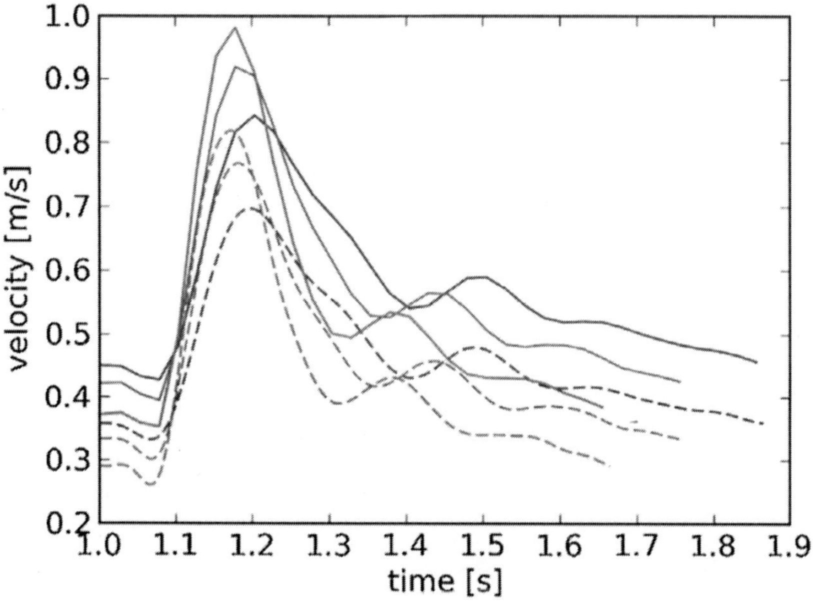

Figure. 3. Velocity time-series for the configurations 1 (bottom lines) to 3 (top lines). Dashed lines indicate the velocity time-series generated for the inflow boundary by pyNS, and solid lines correspond to the maximum velocity measured at the 49 mm plane in HemeLB.

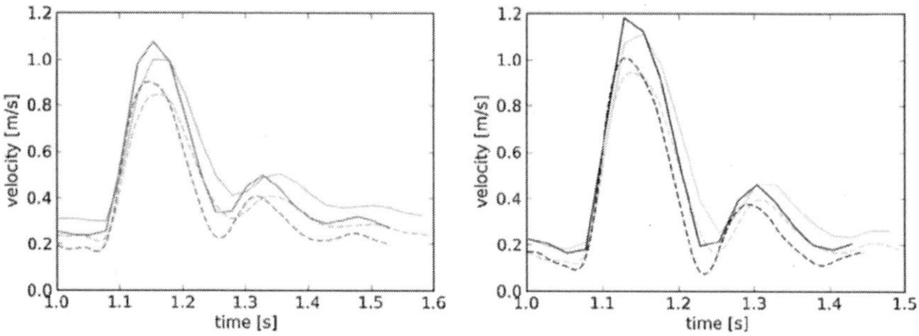

Figure. 4. Left: Velocity time-series as in Fig. 3 but for configurations 4 (bottom line) and 5 (top line). Right: Velocity time-series as in Fig. 3 but for configurations 6 (bottom line) and 7 (top line).

Within the ensemble, the frequency of the cardiac cycles increases with exercise intensity as the heart rate increases. While this seems trivial, it indicates that the time-resolution chosen to couple pyNS and HemeLB is sufficient to reproduce this effect. The good match between the input and the measurement gives confidence that the LB simulations have converged and the cardiac cycles are stable.

4.1. Wall shear stress

Fig. 5 shows snapshots of the WSS for four configurations of the ensemble at the peak systole, when the flow velocity is highest. The colour scale allows us to identify regions of high WSS which are located predominantly at constrictions of the MCA and around the inlet in an assymetric pattern. Based on the WSS values near the inlet, we conclude that the standard circular-shaped Poisseuille velocity profile used in HemeLB is ill suited for patient-specific geometries, as they usually feature non-circular inlets. As a result, we are now developing a method for a modified velocity profile which takes non-circular inlet shapes into account.

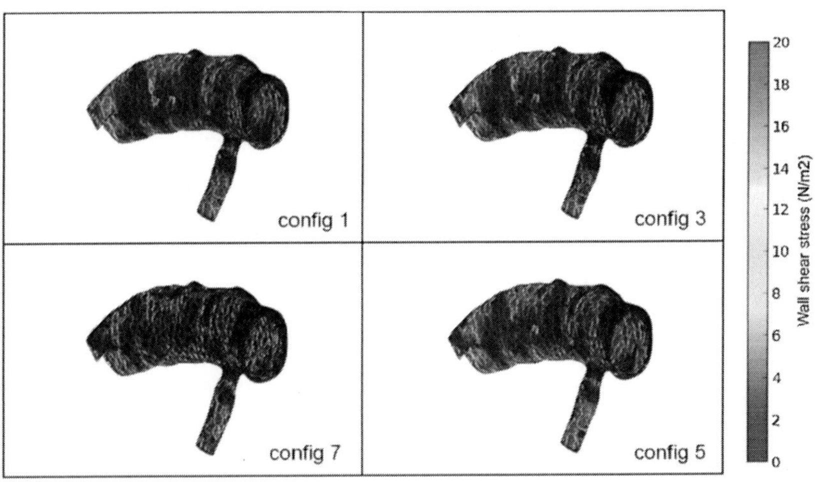

Figure. 5. A snapshot of the wall shear stress for the configurations 1, 3, 5 and 7 at the peak systole. Red regions indicate high wall shear stress while blue indicates low wall shear stress. Exercise intensity is increasing clockwise for the configurations shown. (For interpretation of the references to color in this figure legend, the reader is referred to the web version of the article.)

At the point of interest indicated in Fig. 2, we observe much higher wall shear stress at higher exercise intensity. We present a cross-instance analysis of the WSS at this point in Fig. 6. Here we provide for instance the mean WSS, which decreases linearly with the mean velocity at the inlet (see Table 2, both values are reduced to ~0.75 times the magnitude at rest, when measured at 130% VT exercise intensity). This matches the theoretically expected linear scaling of the WSS with the velocity parallel to the wall. This parallel velocity is expected to be proportional to the velocity at the inlet for simple geometries and flow regimes. Our WSS results are in line with related literature, which report maximum values in MCAs in the range of 14–40 N m^{-2} for unruptured aneurysm geometries [7,31].

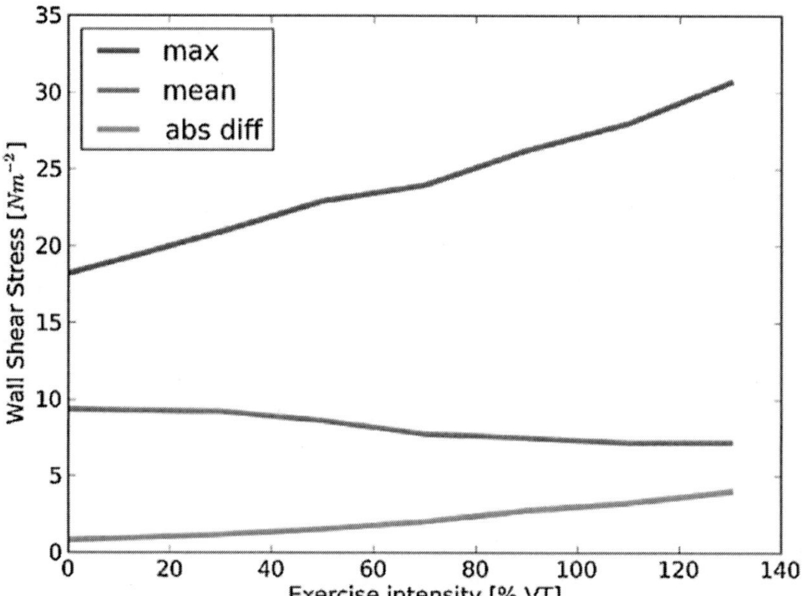

Figure. 6. Cross-instance analysis of the WSS for the location of interest indicated in Fig. 2. We extracted WSS values in the simulations for a range of 3 cardiac cycles (when at rest) up to 6 cardiac cycles (at 130% VT). We present the maximum measured WSS (blue line), the average WSS (green), and the time average of the WSS slope, calculated over intervals of 0.025 s (red). The WSS values at rest can be found on the left side (0% VT). (For interpretation of the references to color in this figure legend, the reader is referred to the web version of the article.)

We also present the maximum WSS in time, which is ~18 N m^{-2} for the run at rest, with a heart rate of 68 bpm and a maximum velocity of 0.84 m s^{-1}. At full exercise intensity, the maximum WSS is much higher, at ~31 N m^{-2} for a heart rate of 134 bpm and a maximum velocity of 1.19 m s^{-1}. Here, while the mean WSS and velocity decrease with exercise intensity, the maximum WSS and velocity increase. Between these two cases, the difference in maximum WSS (a factor of 1.77) cannot be justified solely by the difference in maximum velocity (a factor of 1.45). Indeed, the (much larger) difference in heart rate (a factor of 1.97) may be an important contributor to the magnitude of the maximum WSS. Further investigations are required to explore the exact nature of these relations. The variability of the WSS, here measured as the averaged absolute difference between consecutive extractions at a 0.025 s interval, increases as expected with higher exercise intensity.

5. DISCUSSION AND CONCLUSIONS

We present an automated ensemble simulation framework and its application to model blood flow in the middle cerebral artery under a range of patient-specific cardiac parameters, using a multiscale ensemble approach. We show good agreement of velocity profiles at the inlet with those close to the outlet, and that our non-lattice aligned inflow conditions require further enhancement. FabHemeLB allows us to run the whole workflow for the relatively complicated setup in one tool, including the execution and analysis of the ensemble simulations. It reduces the human effort required for doing these tasks, and by automatically scheduling the ensemble instances in parallel it also allows for efficient use of large core counts and a reduced time to solution. The systematic execution and analysis patterns offered by FabHemeLB allow us to easily identify shortcomings in our existing approach. Not only does this feature in FabHemeLB boost our ongoing research, it also provides the level of data curation required to do future, more extensive, validation studies.

In our case study, we investigate the wall shear stress (WSS) properties in a middle cerebral artery at a location of interest close to the outlet. We find that the mean WSS correlates as expected linearly with the average flow velocity at the inlet. However, in addition we find evidence that the maximum WSS is dependent on the heart rate as well as the average flow velocity. This implies that these relations are non-trivial, and that a comprehensive analysis of flow dynamics in cerebral arteries should not only include the presence of pulsatile flow, but also the presence of these flows over a range of heart rates.

ACKNOWLEDGEMENTS

We are grateful to Rupert Nash for his efforts on enabling property extraction for HemeLB, and to Aditya Jitta for performing the segmentation. This work has received funding from the CRESTA project within the EC-FP7 (ICT-2011.9.13) under Grant Agreements no. 287703, and from EPSRC Grants EP/I017909/1(www.2020science.net) and EP/ I034602 /1. This work made use of the ARCHER supercomputer at EPCC in Edinburgh, via EPSRC and the UK Consortium on Mesoscopic Engineering Sciences (EP/L00030X/1). We have also used the Oppenheimer cluster, administered at the Chemistry Department at University College London.

REFERENCES

1. T. Becske, G.I. Jallo**Subarachnoid Hemorrhage** (2010) eMedicine: http:// emedicine.medscape.com/article/1164341-overview
2. Brain Aneurysm – NHS choices, 2014. http://www.nhs.uk/ conditions /aneurysm /Pages/ Introduction.aspx.
3. P. Marques-Sanches, E. Spagnuolo, F. Martínez, P. Pereda, A. Tarigo, V. Verdier **Aneurysms of the middle cerebral artery proximal segment (M1). Anatomical and therapeutic considerations. Revision of a series. Analysis of a series of the pre bifurcation segment aneurysms** Asian J. Neurosurg. (2010)
4. L. Grinberg, E. Cheever, T. Anor, J. Madsen, G. Karniadakis**Modeling blood flow circulation in intracranial arterial networks: a comparative 3D/1D simulation study** Ann. Biomed. Eng., 39 (1) (2011)
5. L. Formaggia, D. Lamponi, A. Quarteroni**One dimensional models for blood flow in arteries** J. Eng. Math., 47 (3-4) (2003), pp. 251-276
6. D. Fedosov, H. Noguchi, G. Gompper**Multiscale modeling of blood flow: from single cells to blood rheology** Biomech. Model. Mechanobiol. (2014)
7. M. Shojima, M. Oshima, K. Takagi, R. Torii, M. Hayakawa, K. Katada, A. Morita, T. Kiri no**Magnitude and role of wall shear stress on cerebral aneurysm** Stroke (2004)
8. D. Groen, D. Abou Chacra, R.W. Nash, J. Jaros, M.O. Bernabeu, P.V. Coveney**Weighted decomposition in high-performance lattice-Boltzmann simulations: are some lattice sites more equal than others?** Accepted for EASC 2014 (2014, October) ArXiv1410.4713
9. Nektar. http://www.nektar.info/wiki.
10. H. Baek, M. Jayaraman, G. Karniadakis**Wall shear stress and pressure distribution on aneurysms and infundibulae in the posterior communicating artery bifurcation** Ann. Biomed. Eng., 37 (12) (2009), pp. 2469-2487
11. L. Grinberg, J.A. Insley, D.A. Fedosov, V. Morozov, M.E. Papka, G.E. Karniadakis**Tightl y coupled atomistic-continuum simulations of brain blood flow on petaflop supercomputers** Comput. Sci. Eng., 14 (6) (2012), pp. 58-67
12. Palabos. http://www.palabos.org/.
13. H. Anzai, M. Ohta, J.-L. Falcone, B. Chopard**Optimization of flow diverters for cerebral aneurysms** J. Comput. Sci., 3 (1) (2012), pp. 1-7
14. J. Borgdorff, M. Ben Belgacem, C. Bona-Casas, L. Fazendeiro, D. Groen, O. Hoenen, A.Mizeranschi, J.L. Suter, D. Coster, P.V. Coveney, W. Dubitzky, A.G. Hoekstra, P. Strand, B.Chopard**Performance of distributed multiscale simulations** Philos. Trans. R. Soc. A: Math. Phys. Eng. Sci., 372 (2021) (2014)

15. MUSUBI. https://bitbucket.org/apesteam/musubi.
16. M. Hasert, K. Masilamani, S. Zimnya, H. Klimachb, J. Qi, J. Bernsdorf, S. Roller**Comple x fluid simulations with the parallel tree-based lattice Boltzmann solver musubi** J. Comput. Sci., 5 (5) (2014), pp. 784-794
17. M. Bernaschi, S. Melchionna, S. Succi, M. Fyta, E. Kaxiras, J.K. Sircar**Muphy: a parallel multi physics/scale code for high performance bio-fluidic simulations** Comput. Phys. Commun., 180 (9) (2009), pp. 1495-1502
18. B. Gmeiner, H. Köstler, M. Stürmer, U. Rüde**Parallel multigrid on hierarchical hybrid grids: a performance study on current high performance computing clusters** Concurr. Comput.: Pract. Exp., 26 (1) (2014), pp. 217-240
19. D. Groen, J. Borgdorff, C. Bona-Casas, J. Hetherington, R.W. Nash, S.J. Zasada, I.Saverchenko, M. Mamonski, K. Kurowski, M.O. Bernabeu, A.J. Hoekstra, P.V. Coveney**Flexible composition and execution of high performance, high fidelity multiscale biomedical simulations** Interface Focus, 3 (2) (2013) 20120087
20. M.O. Bernabeu, R.W. Nash, D. Groen, H.B. Carver, J. Hetherington, T. Krüger, P.V.Cove ney**Impact of blood rheology on wall shear stress in a model of the middle cerebral artery** Interface Focus, 3 (2) (2013) 20120094
21. D. Groen, S. Rieder, S. Portegies Zwart**Mpwide: a light-weight library for efficient message passing over wide area networks** J. Open Res. Softw., 1 (1) (2013)
22. HemeLB. http://ccs.chem.ucl.ac.uk/hemelb.
23. D. Groen, J. Hetherington, H.B. Carver, R.W. Nash, M.O. Bernabeu, P.V. Coveney**Analy zing and modeling the performance of the HemeLB lattice-Boltzmann simulation environment** J. Comput. Sci., 4 (5) (2013), pp. 412-422
24. M. Mazzeo, S. Manos, P.V. Coveney**In situ ray tracing and computational steering for interactive blood flow simulation** Comput. Phys. Commun., 181 (2) (2010), pp. 355-370
25. S. Manini, L. Antiga, L. Botti, A. Remuzzi**pyNS: An open-source framework for 0D haemodynamic modelling** Ann. Biomed. Eng. (2014), pp. 1-13
26. Caroli, S. Manini, L. Antiga, K. Passera, B. Ene Iordache, S. Rota, G. Remuzzi, A.Bode, J. Leermakers, F. van de Vosse, R. Vanholder, M. Malovrh, J. Tordoir, A. Remuzzi**Validation of a patient-specific hemodynamic computational model for surgical planning of vascular access in hemodialysis patients** Kidney Int., 84 (2013), pp. 1237-1245
27. FabSim. http://www.github.com/djgroen/FabSim.
28. J. Sugawara, M. Tanabe, T. Miyachi, K. Yamamoto, K. Takahashi, M. Iemitsu, T. Otsuki, S.Homma, S. Maeda, R. Ajisaka, M. Matsuda**Non-invasive assessment of cardiac output during exercise in healthy young humans: comparison between modelflow method and doppler echocardiography method** Acta Physiol. Scand., 179 (4) (2003), pp. 361-366
29. Gimias. http://www.gimias.org/.
30. M. Bouzidi, M. Firdaouss, P. Lallemand**Momentum transfer of a Boltzmann-lattice fluid with boundaries** Phys. Fluids, 13 (11) (2001), pp. 3452-3459
31. J.R. Cebral, F. Mut, J. Weir, C. Putman**Quantitative characterization of the hemodynamic environment in ruptured and unruptured brain aneurysms** Am. J. Neuroradiol., 32 (1) (2011), pp. 145-151

CHAPTER 6

A Super-High-Efficiency Algorithm for the Calculation of the Correlation Integral

Zhuping Gong

School of Business and Administration, South China University of Technology, Guangzhou, China

ABSTRACT

When the chaotic characteristics of manufacturing quality level are studied, it is not practical to use chaotic methods because of the low speed of calculating the correlation integral. The original algorithm used to calculate the correlation integral is studied after a computer hardware upgrade. The result is that calculation of the correlation integral can be sped up only by improving the algorithm. This is accomplished by changing the original algorithm in which a single distance threshold-related correlation integral is obtained from one traversal of all distances between different vectors to a high-efficiency algorithm in which all of the distance threshold-related correlation integrals are obtained from one traversal of all of the distances between different vectors. For a time series with 3000 data points, this high-efficiency algorithm offers a 3.7-fold increase in speed over the original algorithm. Further study of the high-efficiency algorithm leads to the development of a super-high-efficiency algorithm, which is accomplished by changing the original and high-efficiency algorithms, in which the add-one operation of the Heaviside function is executed n times, such that the execution of the add-one operation occurs only once. The super-high-efficiency algorithm results in increases in the calculation speed by up to 109 times compared with the high-efficiency algorithm and by approximately 404 times compared with the original algorithm. The calculation speed of the super-high-efficiency algorithm is suitable for practical use with the chaotic method.

Keywords: Super-High-Efficiency, Algorithm, Correlation Integral, Chaotic Time Series

1. INTRODUCTION

The chaotic characteristics of manufacturing quality level using data containing 588 daily product defect fractions, which were obtained from the HZ company, had been studied [1] - [3] . Further, the manufacturing quality level was proven to be chaotic if the correlation dimension was fractional in a study that used the G-P algorithm [4] in the calculation of the correlation dimension.

In this study, the manufacturing quality level time series was first reconstructed as a multi-dimensional phase space, and then the distances between vectors in the reconstructed phase space were calculated under a certain embedded dimension. Next, the correlation integral was calculated according to the relationship between the distance value and distance threshold value of the vectors. Finally, the correlation dimension was determined by log-log coordinates between the correlation integral and the distance threshold value. The problem with this process was the amount of time wasted in calculating the correlation integral, which was a huge obstacle if research results were to be useful in actual practice. Thus, the process for calculating the correlation integral must be improved.

Improving the speed of calculation first required the computer speed to be accelerated. The computer we used initially had a P4 2.8-GHz CPU with 512 Mb of memory. With this configuration, a time series with 3000 data points that was constructed with the Lorenz system required almost 4 hours to calculate the correlation integral. We then used a new computer configured with a 2.8-GHz Core i5 2300 CPU with 2 Gb of memory. However, calculations on the new computer still took 50 minutes. Although this was a 4-5-fold decrease in calculation time, the improvement in calculation efficiency was still not high enough to apply this technology in practice. Clearly, the time spent in calculating the correlation integral was not related so much to the computer hardware configuration but more to the algorithm used in the calculations.

Therefore, we investigated both the algorithm for calculating the correlation integral and methods to improve its efficiency. The improved algorithm had to be workable in practice and was helpful in the application of chaotic time series analysis to quality control in practical manufacturing situations.

2. G-P ALGORITHM

For a given time series$\{x(t),t=1,2,3,\cdots N\}$, its reconstructed phase space with m dimension embedding is

$\{X_i(x(i),x(i+\tau),\cdots x(i+(m-1)\tau)),i=1,2,\cdots,N-(m-1)\tau\}$. The trajectory of the reconstructed vector in the reconstructed phase space is separated exponentially. For two random points X_i and X_j, although the exponential separation will lead them to be irrelevant, they exist in the same attractor. The relevance of these points in the reconstructed phase space is measured by the correlation integral [4] - [6] :

$$C(r)=\lim_{N\to\infty}\frac{1}{M^2}\sum_{i,j=1}^{M}\theta\left[r-\left\|X_i-X_j\right\|\right] \tag{1}$$

where ⍰(x) is the Heaviside function:

$$\theta(x)=\begin{cases}0, & x\le 0\\ 1, & x>0\end{cases} \tag{2}$$

Formulae (1) and (2) show that the correlation integral is related to the distance threshold value r, and the relationship is given by:

$$\lim_{r\to 0}C(r)\propto r^{\nu} \tag{3}$$

where v is the correlation index, and its expression is:

$$v=\log C(r)/\log r \tag{4}$$

The correlation index v is related not only to the distance threshold r but also to the embedded dimension m. v increases gradually with m and then becomes saturated. The saturated correlation index v is termed the correlation dimension D of the time series under study.

3. ORIGINAL CORRELATION INTEGRAL ALGORITHM

3.1. Range of the Distance Threshold

The range of the distance threshold r must be predetermined before calculating the correlation integral. A smaller or larger r causes the correlation integral results to be meaningless.

The procedure that determines the distance threshold value is as follows:

- Among the distance values between different vectors, set the minimum distance value as $R_{min}$and the maximum distance value as R_{max}, and the interval $[R_{min}, R_{max}]$ defines the distance range.
- The distance thresholds are discrete values in the interval $[R_{min}, R_{max}]$, and the incremental change between contiguous threshold values is Δr, which is defined as:

$$\Delta r = r_{i+1} - r_i = \frac{R_{max} - R_{min}}{C}, r_i \in [R_{min}, R_{max}], i = 1, 2, \cdots, C \tag{5}$$

Thus, there are C distance threshold values.

3.2. Original Correlation Integral Algorithm

First, all of the distances between any pair of reconstructed vectors are calculated for a selected embedded dimension m, and the distances are saved as a two-dimensional array. According to formulae (1 - 2), the correlation integral is a function of the distance threshold value r in embedded dimension m. This means that one given r value corresponds to one certain correlation integral. Based on formulae (1 - 2), the correlation integral algorithm, termed the original correlation integral algorithm, is as follows:

- First, set the distance threshold value r as R_{min}, which is the minimum of interval $[R_{min}, R_{max}]$.
- Traversing the distance array that is calculated from the m dimensional reconstructed phase space, if the distance value d_{ij} between the ith and the j th vector is less than r, the Heaviside function value ? adds one.
- Calculate the correlation integral of distance threshold value r according to formula (1).

- The distance threshold value adds Δr according to formula (5), and if the distance threshold value is still located within the interval [R_{min}, R_{max}], the calculation jumps to procedures *b* and *c* and calculates the correlation integral of the current distance threshold value. If the distance threshold value is greater than R_{max}, the calculation finishes.

In this original algorithm, the calculating complexity is defined as:

$$T(N,C,T_a)=\sum_{m=1}^{mt}\left[C\left(N-1-(m-1)\tau\right)\left(N-2-(m-1)\tau\right)t_v/2+T_a t_a\right] \quad (6)$$

Here, t_v is the time spent in one visit of one distance array unit, t_a is the time spent in the operation of adding one, and t_a and t_v are decided by the hardware configuration of the computer used. t_a is the frequency of operations performed to add one in each embedding, and *mt* is the largest embedding dimension.

This algorithm took almost 4 hours to calculate a correlation integral for a time series with 3000 data points and 1 to 13 dimensional embedding. Further, upgrading to the faster computer configuration was not an adequate solution, although the new computer increased the calculation speed by a factor of four. Thus, the algorithm for computing the correlation integral had to be improved to achieve an acceptable calculation time.

4. IMPROVEMENT OF THE CORRELATION INTEGRAL CALCULATION

To improve the calculation time, we conducted an intense study of the original algorithm. Our investigation showed that the low efficiency of the original algorithm occurred because one traverse of the distances array obtains only one correlation integral of one certain distance threshold value. According to formula (5), there are *C* distance threshold values, so to obtain all of these correlation integral values must take C times to traverse the distance array. Because this is a two-dimensional array, a very long time is required to traverse the array.

From this analysis, the first improvement comes by determining whether the traverse times can be reduced, and furthermore, whether one single traverse of the distance value array will be adequate for the calculation of the correlation integral under a certain embedded reconstructed phase space.

4.1. First Improvement: A High-Efficiency Algorithm

If one traverse can obtain C correlation integrals under C distance threshold values, the calculating efficiency will be enhanced tremendously. Along these lines, the original algorithm was studied, and the following relationship between distances and the distance threshold value was found:

$$d_{ij} < r_p \Rightarrow d_{ij} < r_q$$
$$p = 1, 2, \cdots, C; q = p, p+1, \cdots, C \quad (7)$$

where d_{ij} denotes the distance between the ith and j th vectors, and r_p and r_q are the pth and qth distance threshold values. This relationship tells us that if d_{ij} is less than r_p, then d_{ij} must be less than all r_q that are equal to or greater than r_p. This relationship is obvious because the distance threshold value is set as an incremental interval Δr in formula (5).

According to this relationship, when we are certain that the distance value d_{ij} is less than one certain distance threshold value r_p, it is not efficient to perform the add one operation only for the Heaviside function θ of r_p. Rather, it will be much more efficient to perform the add one operation for all of the Heaviside functions θ of all r_q: q=p,p+1,---C. This means that all of the correlation integrals under all threshold values will be obtained by a single traverse of the distance array.

According to this analysis, we improved the correlation integral calculation under a certain embedded dimension m. The improved high-efficiency algorithm is as follows:

- Set all of the distance threshold values as r_i: i=1,2,---C increasingly from R_{min} to R_{max}.
- In all of the distances between the vectors under m dimensional embedding, set the first distance as the working distance d_{ij} based on the sorting order of the distance array.
- Compare d_{ij} with r_i: : i=1,2,---Cuntil $d_{ij} > r_{p-1}$ and $d_{ij} < r_p$, and then add one to all of the Heaviside functions θ(x) for r_q: q=p,p+1,---C
- Set the next distance value as the working distance d_{ij} based on the sorting order of the distance array and return to processes c and d until all of the distance values are traversed.
- Calculate the correlation integral of each distance threshold r_i: i=1,2,---C

The original correlation integral algorithm traverses the distance array once and calculates one correlation integral of one distance threshold; thus, the calculation of all of the correlation integrals of the distance thre- sholds leads to a traverse of the distance array of C times. However, the improved high-efficiency algorithm can obtain the correlation integrals of all of the distance thresholds with only one traverse of the distance array.

In this high-efficiency algorithm, the calculating complexity is defined as:

$$T(N,T_a)=\sum_{m=1}^{mt}\left[\left(N-1-(m-1)\tau\right)\left(N-2-(m-1)\tau\right)t_v/2+T_a t_a\right] \tag{8}$$

The definitions of the parameters are the same as those in formula (6).

4.2. Second Improvement: A Super-High-Efficiency Algorithm

Following a deeper analysis of the high-efficiency algorithm, we realized that in process c, every time we ensure that $d_{ij} > r_{p-1}$ and $d_{ij} < r_p$, the add one operation must be performed for all of the Heaviside functions $\theta(x)$ for r_q: $q=p,p+1,---C$. Thus, the number of single add one operations is C-p + 1 only in one unit visit of the distance array, but because the array has a large number of units, the number of add one operations is quite high. If the add one operation does not operate immediately but simply records the number of times until the end of the traverse, then the add one operation is performed for all Heaviside functions $\theta(x)$, and here, the addition operation is adding not one but the total number of recorded times of adding one. Furthermore, the addition operation can be performed for the Heaviside function θ(x) for all r_q: $q=p,p+1,---C$.

From the foregoing analysis, a more efficient algorithm than the high-efficiency algorithm was developed, which we termed the super-high-efficiency algorithm. The procedure for each embedded dimension m is as follows:

- Set all distance threshold values as r_i: $i=1,2,---C$ increasingly from $R_{\min}$ to $R_{\max}$;
- Set a zero array s[1:C] to store the number of add one operations for all of the distance thresholds;
- In the distance array under m dimensional embedding, set the first distance as the working distance d_{ij} based on the location of the distance in the array;
- Compare d_{ij} with r_i: $i=1,2,---C$ until $d_{ij} > r_{p-1}$ and $d_{ij} < r_p$, and then add one to s[p];

- Set the next distance unit in the array as the working distance d_{ij} and return to processes *d* and e until all of the distance values are traversed;
- According to the values of s[p], $p=1,2,\cdots,c$, perform s[p] add one operations for all Heaviside functions θ(x) of r_q,; $q=p,p+1,\cdots C$.
- Calculate the correlation integral of each distance threshold r_i: $i=1,2,\cdots C$.

The core thinking behind the super-high-efficiency algorithm is to reduce the n times of add one operations of the Heaviside function θ(*x*) of corresponding distance thresholds to an add n operation, which is why this algorithm is more efficient than the high-efficiency algorithm.

In the super-high-efficiency algorithm, the calculating complexity is defined as:

$$T(N,C)=\sum_{m=1}^{mt}\left[\left(N-1-(m-1)\tau\right)\left(N-1-(m-1)\tau\right)(t_v+t_a)/2+C(C-1)t_a/2\right] \tag{9}$$

The definitions of the parameters are the same as those in formula (6), and it is assumed that the add one operation takes the same amount of time as the add n operation.

5. CALCULATION EXPERIMENT

The effects of the original, the high-efficiency, and the super-high-efficiency algorithms need to be verified by calculation experiments.

5.1. Chaotic Systems and Parameter Selection

Time series produced from three classical chaotic systems and a product defect rate that comes from the HZ semiconductor company were selected as the data sources. The four data sources and their selected parameters are:

1) Logistic system

$$x_{n+1}=\mu x_n\left(1-x_n\right),\ \left(n=1,2,\cdots;0\le x_n\le 1\right) \tag{10}$$

This system appears as chaotic only if the parameter $\mu > \mu_{?}=3.569945\cdots$ We set $\mu = 4$ and then produced 3000 data points with this system that were used to construct a chaotic time series.

2) Hénon system

$$\begin{cases} x_{n+1} = 1 - ax_n^2 + y_n \\ y_{n+1} = bx_n \end{cases} \tag{11}$$

Here, if the parameters a = 1.4 and b = 0.3 are set, the system appears chaotic. We produced 3000 data points with this chaotic system that were used to construct a chaotic time series.

3) Lorenz system

$$\begin{cases} \dot{x} = -\sigma(x - y) \\ \dot{y} = -xz + rx - y \\ \dot{z} = xy - bz \end{cases} \tag{12}$$

With the Lorenz system parameters set as $r = 28$, $\sigma = 10$, and $b = 8/3$, this system appears chaotic. We produced 3000 data points with this chaotic system that were used to construct a chaotic time series.

4) HZ system

The data of 588 daily product defects of the HZ company from December 28, 2006 to July 20, 2009 were collected as the time series. A previous study has proven that this time series is chaotic [3] .

5.2. Calculation Experiment and Results

In the calculation experiment, Matlab 2009a was selected as the software platform, and a computer with a 2.8- GHz Core i5 2300 CPU and 2 Gb of memory served as the hardware platform. The original, high-efficiency, and super-high-efficiency algorithms served as the working algorithms with the time series data coming from the Logistic, Hénon, Lorenz, and HZ systems. Due to restrictions imposed by the 2 Gb of memory, the calculation experiment could only be performed for 1 to 13 dimensional embedded time series constructed from the 3000 data points of the Logistic, Hénon, and Lorenz systems. Therefore, the embedding dimensions of the four time series were set uniformly as 1 to 13.

The calculation experiment showed that the correlation integrals were the same for each different algorithm used with the same system. This guaranteed the correctness of the different algorithms.

The efficiency of the different algorithms was measured by the timer of Matlab Version 2009a, and the results are shown in Table 1.

5.3. Analysis of the Results

Based on the times shown in Table 1, three indices were defined as follows:

1) The efficiency of the first improvement

$$E_1 = \frac{T_H}{T_O} \tag{13}$$

2) The efficiency of the second improvement

$$E_2 = \frac{T_S}{T_H} \tag{14}$$

3) The efficiency of the overall improvement

$$E_{\mathrm{All}} = \frac{T_S}{T_O} \tag{15}$$

The effects of the algorithm improvements were calculated according to the data in Table 1 and the application of formulae (13 - 15) and are shown in Table 2.

The following results can be observed from Table 2:

1) The effects of the LHL systems are of the same order of magnitude, and the effect of the HZ system is of a lower magnitude.

2) The main effect of the improvements comes from the second improvement.

6. DISCUSSION

The time spent in the process of calculating the correlation integral differed among the original, high-efficiency, and super-high-efficiency algorithms. This difference was obtained by analyzing formulae (6), (8), and (9). Compared with the original algorithm, the high-efficiency algorithm performed the same number of add one operations, but the traverse time decreased by C-1 times. Compared with the high-efficiency algorithm, the traverse time of the super-high-efficiency algorithm was the same, but the number of add one operations performed decreased substantially, which is the source of the increased efficiency.

A more detailed analysis is given in Table 3.

Table 1. Calculation time of the three algorithms for the four time series.

Algorithm	Logistic	Hénon	Lorenz	HZ
TO (sec)	2748	2776.5	2990.6	103.89
TH (sec)	642.44	670.24	986.17	38.759
TS (sec)	7.00	7.00	7.09	0.385

TO: time consumed by the original algorithm; TH: time consumed by the high-efficiency algorithm; TS: time consumed by the super-high-efficiency algorithm.

Table 2. Effects of algorithm improvement.

Index	Logistic	Hénon	Lorenz	Average of the LHL systems	HZ
E_1	4.277	4.143	3.033	3.7	2.680
E_2	91.777	95.749	139.093	109	100.673
E_{All}	392.571	396.643	421.805	404	269.844

LHL: Logistic, Hénon and Lorenz systems.

Table 3. Degree of complexity of the three algorithms in relation to N, C, and Ta.

Algorithm	N	C	Ta
Original	√	√	√
High-efficiency	√	×	√
Super-high-efficiency	√	√	×

Here, the time complexity of the three algorithms depends entirely on the number of data points N, so time series of different lengths cannot be compared with one other. Because the time series scale comprises 3000 data points in each of the Logistic, Hénon, and Lorenz systems, these systems are comparable in calculating time complexity. The HZ system cannot be compared with these three systems because it used only 588 data points.

Ta is the number of times the add one operation was performed in both the original and high-efficiency algorithms, and depends on the distribution of the distance values and the distance threshold interval number *C*. Different systems have different distance distributions, so different systems will have different values of Ta. This is why different systems with the same data length and same *C* value have different calculation times.

C is the distance threshold interval number. Although the high-efficiency algorithm is not directly related to C, it is indirectly related to *C* through Ta. Therefore, the value of C can affect the time complexity in all three algorithms.

We know from Table 2 that the super-high-efficiency algorithm can produce an approximately 400-fold improvement in calculation speed in a time series with 3000 data points. Based on the analysis in Table 2, three conclusions are obtained:

1) The effect of the first improvement was to increase the calculation speed by only 3.7 times that of the original algorithm for a time series with 3000 data points. This effect results from a change in the algorithm in which the C times traversing of the distance array is reduced to one time.

2) The effect of the second improvement was to increase the calculation speed by about 109 times that of the high-efficiency algorithm. This effect comes from a change to the algorithm in which the performance of the add one operation n times is reduced to the performance of the operation only once.

3) Compared with the original algorithm, the calculation efficiency of the super-high-efficiency algorithm was increased by about 404 fold for a time series with 3000 data points.

7. CONCLUSIONS

The super-high-efficiency algorithm for the calculation of the correlation integral produced an approximately 400-fold reduction in calculation time. This effect was due to a reduction in both the traverse time of the distance array from *C* times to once, and in the number of add one operations performed to a single add operation.

The super-high-efficiency algorithm decreased the calculation speed from almost 1 hour (approximately 2800 seconds) to 7 seconds for a time series with 3000 data points. This 7-second speed was an acceptable calculation speed that rendered chaotic time series analysis usable in practice.

ACKNOWLEDGEMENTS

This paper is funded by the national nature science fund of China (51465020); and scientific research project (social science) of the special fund of the subject and professional building of institution of higher learning in Guangdong province in 2013 (2013WYXM0006) And the soft science research plan of Guangdong province (2013B070206017).

REFERENCES

1. Gong, Z.P. (2010) The Calculating Method of the Average Period of Chaotic Time Series. Systems Engineering, 28, 111-113. (In Chinese)
2. Gong, Z.P. (2011) Comparison of the Calculating Method of Delay Time in the Reconstructed Phase Space of Manufacturing Quality Information System. Systems Engineering, 29, 81-85. (In Chinese)
3. Gong, Z.P. (2012) The Chaotic Characteristic of the Variation of Manufacturing Quality Level Based on Time Series Analysis. Systems Engineering, 30, 38-42. (In Chinese)
4. Grassberger, P. and Procaccia, I. (1983) Measuring the Strangeness of Strange Attractors. Physica D: Nonlinear Phenomena, 9, 189-208. http://dx.doi.org/10.1016/0167-2789(83)90298-1
5. Kantz, H. and Schreiber, T. (1997) Nonlinear Time Series Analysis. Cambridge University Press, Cambridge
6. Kim, H.S., Eykholt, R. and Salas, J.D. (1999) Nonlinear Dynamics, Delay Times, and Embedding Windows. Physica D: Nonlinear Phenomena, 127, 48-60. http://dx.doi.org/10.1016/S0167-2789(98)00240-1

CHAPTER 7

OpenSBLI: A Framework for the Automated Derivation and Parallel Execution of Finite Difference Solvers on a Range of Computer Architectures

Christian T. Jacobs*, Satya P. Jammy, Neil D. Sandham

Aerodynamics and Flight Mechanics Group, Faculty of Engineering and the Environment, University of Southampton, University Road, Southampton, SO17 1BJ, United Kingdom

ABSTRACT

Exascale computing will feature novel and potentially disruptive hardware architectures. Exploiting these to their full potential is non-trivial. Numerical modelling frameworks involving finite difference methods are currently limited by the 'static' nature of the hand-coded discretisation schemes and repeatedly may have to be re-written to run efficiently on new hardware. In contrast, OpenSBLI uses code generation to derive the model's code from a high-level specification. Users focus on the equations to solve, whilst not concerning themselves with the detailed implementation. Source-to-source translation is used to tailor the code and enable its execution on a variety of hardware.

Keywords: High-Performance Computing; Code generation; Computational Fluid Dynamics; Finite difference methods; Graphics processing units

1. INTRODUCTION

High Performance Computing (HPC) systems and architectures are evolving rapidly. Traditional single processor-based CPU clusters are moving towards multi-core/multi-threaded CPUs. At the same time new architectures based on many-core processors such as graphics processing units (GPUs) and Intel's Xeon Phi are emerging as important systems and further developments are expected with energy-efficient designs from ARM and IBM. According to the IT industry, such advances are expected to deliver compute hardware capable of exascale-performance (i.e. 10^{18} floating-point operations per second) by 2018 [1]. Yet many frameworks aimed at computational/numerical modelling are currently not ready to exploit such new and potentially disruptive technologies.

Traditional approaches to numerical model development involve the production of static, hand-written code to perform the numerical discretisation and solution of the governing equations. Normally this is written in a language such as C or Fortran that is considerably less abstract when compared to a near-mathematical domain specific language. Explicitly inserting the necessary calls to MPI or OpenMP libraries enables the execution of the code on multi-core or multi-thread hardware. However, should a user wish to run the code on alternative platforms such as GPUs, they would likely need to re-write large sections of the code, including calls to new libraries such as CUDA or OpenCL, and optimise it for that particular hardware backend [2]. As HPC hardware evolves, an increasing burden faced by computational scientists becomes apparent; in order to keep up with trends in HPC, not only must a model developer be a domain specialist in their area of study, but also an expert in numerical algorithms, software engineering, and parallel computing paradigms [3] ; [4].

One way to address this issue is to introduce a separation of concerns using high level abstractions, such as domain specific languages (DSLs) and active libraries [4]; [5]; [6]; [7] ; [8]. This paradigm shift allows a domain specialist to describe their problem as a high-level, near-mathematical specification. The task of taking this specification and transforming it into executable computer code can then be handled in the subsequent abstraction layer; unlike the traditional approach of hand-writing the C/Fortran code that discretises the governing equations, this layer generates the code automatically from the problem specification. Finally, the generated code can be readily targetted towards a specific hardware platform through source-to-source translation. Hence, domain specialists focus on the equations they wish to solve and the setup of their problem, whilst the parallel computing experts can introduce support for new backends as they become available. At no point does the code have to undergo a fundamental re-write if the desired backend changes. Use of such strategies can have significant benefits for the productivity of both the user and developer, by removing the need to spend time re-writing code and/or the problem specification [5].

Given the motivation for the use of automated solution techniques, in this paper we present a new framework, OpenSBLI, for the automated derivation and parallel execution of finite difference-based models. This is an open-source release of the recent developments in the SBLI codebase developed at the University of Southampton, involving the replacement of SBLI's Fortran-based core with flexible Python-based code generation capabilities, and the coupling of SBLI to the OPS active library [9]; [10]; [11] ; [12] which targets the generated code towards a particular backend using source-to-source translation. Currently, OpenSBLI can generate OPS-compliant C code to discretise and solve the governing equations, using arbitrary-order central finite difference schemes and a choice of either the forward Euler scheme or a third-order Runge-Kutta time-stepping scheme. OpenSBLI then uses OPS to produce code targetted towards different backends. It is worth noting that backend APIs such as OpenMP (version 4.0 and above) are also capable of running on CPU, GPU and Intel Xeon Phi architectures, for example. However, currently OPS has no support for OpenMP version 4.0 and above. Moreover, codes that are written by hand in OpenMP would still potentially need to be re-written if different algorithms or equations were to be considered. Thus, the benefits of code generation still play a crucial role here, regardless of which backend is chosen.

The application of SBLI has so-far concentrated on problems in aeronautics and aeroacoustics, in particular looking at shock-boundary layer interactions (see e.g. [13]; [14]; [15] ; [16] and the references therein for more details). While such applications entail solving the 3D compressible Navier-Stokes equations, in principle other equations expressible in Einstein notation and solved using finite differences are also supported by the new code generation functionality, highlighting another advantage of such a flexible approach to numerical model development. Note also that while OpenSBLI does not yet feature shock-capturing schemes and Large Eddy Simulation models (unlike the legacy SBLI code), these will be implemented in the future as part of the project's roadmap. The main purposes of this initial release is the algorithmic changes to legacy SBLI's core.

Details the abstraction and design principles employed by OpenSBLI are given in Section 2. Section 3 details three verification and validation test cases that were used to check the correctness of the implementation. The paper finishes with some concluding remarks in Section 4.

2. DESIGN

Legacy versions of SBLI comprise static hand-written Fortran code, parallelised with MPI, that implements a fourth-order central differencing scheme and a low-storage, third or fourth-order Runge-Kutta timestepping routine. It is capable of solving the compressible Navier-Stokes equations coupled with various turbulence parameterisations (e.g. Large Eddy Simulation models) and diagnostic routines. In contrast, OpenSBLI is written in Python, and by replacing the legacy core with modern code generation techniques, the existing functionality of SBLI is enriched with new flexibility; the compressible Navier-Stokes equations can still be solved in OpenSBLI for the sake of continuity, but the set of equations that can be readily solved essentially becomes a superset of that of the legacy code. Furthermore, the use of the OPS library allows the generated code to easily be targetted towards sequential, MPI, or an MPI+OpenMP hybrid backend (for CPU parallel execution), CUDA and OpenCL (for GPU parallel execution), and OpenACC (for parallel execution on accelerators), without the need to re-write the model code. OPS is readily extensible in terms of new backends, making the code generation technique an attractive way of future-proofing the codebase and preparing the framework for exascale-capable hardware when it arrives. The main achievement of OpenSBLI is the ability to express model equations at a high-level with the help of the SymPy library [17], expanding the equations based on the index notation, and coupling this functionality with the generation of OPSC-based model code and also with the OPS library which performs code targetting. OpenSBLI's focus on the generation of computational kernels essentially forms a bridge between the high-level equations and the computational parallel loops ('parloops') that iterate over the grid points to solve the governing equations.

For any given simulation that is to be performed with OpenSBLI, the problem (comprising the equations to be solved, the grid to solve them on, their associated boundary and initial conditions, etc) must be defined in a setup file, which is nothing but a Python file which instantiates the various relevant components of the OpenSBLI framework. All components follow the principle of object-oriented design, and each class is explained in detail throughout the subsections that follow. An overview of the class relationships is also provided in Fig. 1.

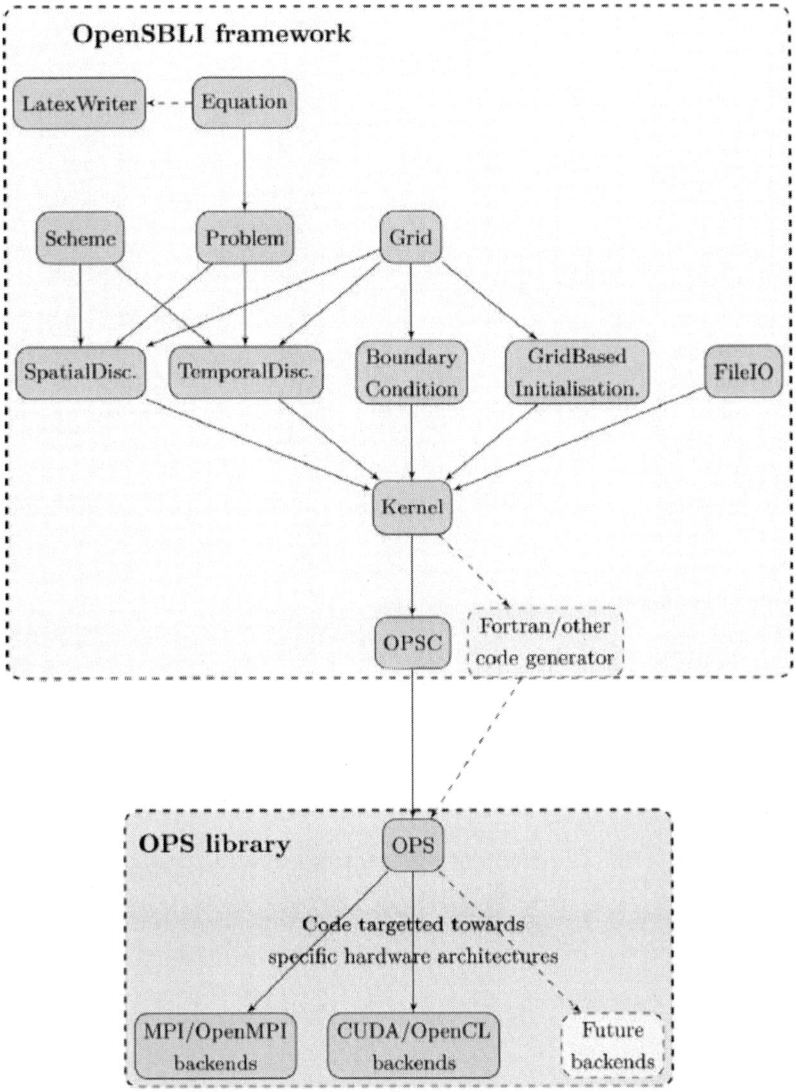

Figure. 1. The overall design of the OpenSBLI framework with respect to the core classes. The code targetting happens within the OPS library. The CPU backends include MPI, OpenMP, hybrid MPI+OpenMP, as well as a sequential version of the code. The GPU backends include CUDA and OpenCL, which can also be combined with MPI to run the code on multiple GPUs in parallel. The only accelerator backend available is OpenACC.

2.1. Equation specification

In a similar fashion to other problem solving environments such as OpenFOAM [18], Firedrake [4], FEniCS [5] ; [6], OPESCI-FD [19], Devito [20] ; [21], deal.II [22] and FreeFEM++ [23], OpenSBLI comprises a high-level interface for specifying the differential equations that are to be solved. These equations (and any accompanying formulas for temperature-dependent viscosity, for example) can be expressed in Einstein notation, also known as index notation. The adoption of such an abstraction is advantageous since it removes the need for the user to expand the equations by hand which can be an error-prone task. Furthermore, much like the Devito domain specific language (DSL) [20] ; [21] for finite difference stencil compilation, OpenSBLI makes use of the SymPy symbolic algebra library that supplies the basic components required for the modelling functionality that has been implemented in the present work. This functionality includes the automatic expansion of indices based on their contraction structure, such that repeated indices are expanded into a sum about that index, and the implementation of various types of differential operator.

2.1.1. Expressing

Consider the conservation of mass equation

$$\frac{\partial \rho}{\partial t} + \frac{\partial}{\partial x_j}[\rho u_j] = 0, \tag{1}$$

where u_j is the *j*th component of the velocity vector $\mathbf{u}$, ρ is the density field, and *xj* is the coordinate field in the *j*th dimension. In an OpenSBLI problem setup file, the user would specify this as a string, giving the left-hand side and right-hand side of the equation in the following format:

*mass="Eq(Der(rho,t),-Conservative(rho*u_j,x_j))"*

The functions Der and Conservative here are OpenSBLI-specific derivative operators, each defined in their own class derived from SymPy's Function class. Other high-level interfaces such as OpenFOAM offer similar differential operators such as div and grad, for example [18]. General derivatives are represented using the Der operator, whereas the Conservativeoperator ensures that the derivative will not be expanded using the product rule. A skew-symmetric form of the derivative is also available using the Skew function, discussed later in Section 3.3. All of these are essentially 'handler'/placeholder objects that OpenSBLI uses for spatial/temporal discretisation after parsing and expanding the equations about the Einstein indices. Special functions such as the Kronecker delta function and the Levi-Civita symbol are also available, derived from SymPy's LeviCivita and KroneckerDelta classes in order to handle Einstein expansion; these too are expanded later by OpenSBLI.

2.1.2. Parsing

Once all of the governing equations have been expressed by the user in string format, they are collected together in OpenSBLI's Problem class (see Appendix A). This class also accepts *substitutions, formulas*, and *constants*. For long equations, such optional *substitutions* (such as the definition of the stress tensor) can be written as a separate string (in the same way as the governing equations) to allow better equation readability, and then automatically substituted into the equations (such as the conservation of momentum and energy equations) at expansion-time instead of performing such error-prone manipulations by hand. The constitutive equations which define a relationship between the prognostic and non-prognostic variables are given as *formulas*, for example temperature-dependent viscosity relations, and an equation of state for pressure. The *constants* are the spatially and temporally independent variables which are represented as strings. Upon instantiation of the Problem class, the process is invoked to transform the equations into their final expanded form.

For each equation in string form, a new OpenSBLI Equation object is created. During its initialisation, SymPy's parse_expr function converts the equation string into a SymPy Eqdata type. Any of the OpenSBLI derivative operators such as Der and Conservative (currently in string format) are replaced by actual instances of the Der and Conservative classes. Similarly, any substitutions given in the Problem are parsed and substituted directly into the expression using SymPy's xreplace function. All other terms in the parsed expression are represented by OpenSBLI's EinsteinTerm class, derived from SymPy's Symbol class, which contains its own methods and attributes for determining/expanding Einstein indices. For example, the class's initialisation method __init__ splits up the term u_j where there are underscore markers, and stores the Einstein index j in a list as a SymPy Idx object. The get_expanded method later replaces the alphabetical Einstein indices with actual numerical indices, replacing _j with 0 and 1, in the 2D case. Finally, any constants in the Problem object are also represented as an EinsteinTerm object, but are flagged as constant terms in OpenSBLI, so that they are not spatially or temporally-dependent. The coordinate vector components *xj* (and the time term *t*) are a special case of an EinsteinTerm; these are marked with a is_coordinate flag so that, during the expansion phase, the EinsteinTerms are made dependent on the coordinate field (and time, if appropriate) to ensure that differentiation is performed correctly.

2.1.3. Expanding

After the parsing and substitution stage, the equations are expanded about repeated indices. Note that this process is performed by OpenSBLI, although various SymPy classes underpin the functionality. Following the example, (1) would be expanded as

$$\frac{\partial \rho}{\partial t} + \frac{\partial}{\partial x_0}[\rho u_0] + \frac{\partial}{\partial x_1}[\rho u_1] = 0. \qquad (2)$$

OpenSBLI loops over each EinsteinTerm stored in the parsed Equation object, and maps it to a SymPy Indexed object. For example, the term u_k would first be mapped to u[k]. The index kin the term is then expanded over 0, ..., $d-1$ (where d is the dimension of the problem) by replacing it with each integer dimension, yielding a SymPy Mutable Dense NDim Array array of size d (for a vector function, or $d \times d$ for a tensor of rank 2) of expanded variables which is stored as a class attribute. For example, expanding the vector u[k] yields the expansion array [u0, u1] in 2D. Upon expansion, the terms are also made spatially-dependent (i.e. indexed by x_0, x_1, x_2 coordinates, depending on the dimension) and, if applicable, temporally-dependent (i.e. indexed also by t). The only exceptions to this are constants such as the Reynolds number *Re*. The expansion array from the previous example then becomes [u0[x0, x1, t], u1[x0, x1, t]] (and [x0, x1] for the constant coordinate field).

Each equation is expanded by locating any repeated indices and then summing over them as appropriate. For example, after mapping each EinsteinTerm (e.g. u_k) to an Indexedobject (e.g. u[k]), the mass equation is represented internally as

*Eq(Der(rho,t),-Conservative(rho*u[k],x[k]))*

Since the index k is repeated, the expansion arrays are used to expand this expression to

Eq(Der(rho[x0,x1,t],t),

*-Conservative(rho[x0,x1,t]*u0[x0,x1,t],x0[x0,x1,t])*

*-Conservative(rho[x0,x1,t]*u1[x0,x1,t]),x1[x0,x1,t]))*

Finally, the Der and Conservative functions are applied, with the expression becoming

Eq(Derivative(rho[x0,x1,t],t),

*-Derivative(rho[x0,x1,t]*u0[x0,x1,t],x0)*

*-Derivative(rho[x0,x1,t]*u1[x0,x1,t],x1))*

which is equivalent to (2). Similar expansion can also be applied for any other equations involving e.g. diagnostic fields. Note how the calls to Der and Conservative have been replaced by calls to SymPy's Derivative class (which in turn uses SymPy's difffunction); while it is SymPy that handles the differentiation, it is OpenSBLI that handles the exact formulation of the derivative (i.e. OpenSBLI has ensured that the derivative has not been expanded using the product rule here).

Any nested derivatives are also handled here. It is not currently possible to specify, for example, diff(diff(u_j, x_i), x_j) using SymPy's diff function directly because the fact that u_j is dependent on x_i and x_j is not taken into account. In contrast, the use of Der and EinsteinTerms like u_j in OpenSBLI allows the derivative to be computed correctly since the terms are made dependent through the use of Indexed objects as previously described. OpenSBLI users must instead use the Der function Der(Der(u_j, x_i), x_j). For each nested derivative (or nested function in general), the inner function is evaluated first along with all other non-nested functions. Only then is the outer function applied.

For the purposes of debugging, OpenSBLI includes a LatexWriter class that takes the expanded equations as input and writes them out in LaTeX format so developers can more easily spot errors, for example where indices have been expanded incorrectly.

2.2. Grid

The governing equations are discretised on a regular grid of solution points that span the domain of interest; an example is provided in Fig. 2. All grid-related functionality is handled by the Grid class, which must be instantiated by the user in the problem setup file. The dimensionality of the problem *d*, the number of points in each dimension, and the grid spacing must all be supplied. A problem of dimension *d* would generate a grid of $\mathrm{N}_{x0} \times \cdots \times \mathrm{N}_{xd-1}$ solution points in total, where N_{xi} represents the user-defined number of grid points in direction *xi*.

For the sake of looping over each solution point and computing the necessary derivatives via the finite difference method, each (non-constant) term is processed further by OpenSBLI; the index of each spatial coordinate (e.g. *x*0) is mapped onto an index over the grid points in that spatial direction (e.g. i0) which will iterate from 0 to $N_{xi}-1$ (for a given direction *xi*) when the computational kernel is eventually generated.

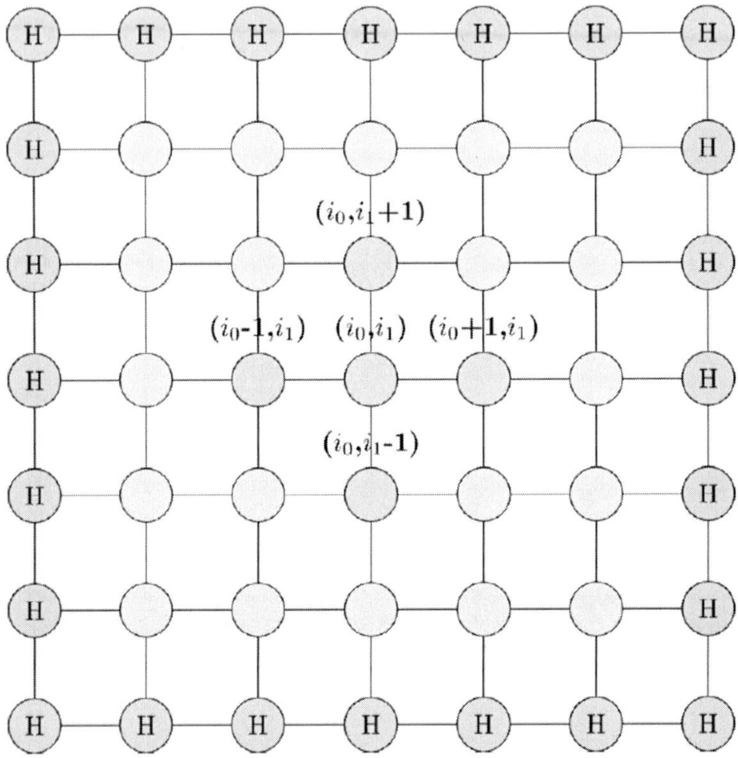

Figure. 2. The regular grid of solution points upon which the governing equations are solved. The grid point indices in the x and y directions are denoted i_0 and i_1, respectively. The halo points that surround the outer-most points of the domain are labelled 'H'. A computational stencil used for second-order central differencing is highlighted red in the centre, with the relative grid coordinates of each point.

In addition to the solution points within the physical domain, a set of halo points (or 'ghost' points), which border the outer-most grid points, are also created automatically depending on the boundary conditions and the spatial order of accuracy. These halo points are necessary to ensure that the derivatives near the boundary can be computed with the same stencil as the 'inner' points. The exact number of halo points required therefore depends on the number of stencil points; for example, in Fig. 2 the stencil for a second-order central difference (using 3 points in each direction) would require one halo point at each end of the domain. The values that these halo points hold depend on the type of boundary condition applied, and this is discussed in more detail in Section 2.6.

Every field/term in the governing equations that is represented by the grid indices holds a so-called 'work array' which essentially contains the field's numerical value at each of the grid points, including the halos. The implementation of initial and boundary conditions is done by accessing and modifying this work array, as will be described in Sections 2.5 and 2.6.

2.3. Computational kernels

The Kernel class defines a sequence of computational steps that should be performed to solve the governing equations. For instance, one kernel may be created to compute the spatial derivative of a field, while another kernel handles the initialisation of the field values based on a given initial condition, and another handles the enforcement of boundary conditions that involve computations. During the instantiation of a kernel, the relevant variables and fields are classified as inputs, outputs and input/outputs (i.e. both an input and an output), and the kernel's range of evaluation (i.e. the range of grid indices over which the kernel is applied). This helps to minimise data transfer, since only those variables/fields required to perform the computation are passed to the generated kernel code.

2.4. Discretisation schemes

Once a grid is created, the equations are discretised upon that grid. For spatial discretisation purposes OpenSBLI offers a central differencing scheme for first and second-order derivatives; all the stencil coefficients are computed using SymPy, which allows stencils of an arbitrary order of accuracy to be created. For temporal discretisation purposes, OpenSBLI features the (first-order) forward Euler scheme as well as the same low-storage, third-order Runge-Kutta timestepping scheme [24] present in the legacy SBLI code.

To use a particular scheme, one should instantiate a discretisation scheme derived from the generic base class called Scheme, which essentially stores the finite difference stencil coefficients or the weights used in a particular time-stepping scheme. Spatial and temporal schemes should be instantiated separately.

For the purpose of spatial discretisation, handled by the Open SBLI SpatialDiscretisationclass, an Evaluations object is created for each of the formulas, and the derivatives in the equations. Each Evaluations object automatically finds and stores the dependencies of a given term (e.g. $\partial(A + B)/\partial x_0$ requires the dependencies A and B). Once all the Evaluationshave been created, they are sorted with respect to their dependencies being evaluated (e.g. if B depends on A, then A should be evaluated first). The next step involves defining the range of grid point indices over which each evaluation should be performed, and also assigning a temporary work array for each evaluation. All of the evaluations are then described by a Kernel object (see Section 2.3). It is here, while creating the kernels, that the (continuous) spatial derivatives are automatically replaced by their discrete counterparts. It should be noted that, for the evaluation of formulas, these kernels are fused together if they have no inter-dependencies to avoid race conditions when running on threaded architectures. Finally, to evaluate the residual for the purposes of temporal discretisation, the derivatives in the expanded equations (represented by an Evaluationsobject) are substituted by their temporary work arrays, and a Kernel is created for evaluating the residual of each equation.

The temporal discretisation, handled by the TemporalDiscretisation class, involves applying the various stages of the time-stepping scheme supplied using the residuals computed by the spatial discretisation process. Similarly, a Kernel object is created for the evaluations in the time-stepping scheme.

2.5. Initial conditions

In order for the prognostic fields to be advanced forward in time, initial conditions can be applied using the GridBasedInitialisation class. This is accomplished in much the same way as specifying equations, but involves assignment of grid variables and work arrays of grid point values. For example, in the simulation setup file the x_0 coordinate can be defined using the grid point index and Δx_0:

$$x0 = \text{“}Eq(grid.grid\_variable(x0), grid.Idx[0]*grid.deltas[0])\text{”},$$

which in turn defines the initial value for each prognostic variable, by assigning this to the array of values at each grid point (also known as the variable's work array), e.g.:

$$rho = \text{“}Eq(grid.work\_array(rho), 2.0*sin(x0))\text{”}.$$

2.6. Boundary conditions

OpenSBLI currently comprises two types of boundary condition, implemented in the classes Periodic Boundary Condition and Symmetry BoundaryCondition. Users may apply different boundary conditions in different directions if they so wish. Periodic boundaries are defined such that, for each prognostic field ϕ, $\phi(x_0) = \phi(xN)$ where N is the number of points in the domain. This condition is achieved via the exchange of halo point data at each end of the domain. Symmetry boundary conditions enforce the condition that $\phi(xN) = \phi(xN_{-1})$ for scalar fields and $\phi i(xN) =- \phi i(xN_{-1})$ for vector fields (in the direction i), which is achieved using a computational kernel.

2.7. Input and output

The state of the prognostic fields can be written to disk every n iterations as defined by the user, or only at the end of the simulation. This functionality is handled by the FileIOclass. OpenSBLI adopts the HDF5 format [25] ; [26] as it features parallel read/write capabilities and therefore has the potential to overcome the serial input/output bottleneck currently plaguing many large-scale parallel applications [27] ; [28]. Future releases of OpenSBLI will come with the ability to read in mesh files and the state fields from an HDF5 file, enabling the restarting of simulations from 'checkpoints' as well as the assignment of initial conditions that cannot be simply defined by a formula.

2.8. Code generation

OpenSBLI currently generates code in the OPSC language which performs the simulation; this is essentially standard C++ code that includes calls to the OPS library. Such functionality is accomplished using the Open SBLI OPSCCode Printer class (derived from SymPy's CCodePrinter class, used to perform the generation of OPSC code statements) and the OPSC class (which agglomerates the literal strings of OPSC statements and kernel functions and writes them to file). The generated code's structure follows a generic template that maps out the order in which the simulation steps/computations are to be called. The template is represented as a multi-line Python string template, with each line containing a place-holder for the code that performs a particular step. Examples include $header which is replaced by any generic boilerplate header code (e.g. #include <stdlib.h> and kernel function prototypes), $initialisation which is replaced by the grid and field setup (e.g. by declaring an OPS block using the ops_decl_block function), and $bc_calls which is replaced by calls to the boundary condition kernel(s). This template can be readily changed to incorporate additional functionality, such as the inclusion of turbulence models. Once all component place-holders have been replaced by OPSC code, the code is written out to disk. For the case of the OPSC language, two files are written; one is a C++ header file containing the computational kernels, and the other is the C++ source file containing various constant definitions (e.g. the timestep size delta_t, and the constants of the Butcher tableau for the time-stepping scheme), OPS data structures, and calls to the kernels specified in the header file.

OpenSBLI's local Python objects (most pertinently, the kernel objects that describe the computations to be performed on the grid) are essentially translated to OPSC data structures and function calls during the preparation of the code. For instance, when declaring computational stencils that define a particular central differencing scheme, the local grid indices stored in the Central scheme object are used to write out an ops_stencildefinition during code generation. Similarly, ops_halo structures and calls to ops_halo_transferare produced to facilitate the implementation of the periodic boundary conditions. All fields are declared as ops_dat datasets; for an example of where these are used, see the function ops_argument_call in the file opsc.py which generates/accumulates calls to the OPS function ops_arg_dat through the use of 'printf'-style string formatting, filling in the 'placeholder' arguments (e.g. %s in Python) with values from the local OpenSBLI objects. Finally, calls to OpenSBLI Kernel objects are represented in OPSC as regular C++ functions (see Fig. 3) which are passed to the ops_par_loop function (see Fig. 4), which executes the function efficiently over the range of grid points within the desired block; OpenSBLI is currently a single-block code so only one block, containing all the grid points, is used. Further details on the OPS data structures and functionality can be found in the work by Reguly et al. [10].

```
#ifndef block_0_KERNEL_H
#define block_0_KERNEL_H

void mms_2_0_block0_0_kernel(const double *phi , double *wk0)
{
wk0[OPS_ACC1(0,0)] = (-2*phi[OPS_ACC0(0,0)] + phi[OPS_ACC0(0,-1)] + phi[OPS_ACC0(0,1)])/pow(deltai1, 2);
}

void mms_2_0_block0_1_kernel(const double *phi , double *wk1)
{
wk1[OPS_ACC1(0,0)] = (-rc0*phi[OPS_ACC0(0,-1)] + (rc0)*phi[OPS_ACC0(0,1)])/deltai1;
}
```

Figure. 3.\ Code snippit showing two kernels from a 2D 'method of manufactured solutions' (MMS) simulation (see Section 3.2) using second-order central differences. The first kernel computes $\frac{\partial^2 \phi}{\partial x_1^2}$ and stores it in a new work array called *wk0*. Similarly, the second kernel computes the first derivative $\frac{\partial \phi}{\partial x_1}$. The constant deltai1represents the grid spacing in the x_1 direction, and *rc*0 holds the constant value of 0.5. Calls to OPS_ACCare used to access the finite difference stencil structure.

```
for (int iteration=0; iteration<5893; iteration++){

   int iter_range7[] = {-1, nx0 + 1, -1, nx1 + 1};
   ops_par_loop_mms_2_0_block0_7_kernel("Save equations", mms_2_0_block, 2, iter_range7,
               ops_arg_dat(phi, 1, stencil1, "double", OPS_READ),
               ops_arg_dat(phi_old, 1, stencil1, "double", OPS_WRITE));


   for (int stage=0; stage<3; stage++){

      int iter_range0[] = {0, nx0, 0, nx1};
      ops_par_loop_mms_2_0_block0_0_kernel("D(phi[x0 x1 t] x1 x1)", mms_2_0_block, 2, iter_range0,
                  ops_arg_dat(phi, 1, stencil0, "double", OPS_READ),
                  ops_arg_dat(wk0, 1, stencil1, "double", OPS_WRITE));

      int iter_range1[] = {0, nx0, 0, nx1};
      ops_par_loop_mms_2_0_block0_1_kernel("D(phi[x0 x1 t] x1)", mms_2_0_block, 2, iter_range1,
                  ops_arg_dat(phi, 1, stencil2, "double", OPS_READ),
                  ops_arg_dat(wk1, 1, stencil1, "double", OPS_WRITE));
```

Figure. 4. Code snippit showing the calls to the two kernels in Fig. 3 inside the inner for-loop. The outer loop is the time-stepping loop which iterates for a user-defined number of iterations. The inner loop iterates over the 3-stages of the low-storage, third-order Runge-Kutta scheme. Note that each kernel is actually executed as an ops_par_loop over all the grid points.

Some optimisations are performed during the code generation stage by OpenSBLI to avoid unnecessary and expensive division operations in the kernels; rational numbers (e.g. finite difference stencil weights that are rational) and constant EinsteinTerms raised to negative powers (e.g. Re^{-1}) are evaluated and stored (e.g. by over-riding the _print_Rationalmethod in the OPSCCodePrinter class).

Once the code generation process is complete, the OPS library is called to target the code towards various backends. These include the sequential code, MPI and hybrid MPI+OpenMP parallellised versions of the code for CPUs, CUDA and OpenCL versions of the code for GPUs, and an OpenACC version for accelerators. The test cases presented in this paper (see Section 3) consider the sequential, MPI, and CUDA backends. Targetting 'hand-written'/manually-generated model code towards a particular architecture is something that is well-known as a time-consuming, error-prone and often unsustainable activity; often numerical models have to be completely re-written, involving many if-else statements and #ifdef-style pragmas to ensure that the correct branch of the code is followed for a given backend. As the number of backends grows, the code becomes unsustainable. In contrast, with the abstraction introduced here through code generation, support for a new backend only needs to be added to the OPS library; the top-level, abstract definition of the equations and their implementation need not be modified due to the separation of concerns, thereby highlighting one of the key advantages of automated model development.

When comparing the number of lines and the complexity of the code that gets generated by OpenSBLI, another advantage of automated model development becomes clear; in the case of the 3D Taylor-Green vortex test case, the problem specification file containing ~100 lines generates OPSC code that is approximately 1500 lines long (excluding blank lines and comments). As more parameterisations (e.g. Large Eddy Simulation turbulence models) and diagnostic field computations are added, it is expected that this number would grow even further relative to the number of lines required in the setup file.

3. VERIFICATION AND VALIDATION

In order to verify the correctness of OpenSBLI and be confident in the ability of the solution algorithms to accurately represent the underlying physics, three representative test cases covering 1, 2 and 3 dimensions were created and are presented here.

3.1. Propagation of a wave

This 1D test case considers the first-order wave equation, given by

$$\frac{\partial \phi}{\partial t} + c\frac{\partial \phi}{\partial x} = 0, \tag{3}$$

where ϕ is the quantity that is transported at constant speed c. The expected behaviour is that an arbitrary initial profile at time $t = 0$ is displaced by a distance $dt = ct$, such that $\phi(x, t = 0) = \phi(x = dT, t = T)$ for some finish time T. The constant c was set to 0.5 m s^{-1} in this case, and the equation was solved on the line $0 \leq x \leq 1$ m. Eighth-order central differencing was used to discretise the domain in space in conjunction with a third-order Runge-Kutta scheme for temporal discretisation. The grid spacing Δx was set to 0.001 m, and the timestep size Δt was set to 4×10^{-4} s, yielding a Courant number of 0.2. A smooth, periodic initial condition $\phi(x, t = 0) = \sin(2\pi x)$ was used, and periodic boundary conditions were enforced at both ends of the domain.

The simulation was run in serial (on an Intel® Core™i7-4790 CPU) until a finish time of $t = 1$ s. The initial and final states of the solution field ϕ are shown in Fig. 5. As desired, the error in the solution is very small at $O(10^{-10})$, and provides some confidence in the implementation of the solution method and the periodic boundary conditions.

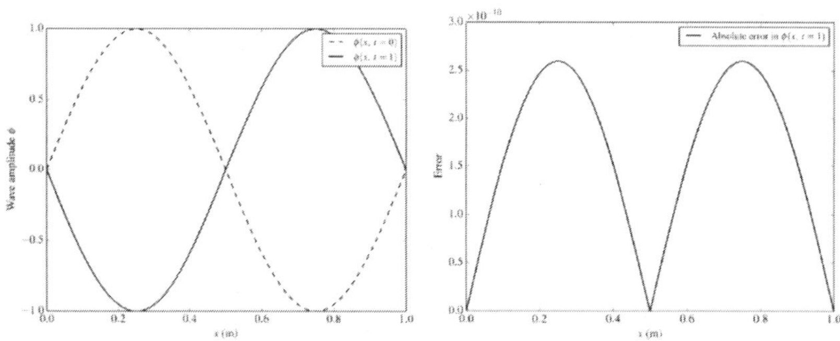

Figure. 5. Results from the 1D wave propagation simulation. Left: The solution field ϕ at time $t = 0$ s and $t = 1$ s. Right: The error between the analytical solution and the numerical solution at time $t = 1$ s.

3.2. Method of manufactured solutions

The method of manufactured solutions (MMS) is a rigorous way to check the correctness of a numerical method's implementation [29]; [30] ; [31]. The overall algorithm involves constructing a manufactured solution ϕm for the prognostic variable(s) ϕ and substituting this into the governing equation. Since the manufactured solution will not, in general, be the exact solution to the equation, a non-zero residual term will be present. This residual term is then subtracted from the RHS such that the manufactured solution essentially becomes the exact/analytical solution of the modified equation (i.e. the one with the source term). A suite of simulations can then be performed using increasingly fine grids to check that the numerical solution converges to the manufactured solution at the expected rate determined by the discretisation scheme.

For this test, the 2D advection-diffusion equation (with a source term S) given by

$$\frac{\partial \phi}{\partial t} + \frac{\partial}{\partial x_j}\left[\phi u_j - k\frac{\partial \phi}{\partial x_j}\right] + S = 0, \tag{4}$$

is considered.

The constant k is the diffusivity coefficient which is set to 0.75 $m^2\ s^{-1}$ here. The prescribed field ui is the ith velocity component, with u_0 = 1.0 m s^{-1} and u_1 =−0.5 m s^{-1}. The prognostic field ϕ is to be determined and has an initial condition of $\phi(x, t = 0) = 0$. In a similar fashion to the works of [29]; [30] ; [31], the manufactured/'analytical' solution $\phi m = \sin(x_0)\cos(x_1)$ employs a mixture of sine and cosine functions since these are continuous and infinitely differentiable. The SAGE framework [32] was used to symbolically determine the residual/source term S.

The domain is a 2D square with dimensions $0 \leq x_0 \leq 2\pi$ m and $0 \leq x_1 \leq 2\pi$ m such that the manufactured solution is periodic. Furthermore, periodic boundary conditions are applied on all sides of the domain. Six central differencing schemes of order 2, 4, 6, 8, 10 and 12 are considered for the spatial discretisation, and a third-order Runge-Kutta scheme is used throughout to advance the equation in time. To perform the convergence analysis, the grid spacing was halved for each successive case such that $\Delta x = \Delta y = \frac{\pi}{2}, \frac{\pi}{4}, \frac{\pi}{8}, \frac{\pi}{16}$,and $\frac{\pi}{32}$. The timestep size Δt was also halved for each case to maintain a maximum bound of 0.025 on the Courant number; this was purposefully kept small and near-constant to minimise the influence of temporal discretisation error [33]. All simulations were run in serial (on an Intel® Core™i7-4790 CPU) until a finish time of $T = 100$ s to ensure that a steady-state solution was attained.

Fig. 6 demonstrates how ϕ converges towards the manufactured solution ϕm as the grid is refined. The convergence rate for each order of the central difference scheme is illustrated in Fig. 7. The anomaly in the twelfth-order convergence plot was likely caused by reaching the limit of machine precision. Overall, these results provide confidence in the correctness of the automatically-generated code/model.

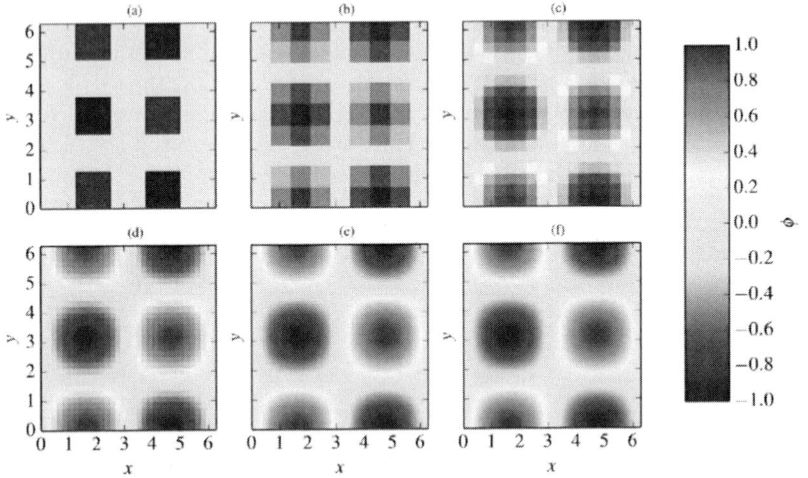

Figure. 6. (a–e) The numerical solution field ϕ at the finish time $t = T$. (f) The manufactured solution ϕm. All results are from the twelfth-order MMS simulation set.

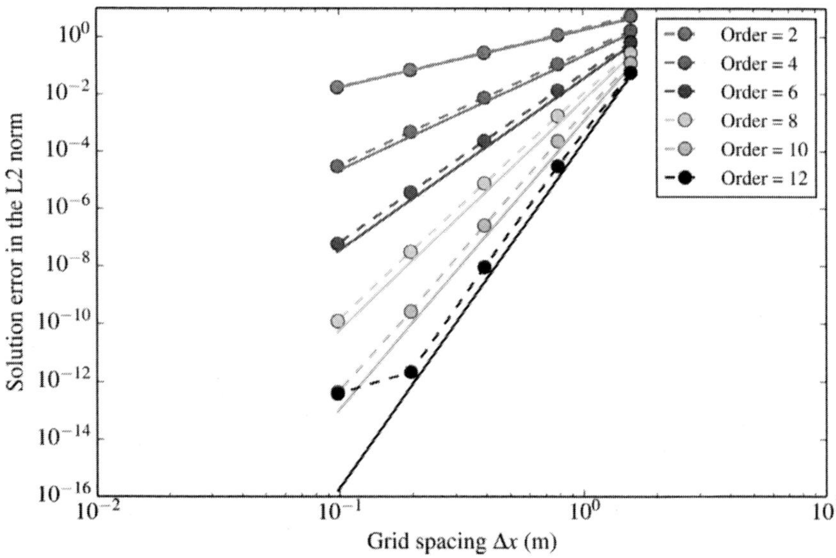

Figure. 7. The absolute error (in the L2 norm) between the numerical solution ϕ and the exact/manufactured solution ϕm, from the suite of MMS simulations. The solid lines represent the expected convergence rate for each order.

3.3. 3D Taylor-Green vortex

The Taylor-Green vortex is a well-known hydrodynamic problem [34]; [35] ; [36]characterised by transition to turbulence, decay of turbulence, and the energy dissipation during its evolution. It is frequently used to evaluate the ability of a numerical method to capture the underlying physical processes. During the initial stages of evolution, the dynamics display structural changes (rolling up, stretching and interaction of the vortices). This process is inviscid in nature. Later the vortices break down and transition into fully-turbulent dynamics. As there are no external forces or turbulence-generating mechanisms, the small-scale structures dissipate all the energy, and the fluid eventually comes to rest [34]. The numerical method employed should be able to capture each of these stages accurately.

The 3D compressible Navier-Stokes equations were solved in non-dimensional form, written in Einstein notation as

$$\frac{\partial \rho}{\partial t} + \frac{\partial}{\partial x_j}[\rho u_j] = 0, \tag{5}$$

$$\frac{\partial \rho u_i}{\partial t} + \frac{\partial}{\partial x_j}[\rho u_i u_j + p\delta_{ij} - \tau_{ij}] = 0, \tag{6}$$

and

$$\frac{\partial \rho E}{\partial t} + \frac{\partial}{\partial x_j}[\rho E u_j + u_j p - q_j - u_i \tau_{ij}] = 0 \tag{7}$$

for the conservation of mass, momentum and energy, respectively. The (dimensionless) quantity ρ is the fluid density, ui is the ith (scalar) component of the velocity vector $\mathbf{u}$, p is the pressure field, E is the total energy. The components of the stress tensor τ are given by

$$\tau_{ij} = \frac{1}{Re}\left(\frac{\partial u_i}{\partial x_j} + \frac{\partial u_j}{\partial x_i} - \frac{2}{3}\delta_{ij}\frac{\partial u_k}{\partial x_k}\right), \tag{8}$$

where δij is the Kronecker Delta function and Re is the Reynolds number. The components of the heat flux term q are given by

$$q_j = \frac{\mu}{(\gamma - 1)M^2 PrRe}\frac{\partial T}{\partial x_j}, \tag{9}$$

where T is the temperature field, γ is the ratio of specific heats, M is the Mach number, and Pr is the Prandtl number. The various quantities are non-dimensionalised using the reference velocity u_{ref}, the reference length L, the reference density ρ_{ref}, and the reference temperature T_{ref}.

The equation of state linking p, ρ and T, is defined by

$$p = \frac{1}{\gamma M^2}\rho T, \tag{10}$$

and the total energy is given by

$$\rho E = \frac{p}{\gamma - 1} + \frac{1}{2}\rho u_j^2. \tag{11}$$

The pressure p is non-dimensionalised by $\rho_{\text{ref}} u_{\text{ref}}^2$.

Central finite difference schemes are non-dissipative and are therefore suitable for accurately capturing turbulent dynamics. However, the lack of dissipation can make the scheme unstable. To improve the stability, a skew-symmetric formulation [37]; [38]; [39] ; [40] was applied to the convective terms in (5)–(7); the convective term then becomes

$$\frac{\partial}{\partial x_j}[\rho\phi u_j] = \frac{1}{2}\left(\frac{\partial}{\partial x_j}\rho\phi u_j + u_j\frac{\partial}{\partial x_j}\rho\phi + \rho\phi\frac{\partial}{\partial x_j}u_j\right), \tag{12}$$

where ϕ should be set to 1, uj and E for the continuity, momentum and energy equations, respectively. It should also be noted that the both the convective and viscous terms are discretised using the same spatial order. In all of the simulations performed, the Laplacian in the viscous term is expanded using a finite difference representation of the second derivative (i.e. not treated by successive first derivatives).

As per the work of DeBonis [35] and Bull and Jameson [36], the equations were solved in a 3D cube, with $0 \le x_0 \le 2\pi L$, $0 \le x_1 \le 2\pi L$, and $0 \le x_2 \le 2\pi L$. Periodic boundary conditions were applied on all surfaces. The following initial conditions were imposed at time $t = 0$:

$$u_0(x_0, x_1, x_2, t=0) = \sin\left(\frac{x_0}{L}\right)\cos\left(\frac{x_1}{L}\right)\cos\left(\frac{x_2}{L}\right), \tag{13}$$

$$u_1(x_0, x_1, x_2, t=0) = -\cos\left(\frac{x_0}{L}\right)\sin\left(\frac{x_1}{L}\right)\cos\left(\frac{x_2}{L}\right), \tag{14}$$

$$u_2(x_0, x_1, x_2, t=0) = 0, \tag{15}$$

$$p(x_0, x_1, x_2, t=0) = \frac{1}{\gamma M^2} + \frac{1}{16}\left(\cos\left(\frac{2x_0}{L}\right) + \cos\left(\frac{2x_1}{L}\right)\right)\left(2 + \cos\left(\frac{2x_2}{L}\right)\right), \tag{16}$$

In all the simulations, $Re = 1600$, $Pr = 0.71$, $M = 0.1$, and $\gamma = 1.4$. The reference quantities L, u_{ref} and ρ_{ref} were set to 1.0, and the reference temperature T_{ref} was evaluated using the equation of state (10).

A fourth-order accurate central differencing scheme was used to spatially discretise the domain, and a third-order Runge-Kutta timestepping scheme was used to march the equations forward in time. A set of simulations was performed over a range of resolutions, namely 64^3, 128^3, 256^3 and 512^3 uniformly-spaced grid points. For the 64^3case, a non-dimensional time-step size Δt of 3.385×10^{-3}[35] was used. Each time the number of grid points was doubled, the time-step size was halved to maintain a constant upper bound on the Courant number. The generated code was

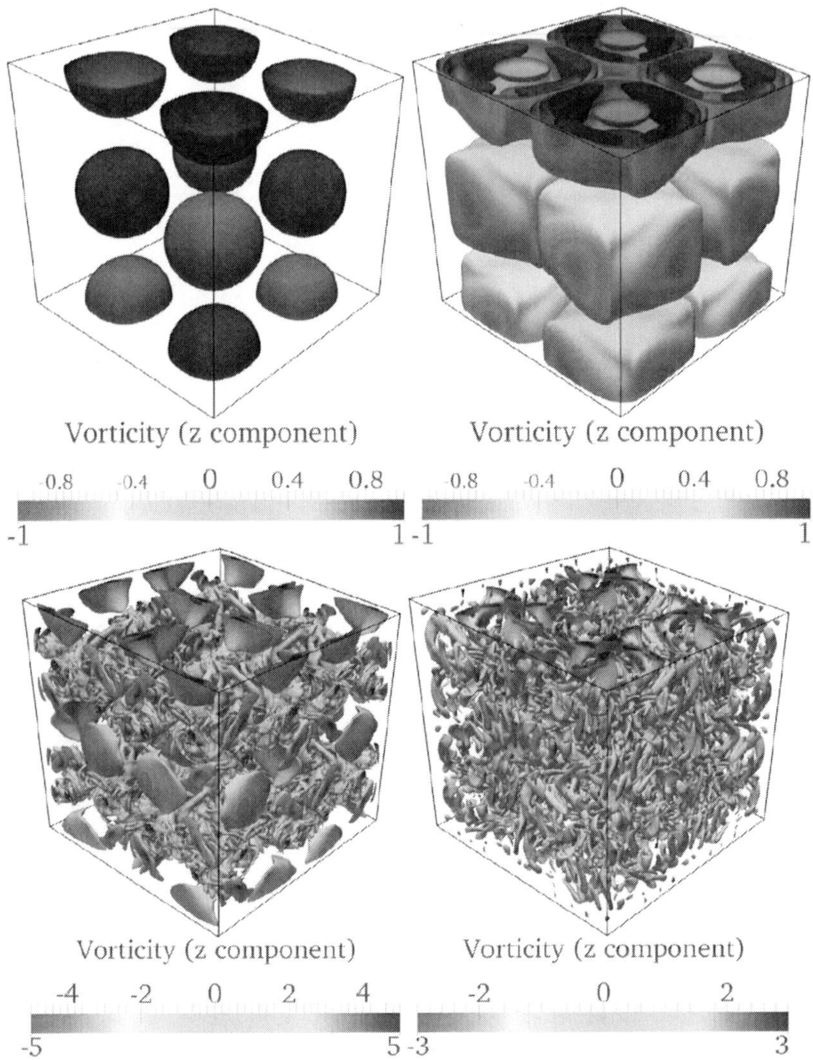

Figure. 8. Visualisations of the non-dimensional vorticity (z-component) iso-contours, from the Taylor-Green vortex test case with a 256^3 grid, at various non-dimensional times. Top left to bottom right: non-dimensional time t = 0, 2.5, 10, 20.

targetted towards the CUDA backend using OPS and executed on an NVIDIA Tesla K40 GPU until a non-dimensional time of $t = 20$, except for the 512^3 case; this was targetted towards the MPI backend and run in parallel over 1440 processes on the UK National Supercomputing Service (ARCHER)

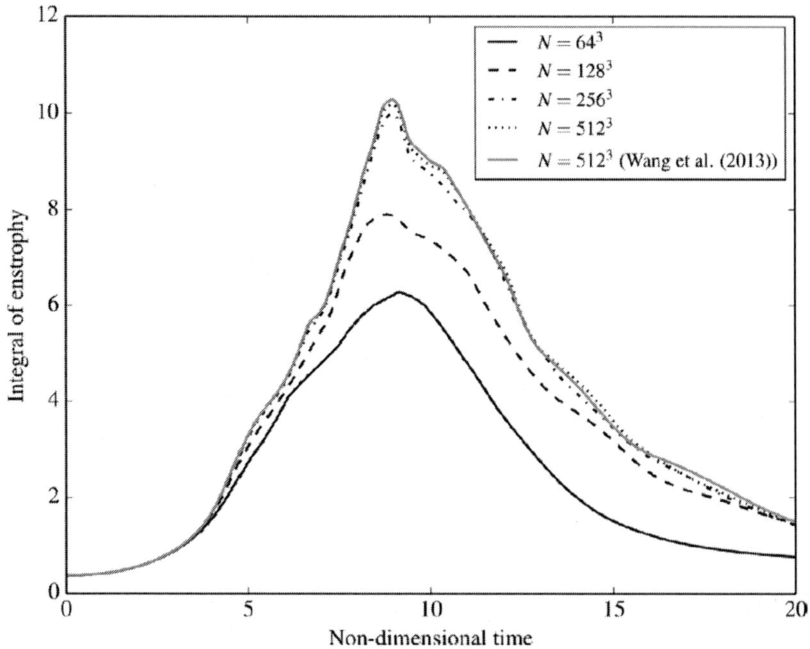

Figure. 9. The integral of the enstrophy in the domain until non-dimensional time $t = 20$, from the Taylor-Green vortex test case. The reference data from [41] is also shown for comparison.

due to lack of available memory on the GPU, and provided a good example of how the backend can be readily changed.

The z-component of the vorticity field at various times can be found in Fig. 8. At non-dimensional time $t = 2.5$ vortex evolution and stretching are clearly visible, progressing onto highly turbulent dynamics where the relatively smooth structures roll-up and eventually breakdown at around $t = 9$. This point is characterised by peak enstrophy in the system. The final stage of the simulation features the decay of the turbulent structures such that the enstrophy tends towards its initial value.

Following the definitions of DeBonis [35], the integrals of the kinetic energy

$$E_k = \frac{1}{\rho_{\mathrm{ref}}\Omega}\int_{\Omega}\frac{1}{2}\rho u_j u_j \mathrm{d}\Omega, \tag{17}$$

and enstrophy

$$\varepsilon = \frac{1}{\rho_{\text{ref}}\Omega}\int_{\Omega}\frac{1}{2}\rho\left(\epsilon_{ijk}\frac{\partial u_k}{\partial x_j}\right)^2 \mathrm{d}\Omega, \tag{18}$$

were computed throughout the simulations. Note that Ω is the whole domain and ϵijk is the Levi-Civita function. These quantities are shown in Figs. 9 and 10 for the various grid resolutions, and are plotted against the reference data from a spectral element simulation by Wang et al. [41] using a 512^3 grid for comparison. Fig. 10 highlights the inviscid nature of the Taylor-Green vortex problem for $t < \sim 3$–4. The transition to turbulence occurs from $\sim 3 < t < 9$ (which is associated with the peak in enstrophy in Fig. 9). Finally, dissipation occurs at $t > 9$. The results show a clear agreement with the reference data, and represents a solid first step towards the validation of OpenSBLI.

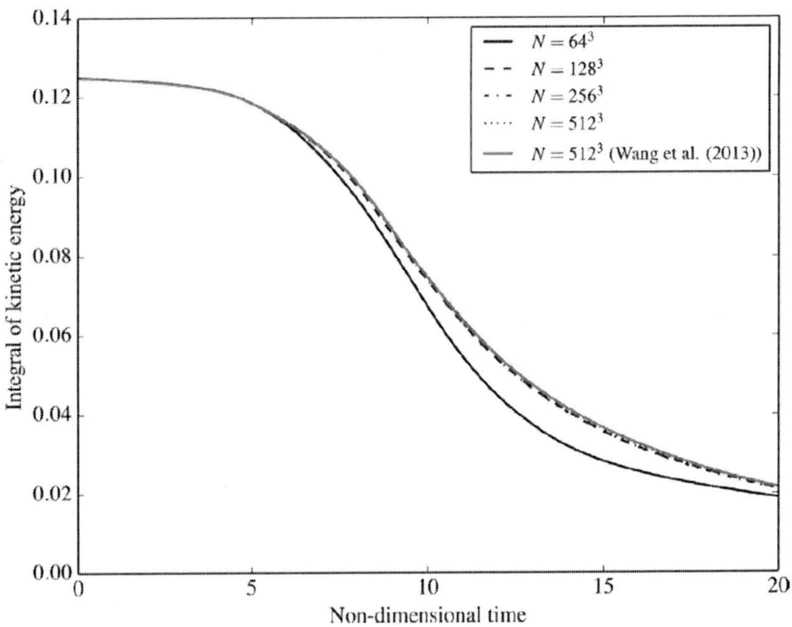

Figure. 10. The integral of the kinetic energy in the domain until non-dimensional time $t = 20$, from the Taylor-Green vortex test case. The reference data from [41] is also shown for comparison.

4. CONCLUSION

Advances in compute hardware are driving a need to change the current state of numerical model development. By developing a new modelling framework based on automated solution techniques, we have effectively future-proofed the core of the SBLI codebase; no longer does a computational scientist need to re-write significant portions of code in order to get it up and running on a new piece of hardware. Instead, the model is derived from a high-level specification independent of the architecture that it will run on, and the underlying code is automatically generated and tailored to a particular backend, the responsibility for which would rest with computer scientists who are experts in parallel programming paradigms. Furthermore, the ease at which the governing equations can be changed is a fundamental advantage of using such abstract specifications. This was highlighted here by considering three test cases, each of which comprised a different set of equations. The discretisation, code generation and code targetting is performed automatically, thereby reducing development costs and potentially avoiding errors, bugs, and non-performant/non-optimal operations. In addition, code that solves the different variants of the same governing equations can be easily generated. For example, in the compressible Navier-Stokes equations, viscosity can be treated either as a constant or as a spatially-varying term. In static, hand-written codes this flexibility comes at the cost of writing different routines for the various formulations, unlike with automated code generation techniques. This is particularly useful when wanting to switch between Cartesian and generalised coordinates. This particular framework also facilitates the fast and efficient switching between different spatial orders of accuracy, and reduces the development time and effort when wishing to try out new numerical formulations of the equations (or a new spatial/temporal scheme) on a wide variety of test cases.

4.1. Future work

Explicit schemes such as the one implemented here can be readily extendible to a range of application areas such as computational aeroacoustics, aero-thermodynamics, problems involving shocks, and hypersonic flow. Incompressible flows may also be handled with the explicit, compressible solver in OpenSBLI so long as the Mach number is sufficiently small. However, this puts tight restrictions on the time-step size thereby limiting the efficiency of the solver, and thus limits the range of applications that OpenSBLI can handle within the context of CFD in general.

Extending to implicit timestepping schemes requires backend support from OPS for matrix inversion, for example. Once this support is implemented, extending OpenSBLI to handle implicit timestepping schemes is straightforward.

The treatment of incompressible flows can be accomplished using schemes such as pressure projection methods [42] ; [43]. Such a method requires (1) the solution of a tentative velocity field, (2) the solution to a pressure Poisson equation (using either direct or iterative solvers), and (3) the update/correction of the velocity field. The equations defining each step would need to be given by the user in OpenSBLI, in a similar fashion to implementing a projection method in the Unified Form Language (UFL) in FEniCS [44] ; [7]. OpenSBLI would also need to recognise that a projection method has been chosen, possibly via a flag set in the problem definition file. For step (2) of the method, direct solvers can be implemented directly once support for matrix inversion and fast Fourier transforms (for example) are included in OPS (which in turn would need to link to various linear algebra packages such as PETSc [45]). This is similar to how an implicit time-stepping scheme would be implemented in OpenSBLI. On the other hand, explicit solution schemes require an iterative solution to the pressure Poisson equation; this is possible by writing the relevant kernel support with an exit criterion (which exits the kernel once a desired tolerance for the solution residual has been attained) and modifying how the code is generated with this in mind. Other methods for incompressible flows such as artificial compressibility methods [42]; [46] ; [47] can be implemented with the current functionality by modifying the input equations in the problem setup file accordingly.

The work considered here only focussed on the MPI and CUDA backends for CPU and GPU execution, respectively. Future work will consider the CPU and GPU performance on other backends, such as OpenMP. For problems such as Mandelbrot Set computation and matrix multiplication, OpenMP has been demonstrated to perform well against other APIs such as CUDA and OpenACC [48]. Future work will also look at comparing the performance of the legacy Fortran-based SBLI code against the OpenSBLI-generated code in order to demonstrate the potential speed-ups that can be obtained.

5. CODE AVAILABILITY

OpenSBLI is an open-source release of the original SBLI code developed at the University of Southampton, and is available under the GNU General Public Licence (version 3). Prospective users can download the source code from the project's Git repository: https://github.com/opensbli/opensbli.

ACKNOWLEDGEMENTS

CTJ was supported by a European Commission Horizon 2020 project grant entitled "ExaFLOW: Enabling Exascale Fluid Dynamics Simulations" (grant reference 671571). SPJ was supported by an EPSRC grant entitled "Future-proof massively-parallel execution of multi-block applications" (EP/K038567/1). The data behind the results presented in this paper are available from the University of Southampton's institutional repository (see http://doi.org/10.5258/SOTON/402072, http://doi.org/10.5258/SOTON/402073, http://doi.org/10.5258/SOTON/402070, http://doi.org/10.5258/SOTON/401892). The authors acknowledge the use of the UK National Supercomputing Service (ARCHER), with computing time provided by the UK Turbulence Consortium (EPSRC grant EP/L000261/1). The authors would also like to thank the anonymous reviewers whose feedback helped improve the quality of this paper, and the NVIDIA Corporation for donating the Tesla K40 GPU used throughout this research.

Appendix A. Example of a simulation setup file

The code in Fig. A.11 contains the key components of a simulation setup file. Specifically, this is taken from the Taylor-Green vortex simulation. Other examples of setup files can be found in the apps directory of OpenSBLI.

```
# Problem dimension
ndim = 3

# Define the compresible Navier-Stokes equations in Einstein notation.
mass = "Eq(Der(rho,t), - Skew(rho*u_j,x_j))"
momentum = "Eq(Der(rhou_i,t), -Skew(rhou_i*u_j,x_j) - Der(p,x_i) + Der(tau_i_j,x_j) )"
energy = "Eq(Der(rhoE,t), - Skew(rhoE*u_j,x_j) - Conservative(p*u_j,x_j) + Der(q_j,x_j) + Der(u_i*tau_i_j,x_j) )"
equations = [mass, momentum, energy]

# Substitutions and constants
stress_tensor = "Eq(tau_i_j, (1.0/Re)*(Der(u_i,x_j) + Der(u_j,x_i)- (2/3)*KD(_i,_j)*Der(u_k,x_k)))"
heat_flux = "Eq(q_j, (1.0/((gama-1)*Minf*Minf*Pr*Re))*Der(T,x_j))"
substitutions = [stress_tensor, heat_flux]
constants = ["Re", "Pr", "gama", "Minf"]

# Formulas for the variables used in the equations
velocity = "Eq(u_i, rhou_i/rho)"
pressure = "Eq(p, (gama-1)*(rhoE - rho*(1/2)*(u_j*u_j)))"
temperature = "Eq(T, p*gama*Minf*Minf/(rho))"
formulas = [velocity, pressure, temperature]

# Create the TGV problem and expand the equations.
problem = Problem(equations, substitutions, ndim, constants=constants,
                  coordinate_symbol="x", metrics=[], formulas=formulas)
expanded_equations = problem.get_expanded(problem.equations)
expanded_formulas = problem.get_expanded(problem.formulas)

spatial_scheme = Central(4) # Fourth-order central differencing in space.
temporal_scheme = RungeKutta(3) # Third-order Runge-Kutta time-stepping scheme.

# Create a numerical grid of solution points
length = [2.0*pi*1.0]*ndim; np = [64]*ndim; deltas = [length[i]/np[i] for i in range(len(length)) ]
grid = Grid(ndim,{'delta':deltas, 'number_of_points':np})

# Perform the discretisation
sd = SpatialDiscretisation(expanded_equations, expanded_formulas, grid, spatial_scheme)
td = TemporalDiscretisation(temporal_scheme, grid, constant_dt=True, spatial_discretisation=sd)

# Boundary condition
boundary_condition = PeriodicBoundaryCondition(grid)
for dim in range(ndim):
    boundary_condition.apply(arrays=td.prognostic_variables, boundary_direction=dim)

# Initial conditions
x = "Eq(grid.grid_variable(x), grid.Idx[0]*grid.deltas[0])"
y = "Eq(grid.grid_variable(y), grid.Idx[1]*grid.deltas[1])"
z = "Eq(grid.grid_variable(z), grid.Idx[2]*grid.deltas[2])"
u = "Eq(grid.grid_variable(u),sin(x)*cos(y)*cos(z))"
v = "Eq(grid.grid_variable(v),-cos(x)*sin(y)*cos(z))"
w = "Eq(grid.grid_variable(w), 0.0)"
p = "Eq(grid.grid_variable(p), 1.0/(gama*Minf*Minf)+ (1.0/16.0) * (cos(2.0*x)+cos(2.0*y))*(2.0 + cos(2.0*z)))"
r = "Eq(grid.grid_variable(r), gama*Minf*Minf*p)"
initial_conditions = [x,y,z,u,v,w,p,r, "Eq(grid.work_array(rho), r)", "Eq(grid.work_array(rhou0), r*u)",
                      "Eq(grid.work_array(rhou1), r*v)", "Eq(grid.work_array(rhou2), 0.0)",
                      "Eq(grid.work_array(rhoE), p/(gama-1) + 0.5* r *(u**2+ v**2 + w**2))"]
initial_conditions = GridBasedInitialisation(grid, initial_conditions)

# I/O save conservative variables at the end of simulation
io = FileIO(temporal_discretisation.prognostic_variables)

# Simulation parameters
l1 = ['niter', 'Re', 'Pr', 'gama', 'Minf', 'mu', 'precision', 'name', 'deltat']
l2 = [20000, 1600, 0.71, 1.4, 0.1, 1.0, "double", "taylor_green_vortex", 3.385*10**-3]
simulation_parameters = dict(zip(l1,l2))

# Generate the code.
OPSC(grid, sd, td, boundary_condition, initial_conditions, io, simulation_parameters)
```

Figure. A.11. A cut-down version of the 3D Taylor-Green vortex setup/configuration file (67 lines long including whitespace), showing the key components and classes available in OpenSBLI.

REFERENCES

1. P. Thibodeau **Scientists, IT Community Await Exascale Computers** (2009) http: //www.computerworld.com/article/2550451/computer-hardware/scientists-it-community-await-exascale-computers.html
2. F. Rathgeber, G.R. Markall, L. Mitchell, N. Loriant, D.A. Ham, C. Bertolli, P.H. Kelly **PyOP2: a high-level framework for performance-portable simulations on unstructured meshes** High Performance Computing, Networking Storage and Analysis, SC Companion, IEEE Computer Society (2012), pp. 1116–1123
3. C.T. Jacobs, M.D. Piggott **Firedrake-Fluids v0.1: numerical modelling of shallow water flows using an automated solution framework** Geoscientif. Model Dev., 8 (3) (2015), pp. 533–547 http://dx.doi.org/10.5194/gmd-8-533-2015
4. F. Rathgeber, D.A. Ham, L. Mitchell, M. Lange, F. Luporini, A.T.T. McRae, G.-T. Bercea, G.R. Markall, P.H.J. Kelly, Firedrake: automating the finite element method by composing abstractions, ACM Trans. Math. Soft., http://arxiv.org/abs/1501.01809.
5. Logg, G.N. Wells **DOLFIN: automated finite element computing** ACM Trans. Math. Soft., 37 (2) (2010) http://dx.doi.org/10.1145/1731022.1731030
6. Logg, K.-A. Mardal, G.N. Wells, *et al.* **Automated Solution of Differential Equations by the Finite Element Method** Springer (2012) http://dx.doi.org/10.1007/978-3-642-23099-8
7. M.S. Alnæs, A. Logg, K.B. Ølgaard, M.E. Rognes, G.N. Wells **Unified form language: a domain-specific language for weak formulations of partial differential equations** ACM Trans. Math. Soft., 40 (2) (2014)
8. F. Luporini, A.L. Varbanescu, F. Rathgeber, G.-T. Bercea, J. Ramanujam, D.A. Ham, P.H.J. Kelly **Cross-loop optimization of arithmetic intensity for finite element local assembly** ACM Trans. Architec. Code Optim., 11 (4) (2015) http: //dx.doi.org/10.1145/2687415
9. M. Giles, I. Reguly, G. Mudalige **OPS C++ User's Manual** University of Oxford (2015) http://www.oerc.ox.ac.uk/projects/ops
10. I.Z. Reguly, G.R. Mudalige, M.B. Giles, D. Curran, S. McIntosh-Smith **The OPS domain specific abstraction for multi-block structured grid computations** Proceedings of the 2014 Fourth International Workshop on Domain-Specific Languages and High-Level Frameworks for High Performance Computing, IEEE Computer Society (2014), pp. 58–67 http://dx.doi.org/10.1109/WOLFHPC.2014.7
11. G.R. Mudalige, I.Z. Reguly, M.B. Giles, A.C. Mallinson, W.P. Gaudin, J.A. Herdman **Performance analysis of a high-level abstractions-based hydrocode on future computing systems** Proceedings of the 5th International Workshop on Performance Modeling, Benchmarking and Simulation of High Performance Computing Systems (PMBS '14), held in conjunction with IEEE/ACM Supercomputing 2014 (SC'14) (2014)
12. S.P. Jammy, G.R. Mudalige, I.Z. Reguly, N.D. Sandham, M. Giles **Block-structured compressible Navier-Stokes solution using the OPS high-level abstraction** Int. J. Comput. Fluid Dyn., 30 (2016), pp. 450–454 http:// dx.doi.org/ 10.1080 /10618562.2016.1243663
13. E. Touber, N.D. Sandham **Large-eddy simulation of low-frequency unsteadiness in a turbulent shock-induced separation bubble** Theor. Comput. Fluid Dyn., 23 (2) (2009), pp. 79–107 http://dx.doi.org/10.1007/s00162-009-0103-z
14. N. De Tullio, N.D. Sandham **Direct numerical simulation of breakdown to turbulence in a Mach 6 boundary layer over a porous surface** Phys. Fluids, 22 (9) (2010) http://dx.doi.org/10.1063/1.3481147
15. J. Redford, N.D. Sandham, G.T. Roberts **Numerical simulations of turbulent spots in supersonic boundary layers: effects of Mach number and wall temperature** Prog. Aerospace Sci., 52 (2012), pp. 67–79 http://dx. doi.org/ 10.1016 /j.paerosci.2011.08.002

16. Wang, N.D. Sandham, W. Hu, W. Liu **Numerical study of oblique shock-wave/boundary-layer interaction considering sidewall effects** J. Fluid Mech., 767 (2015), pp. 526–561 http://dx.doi.org/10.1017/jfm.2015.58
17. SymPy Development Team **SymPy: Python Library for Symbolic Mathematics** (2016) http://www.sympy.org
18. OpenFOAM **OpenFOAM User Guide, Version 2.3.1**(2014)
19. T. Sun **OPESCI-FD: Automatic Code Generation Package for Finite Difference Models** Master's thesis Imperial College London (2016) https: //arxiv.org /abs/1605.06381
20. N. Kukreja, M. Louboutin, F. Vieira, F. Luporini, M. Lange, G. Gorman **Devito: automated fast finite difference computation** Proceedings of the Sixth International Workshop on Domain-Specific Languages and High-Level Frameworks for High Performance Computing (2016) https://arxiv.org/abs/1608.08658
21. M. Lange, N. Kukreja, M. Louboutin, F. Luporini, F. Vieira, V. Pandolfo, P. Velesko, P. Kazakas, G. Gorman **Devito: towards a generic finite difference DSL using symbolic Python** Proceedings of the PyHPC 2016 Conference (2016) https: //arxiv.org/abs/1609.03361
22. W. Bangerth, R. Hartmann, G. Kanschat **deal.II – a general-purpose object-oriented finite element library** ACM Trans. Math. Soft., 33 (4) (2007) http://dx.doi.org/10.1145/1268776.1268779
23. F. Hecht **New development in FreeFem++** J. Numer. Math., 20 (3–4) (2012), pp. 251–265 http://dx.doi.org/10.1515/jnum-2012-0013
24. M.H. Carpenter, C.A. Kennedy **Fourth-Order 2N-Storage Runge-Kutta Schemes, NASA Technical Memorandum 109112** National Aeronautics and Space Administration, Langley Research Center (1994)
25. M. Folk, E. Pourmal **Balancing Performance and Preservation Lessons Learned with HDF5** Proceedings of the 2010 Roadmap for Digital Preservation Interoperability Framework Workshop, US-DPIF '10 (2010), pp. 11:1–11:8 http://dx.doi.org/10.1145/2039274.2039285
26. Collette **Python and HDF5: Unlocking Scientific Data** O'Reilly Media (2013) http://dx.doi.org/10.1007/978-3-642-23099-8
27. Padua **Encyclopedia of Parallel Computing** vol. 4, Springer US (2011) http://dx.doi.org/10.1007/978-3-642-23099-8
28. J. Fu, M. Misun, R. Latham, C.D. Carothers **Parallel I/O Performance for Application-Level Checkpointing on the Blue Gene/P System** Proceedings of the 2011 IEEE International Conference on Cluster Computing (2011), pp. 465–473
29. K. Salari, P. Knupp **Code verification by the method of manufactured solutions, Tech. Rep. SAND2000-1444** Sandia National Laboratories (2000)
30. P.J. Roache **Code verification by the method of manufactured solutions** J. Fluids Eng., 124 (1) (2002), pp. 4–10 http://dx.doi.org/10.1115/1.1436090
31. C.J. Roy **Review of code and solution verification procedures for computational simulation** J. Comput. Phys., 205 (1) (2005), pp. 131–156 http://dx. doi.org /10.1016/j.jcp.2004.10.036
32. W. Stein, D. Joyner **SAGE: system for algebra and geometry experimentation** ACM SIGSAM Bull., 39 (2) (2005)
33. J.M. Vedovoto, A. da Silveira Neto, A. Mura, L.F.F. da Silva **Application of the method of manufactured solutions to the verification of a pressure-based finite-volume numerical scheme** Comput. Fluids, 51 (1) (2011), pp. 85–99 http://dx.doi.org/10.1016/j.compfluid.2011.07.014
34. M.E. Brachet, D.I. Meiron, S.A. Orszag, B.G. Nickel, R.H. Morf, U. Frisch **Small-scale structure of the Taylor-Green vortex** J. Fluid Mech., 130 (1983), pp. 411–452 http://dx.doi.org/10.1017/S0022112083001159
35. J. DeBonis **Solutions of the Taylor-Green vortex problem using high-resolution explicit finite difference methods** 51st AIAA Aerospace Sciences Meeting including the New Horizons Forum and Aerospace Exposition, Aerospace Sciences Meetings (2013) http://dx.doi.org/10.2514/6.2013-382

36. J.R. Bull, A. Jameson **Simulation of the compressible Taylor Green vortex using high-order flux reconstruction schemes** 7th AIAA Theoretical Fluid Mechanics Conference, AIAA Aviation (2014) http://dx.doi.org/10.2514/6.2014-3210
37. G.A. Blaisdell, N.N. Mansour, W.C. Reynolds **Numerical simulations of homogeneous compressible turbulence, Report TF-50** Department of Mechanical Engineering, Stanford University (1991)
38. G.A. Blaisdell, N.N. Mansour, W.C. Reynolds **Compressibility effects on the growth and structure of homogeneous turbulent shear flow** J. Fluid Mech., 256 (1993), pp. 443–485 http://dx.doi.org/10.1017/S0022112093002848
39. G.A. Blaisdell, E.T. Spyropoulos, J.H. Qin **The effect of the formulation of nonlinear terms on aliasing errors in spectral methods** Appl. Numer. Math., 21 (3) (1996), pp. 207–219 http://dx.doi.org/10.1016/0168-9274(96)00005-0
40. S. Pirozzoli **Numerical methods for high-speed flows** Annu. Rev. Fluid Mech., 43 (1) (2011), pp. 163–194 http://dx.doi.org/10.1146/annurev-fluid-122109-160718
41. Z. Wang, K. Fidkowski, R. Abgrall, F. Bassi, D. Caraeni, A. Cary, H. Deconinck, R. Hartmann, K. Hillewaert, H. Huynh, N. Kroll, G. May, P.-O. Persson, B. van Leer, M. Visbal **High-order CFD methods: current status and perspective**Int. J. Numer. Methods Fluids, 72 (8) (2013), pp. 811–845 http://dx.doi.org/10.1002/fld.3767
42. Chorin **The numerical solution of the Navier-Stokes equations for an incompressible fluid** Bull. Am. Math. Soc., 73 (6) (1967), pp. 928–931
43. Chorin **Numerical solution of the Navier-Stokes equations** Math. Comput., 22 (104) (1968), pp. 745–762
44. The FEniCS Project, Incompressible Navier-Stokes equations, Online, 2011. https://fenicsproject.org/documentation/dolfin/1.0.1/python/demo/pde/navier-stokes/python/documentation.html.
45. S. Balay, S. Abhyankar, M. Adams, J. Brown, P. Brune, K. Buschelman, V. Eijkhout, W. Gropp, D. Kaushik, M. Knepley, L. Curfman McInnes, K. Rupp, B. Smith, H. Zhang **PETSc Users Manual, Tech. Rep. Revision 3.5** Argonne National Laboratory (2014)
46. Chorin **A numerical method for solving incompressible viscous flow problems** J. Comput. Phys., 2 (1) (1967), pp. 12–26 http://dx.doi.org/10.1016/0021-9991(67)90037-X
47. T. Ohwada, P. Asinari **Artificial compressibility method revisited: asymptotic numerical method for incompressible Navier-Stokes equations** J. Comput. Phys., 229 (5) (2010), pp. 1698–1723 http://dx.doi.org/10.1016/j.jcp.2009.11.003
48. C.L. Ledur, C.M.D. Zeve, J.C.S. dos Anjos **Comparative analysis of OpenACC, OpenMP and CUDA using sequential and parallel algorithms** Proceedings of the 11th Workshop on Parallel and Distributed Processing (WSPPD), 2013 (2013)

CHAPTER 8

Statistical Models for Web Pages Usability

Saad Subair, Hussah AlEisa

College of Computer and Information Sciences, Princess Nourah bint Abdulrahman University, Riyadh, KSA

ABSTRACT

The usability of an interface is a fundamental issue to elucidate. Many researchers argued that many usability results and recommendations lack empirical and experimental data. In this research, the usability of the web pages is evaluated using several carefully selected statistical models. Universities web pages are chosen as subjects for this work for ease of comparison and ease of collecting data. A series of experiments has been conducted to investigate into the usability and design of the universities web pages. Prototype web pages have been developed according to the structured methodologies of web pages design and usability. Universities web pages were evaluated together with the prototype web pages using a questionnaire which was designed according to the Human Computer Interactions (HCI) heuristics. Nine (users) respondents' variables and 14 web pages variables (items) were studied. Stringent statistical analysis was adopted to extract the required information to form the data acquired, and augmented interpretation of the statistical results was followed. The results showed that the analysis of variance (ANOVA) procedure showed there were significant differences among the universities web pages regarding most of the 23 items studied. Duncan Multiple Range Test (DMRT) showed that the prototype usability performed significantly better regarding most of the items. The correlation analysis showed significant positive and negative correlations between many items. The regression analysis revealed that the most significant factors (items) that contributed to the best model of the universities web pages design and usability were:

multimedia in the web pages, the web pages icons (alone) organisation and design, and graphics attractiveness. The results showed some of the limitations of some heuristics used in conventional interface systems design and proposed some additional heuristics in web pages design and usability.

Keywords: Usability, HCI, Web Pages Interface, Statistical Models, Ergonomics

1. INTRODUCTION

User interface is a system layer through which the computer and users communicate [1] - [3] . It is an essential part of human computer interaction (HCI). A survey conducted by Myers and Rosson [4] showed that software development devoted almost half of the code to the user interface [4] [5] . Dealing with or achieving goals using the interface is known as the usability of the interface. Usability of a system is how easy the system to use and how easy and efficient to perform task [6] -[8] .

The Web is a huge communication system. From a technical point of view, the Web is a system of exchanging data over computer networks using special software. The interfaces used in the web pages are graphical user interfaces that utilize graphics, colours, and icons. Researches showed that there is still a big gap between the researches of the Human Computer Interaction (HCI) and hypertext systems, essentially the web [9] . Shneiderman [6] [10] argued that many researchers' experience lack empirical data to validate or solidify their conclusions. Web sites can be well categorized by the originator's identity such as individual group, university, corporation, and non-profit organization [10] . Information about users can guide web designers to a better design.

The problems that inspired the research of this study can be listed as follows:

1. There were no empirical studies that categorized the users of the universities web pages, studied their preference, and the problems they were facing when using these web pages.
2. To specify the relative importance of some items or parameters of interface design and usability.

The main objective of this study is to investigate into the relations between the items of web pages design and usability and point out which items or variables contribute significantly to web pages design of the universities.

The hypothesis of this study is that:

1. There are significant differences in the design and development of web pages interfaces among the universities;
2. A prototype of a university sample web pages may stimulate and inspire well-structured procedural web pages interface design;
3. Empirical data from a questionnaire and statistical analysis of these data is expected to lead to conclusions and recommendations about the user preference and design aspects of the universities web pages.

The study was focused on the design and development of user interface for the web pages of the universities. The contents of the web pages (the text and the issues inside) were not subjected to detailed studies.

Buckingham and McKnight [9] and Lee and Kozar, [11] reviewed several articles on the web usability and concluded that the link between the hypertext and the web is getting closer. Bieber et al. [12] predicted the evolution of the Web. They examined an array of hypertext technologies which have yet to be adopted in the web development. Smith et al. [13] discussed the information structuring and searching in the Web and reviewed their common problems. Erskine et al. [14] described the application to web site redesign of an adapted form of scenario-based design through claims analysis. Their approaches were contrasting with other theoretical approaches found in the HCI field. They presented a case study of web site redesign which illustrated how scenarios can serve as tools of building interactivity. In a comparative analysis of book-based text and hypertext, Thimble by [15] argued that comprehensive, iterative design and testing (the conventional system development life cycle) of Web structures soon becomes impractical without good authoring tools. He described a tool for Web authoring which uses a page layout language and database of pages to relieve cognitive overhead for authors. Studying distance learning on the web, Benyon et al. [16] presented a case study of iterative, courseware design. They discussed the different presentation and linking constraints imposed by mark-up tools and browser technology. They concluded that before the web-based learning become a practical and effective option, better authoring tools and clearer pedagogical design principles are needed. The knowledge and experience already gained in other user interface design domains such as menu systems and hypertext [7] [17] - [21] and the results of the research in information retrieval systems [22] [23] can be helpful in web design till empirical data and cases accumulate and give solid and clear methodologies for web pages design. Two of the most important issues in web pages design are length and number of links in web pages. Icon, home, or index page to point to fragments is necessary. Meaningful structures that guide users to the fragments they want is the goal, but excessive fragmentation disrupts those who wish to read or print the full text. Providing users with a visual overview of the web site may reduce the disorientation of the web pages which is a result of adding new materials to the web pages [6] . Staggers [24] suggested that compact vertical design within a page to reduce scrolling of pages can be helpful in web pages design. Some white space can help organize a display of a page. Using excessive horizontal rules or blank lines to separate items is a common mistake in web pages design [25] . Shneiderman [10] suggested that sequencing, clustering, and emphasis for objects within a page are very important issues to consider. The first object in a page is expected to be an important item and likely to be selected (Object/Ac- tion). Clustering related items shows meaningful annotation and relationships between them. Important items can be emphasized with large fonts, color highlights, and surrounding boxes. Questionnaires were used to evaluate the usability of interfaces [26] . However, researchers found problems and weaknesses in the questionnaires developed in the past [10] [27] - [31] . The problems varied from non-represented population sample, to lack of validation and low reliabilities of the questionnaires. Users' acceptance or subjective satisfaction of a system is a critical measure of the system success. Chin et al. [29] developed a measurement instrument which measures the user's subjective rating of the human computer interface called the Questionnaire for User Interface Satisfaction (QUIS). Improvements to the questionnaires for evaluating computer systems were introduced when the sample size and the number of different systems to evaluate were increased [32] . According to the Psychometric Theory the reliability of a questionnaire can be increased by increasing the number of items and scaling steps or points, but much many steps on the scale may affect its reliability adversely (Nunnally [33]).

2. METHODOLOGY

In this study, the problems and the objectives of this research which are stated earlier were attempted to be solved or achieved through:

1. Designing and developing a prototype user interface for a university web page using recent web pages development methodologies.
2. Making the necessary amendments and improvements to the prototype.
3. Evaluating the prototype and other universities web pages interface in a comparative approach through a questionnaire.
4. Using the analysis of variance (ANOVA) procedure to point out the significant differences among the universities web pages with regard to each item (variable) of web pages design.

5. Using correlation analysis to investigate the relationship and association between (items) of universities web pages design variable (each two at a time) as well as multiple relationships
6. Using regression analysis to investigate the causality relationships to find out the most significant independent variables (items) that form the best model or models for the universities web pages design.

Like other computer programs, user interface systems go through stages or phases of development life cycle. In general, the phases that are usually followed in developing user interface systems are similar to those of conventional information system development life cycle with minor modifications. However, Nielsen [2] suggested the usability engineering lifecycle to be particularly applied to interface development and evaluation. The art of prototyping was followed in this study. The first prototype was further improved in sequence till it reached an acceptance level. The first prototype was planned to be a fast design using as little resources as possible. Although HTML is fairly easy programming language, enabling images and adjusting the layouts was somewhat tedious and time consuming. The very first prototype was a scenario where both the different features and functions of the web pages were reduced or cut down. At the next step, this scenario was extended to a horizontal prototype where the surface layer or layout of the web pages (front pages) was in its complete form but the links were not linked to real data or information. In this stage, the concept of prototyping was utilized to the best.

As in conventional system development methodologies, iteration was essential in this design. Most of the steps that were followed needed to be revised. This made the iteration an inevitable process in this design. However, the iteration was done carefully since some iteration might revert the usability of the prototype to some poor usability levels. Shneiderman's Objects/Actions Interface (OAI) Model which follows a hierarchical decomposition of objects and actions in the task and interface domains is considered to decompose complex information problem and building a comprehensible and effective prototype. The benefits from this model were centered in building the required interface objects for the prototype and specifying the required actions on that objects.

2.1. The Experiment

After the prototype was developed, an experiment was conducted to investigate into the usability aspects of four universities web pages and the prototype interface design. The universities are:

1. Univ1;

2. Univ2;
3. Univ3;
4. Univ4;
5. The prototype.

The experiment was a comparative evaluation of the usability problems of these web pages. A sample of 59 students was given a questionnaire to respond to. The students were given enough time to go through and navigate all the five web pages and then asked to attempt answering the questions of the questionnaire. The questions of the questionnaire were mainly designed to point out the usability aspects or attributes of the universities web pages interfaces. The computer laboratory where the experiment was conducted was organized in a way that each user will be independent in his navigation of the web pages and his response to the questions. More than one lab was used for the experiment or evaluation. The evaluation environment was kept to be as normal as possible. Since the test took considerable time (1 to 2 hours), the experiment was conducted in sessions and the users were left free to have a break at any time and respond to the questions in a relax way. The users were informed clearly that all the answers to the questionnaire questions are for the sake of evaluating the web pages and not the users and any preliminary data and personal information will be kept confidential. It is in the guidelines of the questionnaire that, the user may not write his/her name if he/she wishes to.

The answer sheets of the questionnaire were collected after the users finished answering the questions and the data was transferred to the SPSS (Statistical package for social sciences) to be analyzed statistically.

The questionnaire was designed using a rating scale ascending from 1 to 5 which was designed to be administered at controlled experimental conditions. The questions were designed such that each question represents an item of design or heuristic of the interface for the universities web pages. The response of the user (respondent) shows his preference or judgement level of that item. Each item was considered as a variable contributing to the web pages design.

2.2. Items (Variables) Studied

In this study, 23 items or variables were designed as questions in the questionnaire and assigned to the five point scale described before. Nine of the 23 items were concerned with the respondents' knowledge background while the other 14 variables were items of web pages design. The 23 items are listed below.

(a) Respondents Knowledge Background

1. Respondents computer exposure;
2. Respondents internet exposure;
3. Respondents universities web pages exposure;
4. Importance of looking up information on universities in the Internet;
5. Respondents trust of web pages information;
6. Respondents music preference in the universities web pages;
7. Respondents animated pictures preference in the universities web pages;
8. Respondents graphics preference in the universities web pages;
9. Importance of developing web pages for the universities.

(b) Items (Variables) of Web Pages Design

1. Color amount;
2. Contrast against the background;
3. Graphics information provision;
4. Color and Graphics feel of Consistency;
5. Navigation;
6. Location tracking inside the web pages;
7. Hyperlinks (hotspot) semantics;
8. Forms organization and helpfulness;
9. Animated pictures effect on web pages look;
10. Text (alone) organization and design;
11. Icons (alone) organization and design;
12. Icons and text organization and design;
13. Multimedia preference;
14. Universities web pages ranking.

2.3. Statistical Analysis

The data was collected from the questionnaire answer sheets and coded and transferred into SPSS®statistical package to obtain the statistical analysis for the descriptive statistics of the sample, the analysis of variance (ANOVA) of the universities web pages, and the correlation analysis of the items (variables). The data was then converted into text file and uploaded to the university's mainframe to use SAS® statistical package to obtain the regression analysis output which is clearer in SAS®. Interpretations of all these statistical measures would reveal the significant difference among the universities web pages interface design for each item and point out the better items or variables for the universities web pages design.

2.4. Analysis of Variance Procedure

The ANOVA partitions the source of variations into within groups and between groups variation to obtain the F-value which in turn determines whatever to accept or reject the null hypothesis. The analysis of variance (ANOVA) table will then look as illustrated in Table 1.

Duncan's multiple range test (DMRT) was used to test the differences between groups (universities).

2.5. Correlation Analysis

Simple correlation between the nine respondents' computer and Internet level background items (variable) and the 23 items of universities web pages design (independently) were calculated from the original data using SPSS. The correlation coefficient (r) between two characters (X and Y) is usually expressed as follows:

$$r = \frac{\sum xy}{\sqrt{\left(\sum x^2\right)\left(\sum y^2\right)}}$$

where

x = x − X and y = y − Y and

x = Independent variable (item); X is the mean of character X.

y = Independent variable (item); Y is the mean of character Y.

2.6. Regression Analysis

Simple and multiple linear regressions of some items (independent variables) on the ranking of the university web pages item (dependent variable) were estimated. The formula for multiple linear regression is as follows:

$$Y = a + b_1 X_1 + b_2 X_2 + \cdots + b_n X_n$$

where:

Y = dependent variable;

$X_{1\text{-}n}$ = independent variables;

a = intercept of the regression line on the Y axis;

$b_{1\text{-}n}$ = linear regression coefficients;

n = N Univ3ber of independent variables.

Backward, forward, and stepwise selection regression analysis has also been attempted. In stepwise selection variables are added (or removed) one by one into (from) the model. F-statistic is calculated after a variable is added into (or removed from) the model. A variable that did not produce significant F-value at a redefined pro- bability level is then deleted from (or added to) the model. In forward selection, variables already in the model do not necessary stay there and the process ends when none of the variables outside the model is significantly contributing to the model at the specified probability level [34] [35] .

Table 1. Analysis of variance (ANOVA) table based on individual observations[c].

Source of variation	df[a]	Mean square (variance)	$F_{(Observed)}$[b]
Between groups	k − 1		$\text{Variance}_{(\text{Between groups})}$
Within groups (error)	N − k		$\text{Variance}_{(\text{Within groups})}$
Total	N − 1		

[a]k = the n Univ3ber of groups; N = the total n Univ3ber of subjects in the study; [b]Significance level (probability level); [c]Adapted from (Shavelson, 1996).

3. RESULTS AND DISCUSSION

In this article, the nine respondents' knowledge background variables will be discussed first followed by the 14 web pages variables.

3.1. Respondents' Knowledge Background

A correlation analysis study was used to explain the relationship between the nine variables of the respondents' knowledge background. **Table 2** shows that respondents' computer exposure has positive significant correlation with the respondents' internet exposure, respondents' exposure to universities web pages, importance of looking up information about universities in the internet, and the importance of developing web pages for universities. However, respondents' computer exposure showed significant negative correlation with the respondents' preference to animation in universities web pages.

Respondents' internet exposure showed positive significant correlation with the respondents' exposure to uni- versities web pages, respondents' trust of the information of the web, and the importance of developing web pages for universities. Meanwhile, respondents' internet exposure showed significant negative correlation with the respondents' music preference and animated pictures preference in the web pages of universities (**Table 2**). **Table 2** also shows that respondents exposure to universities and importance of looking information about universities in the internet followed almost the same pattern of relation with the other variables as the previous variable respondents' internet exposure.

Table 2 also shows that respondents who trust information of the web believe in the importance of developing web pages for universities. It is also shown that respondents who prefer music in the universities web pages they also prefer animated pictures and graphics in the web pages of

Table 2. Correlation coefficient of the 12 items (variables) of the respondents computers and Internet usage background.

	Respondents computer exposure	Respondents internet exposure	Respondents exposure to universities web pages	Importance of looking information about universities in the internet	Respondents trust of the information of the web	Respondents music preference in the universities web pages	Respondents animation pictures preference in the universities web pages	Respondents graphics preference in the universities web pages
Respondents internet exposure	0.52**							
Respondents exposure to universities web pages	0.40**	0.59**						
Importance of looking information on universities in internet	0.24**	0.13	0.23**					
Respondents trust of the information of the web	0.20**	0.26**	0.18*	0.11				
Respondents music preference in the universities web pages	−0.09	−0.17*	0.02	0.17*	−0.08			
Respondents animation pictures preference in universities web pages	−0.19**	−0.30**	−0.10	-	-0.13	0.55**		
Respondents graphics preference in the universities web pages	0.05	0.05	0.14*	0.07	-	0.36**	0.48**	
Importance of developing web pages for universities	0.33**	0.36**	0.28**	0.38**	0.23**	−0.13	−0.21**	0.20**

universities. However, **Table 2** shows a negative significant correlation between respondents animation preference in universities web pages and the importance of developing web pages for universities. This result reveals that respondents who believe in the importance of developing web pages for universities do not necessarily prefer animated pictures in these web pages. On the other hand, the table shows that the same respondents prefer graphics in these web pages.

3.2. Items (Variables) of Interface Design and Usability

The descriptive statistics of the 14 items of the web pages interface design and usability are shown in **Table 3**. The number of observations (N), means, mean standard errors, standard deviations, and variances is shown in **Table 3**. Low variances in the table show continuous data and hence suggest normal distribution of the data. The analysis of variance (ANOVA) of the variables is shown in **Table 4**. Only variables that showed significant differences among universities web pages are displayed. Navigation of the web pages, location tracking, hyperlinks semantics, animated pictures effect on the web pages, and icons (alone) organization did not show significant difference in the universities web pages with regards to those variables and hence they are not displayed in the table.

However, there are significant differences among the universities web pages regarding the rest of the variables. Duncan's multiple range test (**Table 5**) was used to arrange these variables in significantly different groups. Table 5 shows that the prototype web pages were grouped in the best group in almost all the variables showed in the table. This reveals that the prototype was almost the best among the other universities web pages.

Table 6 shows the correlation coefficients between the 14 variables studied. The web pages color amount has no significant correlation with any of the variables of the web pages studied. The web pages contrast against the background showed significant positive correlation with the web pages icons and text organization, color and graphics consistency, navigation, location tracking, and forms organization and helpfulness.

However, the contrast against the background showed no significant correlation with the web pages animated pictures effect and icons (alone) organization. The web pages graphics information provision has significant positive correlation with all the other variables studied. Specifically, it showed high positive correlation with the web pages hyperlinks semantics (Table 6). Table 6 shows that web pages color and graphics feel of consistency has positive significant correlation with most of the variables studied. However, it has no significant correlation with the text (alone) and icons (alone) organization but has high positive significant correlation with the icons

Table 3. Number of observations, means, standard errors, standard deviations, and variances of the 14 items (variables) of the universities web pages.

Item (variable)	Number of observation	Mean	Standard error	Standard deviation	Variance
Color amount	292	2.90	0.05	0.78	0.60
Contrast against the background	294	3.29	0.06	1.05	1.10
Graphics information provision	290	3.27	0.05	0.84	0.71
Color and graphics feel of consistency	295	3.12	0.05	0.86	0.73
Navigation	295	3.69	0.05	0.94	0.89
Location tracking inside the web pages	294	3.54	0.06	0.97	0.94
Hyperlinks (hotspot) semantics	295	3.56	0.06	1.01	1.02
Forms organization and helpfulness	290	3.38	0.05	0.87	0.75
Animated pictures effect on web pages look	290	3.61	0.05	0.94	0.88
Text (alone) organization and design	290	3.03	0.06	1.00	1.01
Icons (alone) organization and design	293	3.27	0.06	1.00	0.99
Icons and text organization and design	294	3.62	0.05	0.90	0.80
Multimedia preference	294	3.09	0.06	0.99	0.99
Universities web pages ranking	280	2.97	0.08	1.40	1.96

Table 4. Analysis of variance (ANOVA) tables of the five universities for the 35 items (variables) of the web pages studied[a].

Item (variable)		df	Sum of square	Mean square	F-value	$p > F$
Colour amount						
	Between groups	4	7.422	1.856	3.172	0.014***
	Within groups	287	167.893	0.585		
	Total	291	175.315			
Contrast against the background						
	Between groups	4	18.673	4.668	4.435	0.002***
	Within groups	289	304.171	1.052		
	Total	293	322.844			
Graphics information provision						
	Between groups	4	11.977	2.994	4.431	0.002***
	Within groups	285	192.578	0.676		
	Total	289	204.555			

Colour and graphics feel of consistency						
	Between groups	4	14.590	3.647	5.262	0.000***
	Within groups	290	201.017	0.693		
	Total	294	215.607			
Forms organization and helpfulness						
	Between groups	4	5.876	1.469	1.973	0.099*
	Within groups	285	212.155	0.744		
	Total	289	218.031			
Text (alone) organization and design						
	Between groups	4	9.103	2.276	2.296	0.059*
	Within groups	285	282.552	0.991		
	Total	289	291.655			
Icons and text organization and design						
	Between groups	4	9.526	2.382	3.048	0.017**
	Within groups	289	225.807	0.781		
	Total	293	235.333			
Multimedia preference						
	Between groups	4	10.285	2.571	2.671	0.032**
	Within groups	289	278.236	0.963		
	Total	293	288.520			
Universities web pages ranking						
	Between groups	4	27.086	6.771	3.584	0.007***
	Within groups	275	519.625	1.890		
	Total	279	546.711			

***: Significant at 0.01 level of probability. **: Significant at 0.05 level of probability. *: Significant at 0.10 level of probability. ^{a}: Only significant variables are displayed.

Table 5. Grouping of the five universities for the 35 items (variables) studied according to Duncan's multiple range test[d].

	Colour amount	Contrast against the back ground	Graphics information provision	Colour and graphics feel of consistency	Navigation	Hyperlink (hotspot) semantics	Forms organization and helpfulness	Text (alone) organization and design	Icons and text organization and design	Multi-media preference	Universities web pages ranking
Univ1	2.74^{a}	3.12^{a}	3.12^{ab}	3.19^{bc}	3.73^{ab}	3.37^{a}	3.38^{ab}	3.19^{b}	3.36^{a}	3.02^{ab}	2.95^{b}
Univ2	2.91^{ab}	3.58^{b}	3.34^{bc}	3.36^{c}	3.78^{ab}	3.59^{ab}	3.40^{ab}	2.95^{ab}	3.80^{b}	3.29^{b}	3.39^{b}
Univ3	3.15^{b}	3.24^{ab}	3.30^{bc}	2.98^{ab}	3.66^{ab}	3.59^{ab}	3.38^{ab}	3.16^{b}	3.58^{ab}	3.20^{b}	3.00^{b}
Univ4	2.73^{a}	2.95^{a}	2.98^{a}	2.76^{a}	3.44^{a}	3.42^{ab}	3.14^{a}	2.72^{a}	3.53^{ab}	2.76^{a}	2.43^{a}
Prototype	2.98^{ab}	3.59^{b}	3.58^{c}	3.32^{c}	3.83^{b}	3.80^{b}	3.59^{b}	3.16^{b}	3.84^{b}	3.19^{b}	3.07^{b}

[d]Means with the same letters indicate no significant difference from each other at the 0.05 probability level as determined by DMRT.

Table 6. Correlation coefficients of the 14 items (variables) of the universities web pages design and usability[a].

	Colour amount	Contrast against the background	Graphics information provision	Colour and graphics feel of consistency	Navigation	Location Tracking	Hypelinks Semantics	Forms organization	Animated pictures effect	Text (alone) organization	Icons (alone) organization	Icons and text organization	Multimedia preference
Contrast against the background	0.05												
Graphics information provision	0.11	0.28**											
Colour and graphics feel of consistency	0.01	0.33**	0.27**										
Navigation	0.09	0.28**	-	0.14*									
Location Tracking	0.05	0.28**	0.25**	0.26**	0.41**								
Hypelinks Semantics	0.03	0.21**	0.34**	0.21**	0.33**	0.40**							
Forms organization	−0.04	0.27**	0.23**	0.30**	0.33**	0.41**	0.37**						
Animated pictures effect	0.00	0.13	-	0.20**	0.10	0.31**	0.22**	0.31**					
Text (alone) organization	0.01	0.20**	0.16**	0.11	0.22**	0.29**	0.26**	0.41**	0.18**				
Icons (alone) organization	0.07	0.09	0.21**	0.11	0.11	-	0.27**	0.21**	0.24**	0.17*			
Icons and text organization	0.06	0.35**	0.28**	0.33**	0.32**	0.44**	0.26**	0.26**	0.35**	-	0.27**		
Multimedia preference	0.13	0.21**	0.20**	0.20**	0.11	0.18**	0.20**	0.29**	-	0.35**	0.24**	-	
Universities web pages ranking	0.11	0.15*	0.15*	0.19**	-	0.03	0.10	0.17*	0.13	0.09	0.18*	0.18**	0.27**

and text together organization and helpfulness. Navigation of the web pages has significant positive correlation with all the variables studied except the web pages icons (alone) organization and the multimedia preferences in the universities web pages. Location tracking and hyperlinks semantics showed significant positive correlation (separately) with almost all the other variables studied except the ranking of the universities web pages. This result indicated that the ranking of the web pages of the universities did not rely much on these two variables. The web pages forms organization has significant positive correlation with the rest of the variables studied. Forms organization associated significantly with the ranking of the universities web pages. The animated pictures effect on the web pages showed significant positive correlation with the web pages text (alone), icons (alone), and icons and text (together) organization and helpfulness.

However, animated pictures have no significant correlation with the ranking of the web pages. Text (alone) organization has significant positive correlation with the icons alone organization and multimedia preferences in the web pages of the universities. However, text (alone) has no significant correlation with the ranking of the universities web pages. Icons (alone) organization has significant positive correlation with the icons and text (together) organization, multimedia preference, and the web pages ranking. Icons and text (together) organization and multimedia preference have significant positive correlation (separately) with the ranking of the universities pages.

3.3. Regression Analysis

Multiple linear regression results are shown in **Table 7**. The ANOVA (analysis of variance) table for regression shows that the model was significant at the 0.01 probability level, indicating that the contribution of independent variables coefficients were not equal to zero (alternate hypothesis). The table also shows that multimedia pre- ference in the universities web pages contributed highly and significantly to the model. However, some regression analysis procedures were used, including, forward selection, backward elimination, and stepwise selection. Maximum coefficient of determination (R^2) was achieved by including specific number of independent variables.

Table 7. Linear regression analysis between universities web pages ranking and other items (variables) of web pages design and usability.

Source	Degrees of freedom	Sum of square	Mean square	F	$p > F$
Regression	13	70.00086	5.384682	7.26	0.0001
Error	243	180.1159	0.741218		
Total	256	250.1167			

Variable	Parameter estimate	Standard error	Sum of squares	F	$p > F$
Intercept	1.061532	0.440591	4.302695	5.80	0.0167
Color amount	0.027230	0.075147	0.097328	0.13	0.7174
Contrast against the background	−0.030090	0.059111	0.192028	0.26	0.6112
Graphics information provision	0.068228	0.074656	0.619078	0.84	0.3617
Color and graphics feel of consistency	−0.030560	0.071732	0.134551	0.18	0.6704
General navigation	0.069340	0.068237	0.765360	1.03	0.3106
Location tracking	−0.064310	0.069323	0.637798	0.86	0.3545
Hyperlinks (hotspots) semantics	0.121541	0.073557	2.023700	2.73	0.0998

Forms organization and helpfulness	0.026652	0.070631	0.105537	0.14	0.7063
Animated pictures	−0.055670	0.089501	0.286755	0.39	0.5345
Text (alone) organization and design	−0.109380	0.064121	2.157001	2.91	0.0893
Icons (alone) organization and design	0.200612	0.057933	8.888093	11.99	0.0006
Icons and text organization and design	−0.018650	0.060757	0.069868	0.09	0.7591
Multimedia preference	0.416697	0.073722	23.68064	31.95	0.0001

Table 8 shows a summary of forward selection procedure for the dependent variable universities web pages ranking. The procedure showed that only eight independent variables contributed significantly to the model. However, universities web pages multimedia preference and icons (alone) organization and design were contributed significantly to the model. The forward selection begins with the most independent variable contributing highly to the model and advances to the next most contributing one till it reaches a satisfactory model.

Unlike the forward selection, backward elimination begins with the complete model including all the independent variables and then started to eliminate the most non-significant variable in the model considering its F-value, then proceed to the next most non-significant one. **Table 9** shows a summary of the backward elimination procedure. As seen in the table, nine variables were removed from the model. This indicated that these 9 variables did not contribute significantly to the model (in this order) as the rest of the variables did.

The forward selection and backward elimination adds and eliminates (respectively) independent variables to (from) the model regardless of what variables are already outside or existing in the model. Stepwise selection re- examines at every stage the regression of the variables incorporated into the model in previous stages. A variable which might have been the best single variable to enter the model at an early stage, might at later stage, be superfluous because of the relationships between it and other variables now in the regression model [34] . **Table 10** shows a summary of stepwise selection regression procedure model which included only four variables that were significantly contributed to the model at the 0.15 probability level. That is to say, universities web pages multimedia preference, icons (alone) organization, text (alone) organization, and hyperlinks (hotspots) semantics were the four most independent variables contributing to the model significantly.

Coefficient of determination (R^2) measures the proportion of total sum of squares of the variables that is explained by the regression line. It is a measure of how closely the points (observation) fit the least square line. Consequently, the line that has maximum R^2 represents the best fitting line [36] . **Table 11** shows the maxi mum R^2 that is attained with one independent variable, two independent variables, and so on till 13 independent variables. This might indicate the ranking importance of an independent variable as a single or in combination with other independent variables to universities web pages ranking.

Table 8. Summary of forward selection procedure for dependent variable universities web pages ranking.

Step	Variable entered	Number in	Partial R^2	$P > F$
1	Multimedia preference	1	0.2132	0.0001
2	Icons (alone) organization and design	2	0.0401	0.0003
3	Text (alone) organization and design	3	0.0073	0.1151
4	Hyperlinks Semantics	4	0.0094	0.0736
5	Location Tracking	5	0.0021	0.3959
6	Navigation	6	0.0019	0.4177
7	Graphics information provision	7	0.0017	0.4511
8	Animated pictures effect	8	0.0017	0.4466

Table 9. Summary of backward elimination procedure for dependent variable universities web pages ranking.

Step	Variable removed	Number in	Partial R^2	$P > F$
1	Icons and text organization	12	0.0003	0.7591
2	Color amount	11	0.0004	0.7161
3	Forms organization	10	0.0003	0.7426
4	Color and graphics feel of consistency	9	0.0006	0.6585
5	Contrast against the background	8	0.0009	0.5751
6	Animated pictures effect	7	0.0017	0.4466
7	Graphics information provision	6	0.0017	0.4511
8	Navigation	5	0.0019	0.4177
9	Location tracking	4	0.0021	0.3959

Table 10. Summary of stepwise procedure for dependent variable universities web pages ranking[a].

Step	Variable entered/removed	Number in	Partial R^2	Model R^2	$P > F$
1	Multimedia preference	1	0.2132	0.2132	0.0001
2	Icons (alone) organization and design	2	0.0401	0.2534	0.0003
3	Text (alone) organization and design	3	0.0073	0.2607	0.1151
4	Hyperlinks semantic	4	0.0094	0.2700	0.0736

[a]: All variables in the model are significant at 0.15 probability level.

Table 11. R^2 of single and/or combination of different independent variable (s) with dependent variable universities web pages ranking.

Number of variables in the model	Independent variable (s)	R^2
1	Multimedia preference	0.2132491
1	Icons (alone) organization	0.084938
1	Hyperlinks semantics	0.055802
2	Icons (alone) organization and design and multimedia preference	0.253370
2	Hyperlinks semantics and multimedia preference	0.222795
2	Navigation and multimedia preference	0.218554
3	Text (alone) organization and design and icons (alone) organization and design and multimedia preference	0.260674
3	Hyperlinks semantics and icons (alone) organization and design and multimedia preference	0.259198
3	Color amount and icons (alone) organization and design and multimedia preference	0.256156
4	Hyperlinks semantics and text (alone) organization pages and icons (alone) organization and design and multimedia preference	0.270027
4	Navigation and text (alone) organization pages and icons (alone) organization and design and multimedia preference	0.263917
4	Graphics information provision and text (alone) organization pages and icons (alone) organization and design and multimedia preference	0.263123
5	Location tracking and hyperlinks semantics and text (alone) organization and icons (alone) organization and design and multimedia preference	0.272125
5	Graphics information provision and hyperlinks semantics and text (alone) organization pages and icons (alone) organization and design and multimedia preference	0.271217
5	Navigation and hyperlinks semantics and text (alone) organization pages and icons (alone) organization and design and multimedia preference	0.271186

The results showed that the universities web pages location tracking, hyperlinks semantics, text (alone) organization, icons (alone) organization, and design, and multimedia preferences were among the first five variables (items) that contributed significantly to the universities web pages ranking. However, limitation to R^2 is that the addition of an irrelevant independent variable to the regression model will increase R^2 even though the irrelevant variable is not related to the other variables in the model [36] . The assumption of this model is that all the variables (items) have been estimated precisely.

The results of **Tables 7-11** suggest that, some independent variables significantly contributed to the regression model and others did not. For a single independent variable, multimedia preference of the universities web pages gave the most significant contribution to the dependent variable universities web pages ranking (Table 11). For two independent variables, icons (alone) organization and design and multimedia preference of the universities web pages gave the most significant contribution to the dependent variable universities web pages ranking. For three independent variables, the universities web pages text (alone) organization and design, icons (alone) organization and design, and multimedia preference gave the most significant contribution to the dependent variable universities web pages ranking (Table 11).

4. CONCLUSIONS

From the discussion, some important conclusions and findings concerning both users (respondents) and items (variables) of the web pages of the universities [37] were unveiled. These conclusions and findings can be summarized as follows:

1. Respondents who were frequent in using computers did not prefer animated pictures in universities web pages (**Table 2**).
2. Respondents who frequently access the Internet believe that it is important to develop web pages for universities (**Table 2**).
3. Respondents who were well exposed to computer did not prefer music and animation in the universities web pages (**Table 2**).
4. There were significant differences between the universities web pages regarding most of the variables studied (**Table 4**).
5. The prototype which was developed according to a scientific methodology was proved to be better than most of the other universities web pages (**Table 5**).
6. Good contrast against the background of the web pages made the navigation of the web pages more convenient (**Table 6**).
7. Good design of graphics made the icons and hyperlinks meaningful and had metaphoric dimensions (**Table 6**).
8. The usage of the icons and text together in the universities web pages design was proved to be better than the usage of the text alone or the icons alone (**Table 6**).
9. Multimedia in the web pages, the web pages, icons organization and design, the web pages text organization and design, and hyperlinks semantics in the web pages independently and separately contributed significantly (in this order) to the ranking of the universities web pages (**Table 7**).
10. The best five items of web pages design and usability that together formed the most significant model were: location tracking, hyperlinks semantics, text (alone) organization and design, icons (alone) organization and design, and multimedia preference (**Table 11**).

5. RECOMMENDATIONS FOR FUTURE WORK

1. Further studies can be done to increase the validity and reliability of the questionnaire used in this study. This can be achieved by increasing and randomizing the sample of the test users (respondents), increasing the steps of the scale (e.g. 10 points scale), and increasing the number of the questions used in the questionnaire.
2. In further studies, it is suggested that the questions of the questionnaire can be improved by being more specific and adhere to the scale designed for that. General questions can yield confounding results. For one item or variable of interface usability there may be several questions instead of only one or few.
3. In this study, only one sample group was used. In future work, the sample users (respondents) can be divided into two groups: users and experts. More sophisticated scale (e.g. 10 points scale) can be constructed for expert users (respondents). In 10 points scale for instance, expert users can better distinguish the different levels (points) of a single question which can yield a wide range of data which could be better analyzed statistically.
4. Additional statistical models can rigorously analyze the usability of the web pages and hence more valid and verified conclusions can be achieved.

ACKNOWLEDGEMENTS

The authors would like to thank Princess Nourah bint Abdulrahman University Riyadh, KSA, for assistance they rendered concerning this work.

REFERENCES

1. Jacko, J.A. (2012) The human-Computer Interaction Handbook: Fundamentals, Evolving Technologies, and Emerging Applications. CRC Press, Florida, 1518. http:// dx.doi.org/10.1201/b11963
2. Nielsen, J. (1993) Usability Engineering. Academic Press, Inc., Harcourt Brace & Company, San Diego, USA.
3. Landseadel, P. (1994) Methodologies in User Interface Design. IEEE Electronic Library, No. 0-7803-1910-9/94. http://dx.doi.org/10.1109/autest.1994.381557
4. Myers, B.A. and Rosson, M.B. (1992) Survey on User Interface Programming. CHI'92 Proceedings of the SIGCHI Conference on Human Factors in Computing Systems, Monterey, CA, 3-7 May 1992, 195-202. http://dx.doi.org/10.1145/142750.142789

5. Norman, D.A. (2010) Natural User Interfaces Are Not Natural. Interactions, 17, No. 3.http://dx.doi.org/10.1145/1744161.1744163
6. Shneiderman, B., Kent, L. Norman, C.P., Benjamin, B., Bederson, D.A. and Golbeck, J. (2013) 30 Years at the University of Maryland's Human-Computer Interaction Lab (HCIL). Interactions, 20, 50-57. http://dx.doi.org/10.1145/2508061
7. Nielsen, J. (2013) Usability 101: Introduction to Usability. Jakob Nielsen's Alertbox, 25.
8. Davis, P. and Shipman, F. (2011) Learning Usability Assessment Models for Web Sites. Proceedings of the 16th International Conference on Intelligent User Interfaces, Palo Alto, 13-16 February 2011, 195-204. http://dx.doi. org/ .1145 /1943403.1943433
9. Buckingham, S. and McKnight, C. (1997) Word Wide Web Usability: Introduction to This Special Issue. International Journal of Human-Computer Studies, 47, 1-4. http://dx.doi.org/10.1006/ijhc.1997.0132
10. Shneiderman, B. (1997) Designing Information-Abundant Web Sites: Issues and Recommendations. International Journal of Human-Computer Studies, 47, 5-30. http://dx.doi.org/10.1006/ijhc.1997.0127
11. Lee, Y. and Kozar, K. (2012) Understanding of Website Usability: Specifying and Measuring Constructs and Their Relationships. Decision Support Systems, 52, 450-463. http://dx.doi.org/10.1016/j.dss.2011.10.004
12. Bieber, M., Vitali, F., Ashman, H., Balasubramanian, V. and Oinas-Kukkonen, H. (1997) Fourth Generation Hypermedia: Some Missing Links for the World Wide Web. International Journal of Human-Computer Studies, 47, 31-66.http:// dx. doi.org/10.1006/ijhc.1997.0130
13. Smith, P.A., Newman, I.A. and Parks, L.M. (1997) Virtual Hierarchies and Virtual Networks: Some Lessons from Hypermedia Usability Research Applied to the World Wide Web. International Journal of Human-Computer Studies, 47, 67-95. http://dx.doi.org/10.1006/ijhc.1997.0128
14. Erskine, L.E., Carter-Tod, D. and Burton, J.K. (1997) Dialogical Techniques for the Design of Web Sites. International Journal of Human-Computer Studies, 47, 169-196. http://dx.doi.org/10.1006/ijhc.1997.0129
15. Thimbleby, H. (1997) Gentler: A Tool for Systematic Web Authoring. International Journal of Human-Computer Studies, 47, 139-168. http://dx.doi. org/ 10.1006/ijhc.1997.0131
16. Benyon, D., Stone, D. and Woodroffe, M. (1997) Experience with Developing Multimedia Courseware for the World Wide Web: The Need for Better Tools and Clear Pedagogy. International Journal of Human-Computer Studies, 47, 197-218. http://dx.doi.org/10.1006/ijhc.1997.0126
17. Koved, L. and Shneiderman, B. (1986) Embedded Menus: Selecting Items in Context. Communications of the ACM, 29, 312-318. http://dx.doi.org/10.1145/5684.5687
18. Shneiderman, B. and Kearsley, G. (1989) Hypertext Hands-On: An Introduction to a New Way of Organising and Accessing Information. Addison-Wesley, Reading.
19. Norman, K. (1991) The Psychology of Menu Selection: Designing Cognitive Control at the Human/Computer Interface. Ablex, Norwood.
20. Rivlin, E., Rotafogo, R. and Shneiderman, B. (1994) Navigating in Hyperspace: Designs for a Structure-Based Toolbox. Communications of the ACM, 37, 87-96. http://dx.doi.org/10.1145/175235.175242
21. Isakowitz, T., Stohr, E.A. and Balasubramanian, P. (1995) RMM: A Methodology for Hypermedia Design. Communications of the ACM, 38, 34-44. http://dx. doi. org/10.1145/208344.208346
22. Belkin, N.J. and Croft, B.W. (1992) Information Filtering and Information Retrieval: Two Sides of the Same Coin? Communications of the ACM, 35, 29-38. http://dx. doi.org/10.1145/138859.138861
23. Marchionini, G. (1989) Making the Transition from Print to Electronic Encyclopaedia: Adaptation of Mental Models. International Journal of Man-Machine Studies, 30, 591-618. http://dx.doi.org/10.1016/S0020-7373(89)80012-4

24. Staggers, N. (1993) Impact of Screen Density on Clinical Nurses' Computer Task Performance and Subjective Screen Satisfaction. International Journal of Man-Machine Studies, 39, 775-792. http://dx.doi.org/10.1006/imms.1993.1083
25. Horton, W., Taylor, L., Ignacio, A. and Hoft, N.L. (1996) The Web Page Design Cookbook. John Wiley & Sons, Inc., New York.
26. Norman, D. and Nielsen, J. (2013) 10 Usability Heuristics for User Interface Design.http://www.nngroup.com/articles/ten-usability-heuristics/
27. Bailey, J.E. and Person, S.W. (1983) Development of a Tool for Measuring and Analysing Computer User Satisfaction. Management Science, 29, 530-545. http://dx.doi.org/10.1287/mnsc.29.5.530
28. Chin, J.P., Norman, K.L. and Shneiderman, B. (1987) Subjective User Evaluation of CF PASCAL Programming Tools. Technical Report (CAR-TR-304), Human-Computer Interaction Laboratory, University of Maryland, College Park.
29. Chin, J.P., Virginia, A.D. and Norman, K.L. (1988) Development of an Instrument Measuring User Satisfaction of the Human-Computer Interface. Proceedings of the SIGCHI Conference on Human Factors in Computing Systems, Paris, 27 April-2 May 2013, 213-218. http://dx.doi.org/10.1145/57167.57203
30. Ives, B., Olson, M.H. and Baroudi, J.J. (1983) The Measurement of User Information Satisfaction. Communication of the ACM, 26, 785-793. http://dx.doi. org/ 10.1145/358413.358430
31. Larcker, D.F. and Lessig, V.P. (1980) Perceived Usefulness of Information: A Psychometric Examination. Decision Science, 11, 121-134. http://dx. doi.org /10.1111/j.1540-5915.1980.tb01130.x
32. Rushinek, A. and Rushinek, S.F. (1986) What Makes User Happy? Communication of the ACM, 29, 594-598. http://dx.doi.org/10.1145/6138.6140
33. Nunnally, J.C. (1978) Psychometric Theory. McGraw-Hill Book Company, New York.
34. Drapper, N.R. and Smith, H. (1981) Applied Regression Analysis. John Wiley & Sons Inc., New York.
35. SAS Institute Inc. (1987) SAS /STATTM Guide for Personal Computers. Version 6, SAS Institute Inc., Cary.
36. Retherfor, R.D. and Choe, M.K. (1993) Statistical Models for Causal Analysis. John Wiley & Sons, Inc., New York. http://dx.doi.org/10.1002/9781118033135
37. Subair, S. (2014) The Usability Engineering of the Universities Web Pages User Interfaces: A Pragmatic Approach. Princess Nourah bint Abdulrahamn University Printing Press, Riyadh, in Press.

CHAPTER 9

Chaospy: An Open Source Tool for Designing Methods of Uncertainty Quantification

Jonathan Feinberg[a,b],, Hans Petter Langtangen[a,c]

[a]*Center for Biomedical Computing, Simula Research Laboratory, P.O. Box 134, Lysaker, Norway*

[b]*Department of Mathematics, University of Oslo, P.O. Box 1053, Blindern, Oslo, Norway*

[c]*Department of Informatics, University of Oslo, P.O. Box 1080, Blindern, Oslo, Norwaya*

ABSTRACT

The paper describes the philosophy, design, functionality, and usage of the Python software toolbox Chaospy for performing uncertainty quantification via polynomial chaos expansions and Monte Carlo simulation. The paper compares Chaospy to similar packages and demonstrates a stronger focus on defining reusable software building blocks that can easily be assembled to construct new, tailored algorithms for uncertainty quantification. For example, a Chaospy user can in a few lines of high-level computer code define custom distributions, polynomials, integration rules, sampling schemes, and statistical metrics for uncertainty analysis. In addition, the software introduces some novel methodological advances, like a framework for computing Rosenblatt transformations and a new approach for creating polynomial chaos expansions with dependent stochastic variables.

Keywords: Uncertainty quantification Polynomial chaos expansions Monte Carlo simulation Rosenblatt transformations Python package

1. INTRODUCTION

We consider a computational science problem in space $\boldsymbol{x}$ and time t where the aim is to quantify the uncertainty in some response Y, computed by a forward model f, which depends on uncertain input parameters Q:

$$(1) Y=f(\mathbf{x},t,\boldsymbol{Q}).$$

We treat $\boldsymbol{Q}$ as a vector of model parameters, and Y is normally computed as some grid function in space and time. The uncertainty in this problem stems from the parameters $\boldsymbol{Q}$, which are assumed to have a known joint probability density function $p_{\boldsymbol{Q}}$. The challenge is that we want to quantify the uncertainty in Y, but nothing is known about its density p_Y. The goal is then to either build the density p_Y or relevant descriptive properties of Y using the density p_Q and the forward model f. For all practical purposes this must be done by a numerical procedure.

In this paper, we focus on two approaches to numerically quantify uncertainty: Monte Carlo simulation and non-intrusive global polynomial chaos expansions. For a review of the former, there is a very useful book by Rubinstein, Reuven and Kroese [1], while for the latter, we refer to the excellent book by Xiu [2]. Note that other methods for performing uncertainty quantification also exist, such as perturbation methods, moment equations, and operator based methods. These methods are all discussed in [2], but are less general and less widely applicable than the two addressed in this paper.

The number of toolboxes available to perform Monte Carlo simulation is vastly larger than the number of toolboxes for non-intrusive polynomial chaos expansion. As far as the authors know, there are only a few viable options for the latter class of methods: *The Dakota Project*(referred to as Dakota) [3], the *Opus Open Turns library* (referred to as Turns) [4], *Uncertenty Quantification Toolkit*[5], and *MIT Uncertenty Quantification Library*[6]. In this paper we will focus on the former two: Dakota and Turns. Both packages consist of libraries with extensive sets of tools, where Monte Carlo simulation and non-intrusive polynomial chaos expansion are just two tools available among several others.

It is worth noting that both Dakota and Turns can be used from two perspectives: as a user and as a developer. Both packages are open source projects with comprehensive developer manuals. As such, they both allow anyone to extend the software with any functionality one sees fit. However, these extension features are not targeting the common user and require a deeper understanding of both coding practice and the underlying design of the library. In our opinion, the threshold for a common user to extend the library is normally out of reach. Consequently, we are in this paper only considering Dakota and Turns from the point of view of the common user.

Dakota requires the forward model f to be wrapped in a stand-alone callable executable. The common approach is then to link this executable to the analysis software through a configuration file. The technical steps are somewhat cumbersome, but has their advantage in that already built and installed simulation software can be used without writing a line of code.

Alternative to this direct approach is to interact with an application programming interface (API). This approach requires the user to know how to program in the supported languages, but this also has clear benefits as an interface through a programming language allows for a deeper level of integration between the user's model and the UQ tools. Also, exposing the software's internal components through an API allows a higher detailed control over the tools and how they can be combined in statistical algorithms. This feature is attractive to scientists who would like the possibility to experiment with new or non-standard methods in ways not thought of before. This approach is used by the Turns software (using the languages Python or R) and is supported in Dakota through a library mode (using C++).

For example, consider bootstrapping [7], a popular method for measuring the stability of any parameter estimation. Neither Dakota nor Turns support bootstrapping directly. However, since Turns exposes some of the inner components to the user, a programmer can combine these to implement a custom bootstrapping technique.

This paper describes a new, third alternative open source software package called Chaospy [8]. Like Dakota and Turns, it is a toolbox for analysing uncertainty using advanced Monte Carlo simulation and non-intrusive polynomial chaos expansions. However, unlike the others, it aims to assist scientists in constructing tailored statistical methods by combining a lot of fundamental and advanced building blocks. Chaospy builds upon the same philosophy as Turns in that it offers flexibility to the user, but takes it significantly further. In Chaospy, it is possible to gain detailed control and add user defined functionality to all of the following: random variable generation, polynomial construction, sampling schemes, numerical integration rules, response evaluation, and point collocation. The software is designed from the ground up in Python to be modular and easy to experiment with. The number of lines of code to achieve a full uncertainty analysis is amazingly low. It is also very easy to compare a range of methods in a given problem. Standard statistical methods are easily accessible through a few lines of R or Pandas [9] code, and one may think of Chaospy as a tool similar to R or Pandas, just tailored to polynomial chaos expansion and Monte Carlo simulation.

Although Chaospy is designed with a large focus on modularity, flexibility, and customization, the toolbox comes with a wide range of pre-defined statistical methods. Within the scope of Monte Carlo sampling and non-intrusive polynomial chaos expansion, Chaospy has a competitive collection of methods, comparable to both Dakota and Turns. It also offers some novel features regarding statistical methods, first and foremostly a flexible framework for defining and handling input distributions, including *dependent* stochastic variables. Detailed comparisons of features in the three packages appear throughout the paper.

The paper is structured as follows. We start in Section 2 with a quick demonstration of how the software can be used to perform uncertainty quantification in a simple physical problem. Section 3 addresses probability distributions and the theory relevant to perform Monte Carlo simulation. Section 4 concerns non-intrusive polynomial chaos expansions, while conclusions and topics for further work appear in Section 5.

2. A GLIMPSE OF CHAOSPY IN ACTION

To demonstrate how Chaospy is used to solve an uncertainty quantification problem, we consider a simple physical example of (scaled) exponential decay with an uncertain, piecewise constant coefficient:

$$u'(x) = -c(x)u(x), \quad u(0) = u_0, \quad c(x) = \begin{cases} c_0, & x < 0.5 \\ c_1, & 0.5 \leq x < 0.7 \\ c_2, & x \geq 0.7 \end{cases} \tag{2}$$

Such a model arises in many contexts, but we may here think of $u(x)$ as the porosity at depth x in geological layers and c_i as a (scaled) compaction constant in layer number i. For simplicity, we consider only three layers with three uncertain constants c_0, c_1, and c_2.

The model can easily be evaluated by solving the differential equation problem, here by a 2nd-order Runge–Kutta method on a mesh x, coded in Python as:

```
def model(x, u0, c0, c1, c2):
    def c(x):
        if   x < 0.5:                return c0
        elif 0.5 <= x < 0.7:    return c1
        else:                        return c2

    N = len(x)
    u = np.zeros(N)

    u[0] = u0
    for n in xrange(N-1):
        dx = x[n+1] - x[n]
        K1 = -dx*u[n]*c(x[n])
        K2 = -dx*u[n] + K1/2*c(x[n]+dx/2)
        u[n+1] = u[n] + K1 + K2
    return u
```

Alternatively, the model can be implemented in some external software in another programming language. This software can either be run as a stand-alone application, where the Python function model runs the application and communicates with it through input and output files, or the model function can communicate with the external software through function calls if a Python wrapper has been made for the software (there are numerous technologies available for creating Python wrappers for C, C++, and Fortran software).

The Chaospy package may be loaded by

import chaposy as cp

Each of the uncertain parameters must be assigned a probability density, and we assume that c_0, c_1, and c_2 are stochastically independent:

```
c0 = cp.Normal(0.5, 0.15)
c1 = cp.Uniform(0.5, 2.5)
c2 = cp.Uniform(0.03, 0.07)
# Joint probability distribution
distribution = cp.J(c0, c1, c2)
```

The sample points (c_0, c_1, c_2) in probability space, where the model is to be evaluated, can be chosen in many ways. Here we specify a third-order Gaussian Quadrature scheme tailored to the joint distribution:

```
nodes, weights = cp.generate_quadrature(
    order=3, domain=distribution, rule="Gaussian")
```

The next step is to evaluate the computational model at these sample points (object nodes):

```
x = np.linspace(0, 1, 101)
samples = [model(x, u0, node[0], node[1], node[2])
           for node in nodes.T]
```

Now, samples contains a list of arrays, each array containing *u* values at the 101 x values for one combination (c_0, c_1, c_2) of the input parameters.

To create a polynomial chaos expansion, we must generate orthogonal polynomials corresponding to the joint distribution. We choose polynomials of the same order as specified in the quadrature rule, computed by the widely used three-term recurrence relation (ttr):

Polynomials =cp.orth_ttr(order=3,dist=distribution)

To create an approximate solver (or surrogate model), we join the polynomial chaos expansion, the quadrature nodes and weights, and the model samples:

Model_approx=cp.fit_quardrature(polynomials,nodes,weights,samples)

The model_approx object can now cheaply evaluate the model at a point (c_0, c_1, c_2) in probability space for all *x* points in the x array. Built-in tools can be used to derive statistical information about the model response:

Mean= cp.E(model_approx,distribution)

Deviation=cp.std(model_approx,distribution)

The mean and deviation objects are arrays containing the mean value and standard deviation at each point in x. A graphical illustration is shown in Fig. 1.

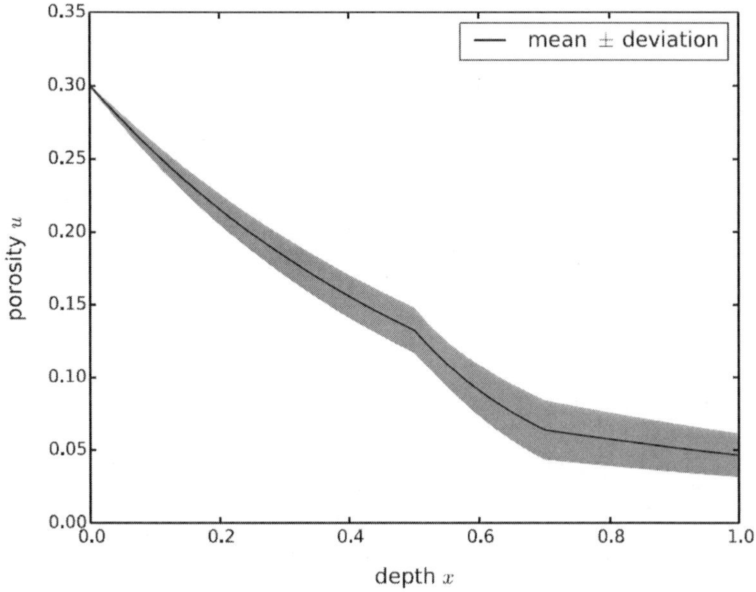

Figure. 1. Solution of a simple stochastic differential equation with uncertain coefficients.

The accuracy of the estimation is comparable to what Dakota and Turns can provide. Fig. 2shows that the estimation error in the three software toolboxes are almost indistinguishable. The error is calculated as the absolute difference between the true value and the estimated value integrated over the depth *x*:

$$\varepsilon_{\mathbb{E}} = \int_0^1 |\mathbb{E}(u) - \mathbb{E}(u_{\text{approx}})| \mathrm{d}x \quad \varepsilon_{\mathbb{V}} = \int_0^1 |\mathbb{V}(u) - \mathbb{V}(u_{\text{approx}})| \mathrm{d}x$$

Both the point collocation method and the pseudo-spectral projection method are included. The former is calculated using two times the random collocation nodes as the number of polynomials, and the latter using Gaussian quadrature integration with quadrature order equal to polynomial order. Note that Turns does not support pseudo-spectral projection, and is therefore only compared using point collocation.

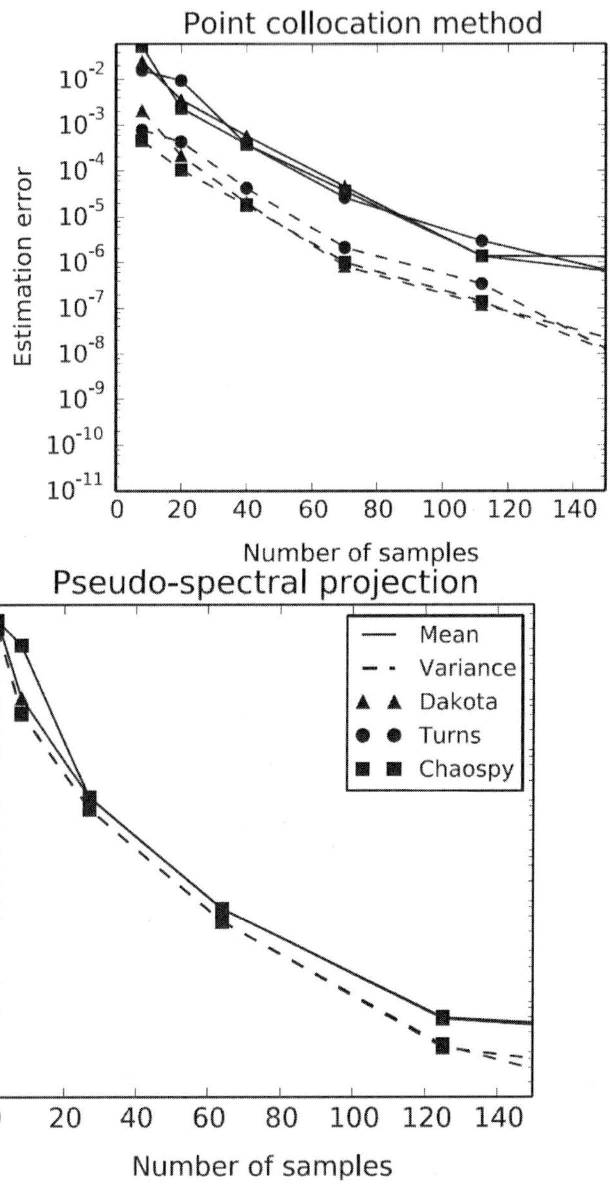

Figure. 2. The error in estimates of the mean and variance, computed by Dakota, Turns, and Chaospy using point collocation and pseudo-spectral projection, is almost identical.

3. MODELLING RANDOM VARIABLES

3.1. Rosenblatt transformation

Numerical methods for uncertainty quantification need to generate pseudo-random realizations

$$\{\mathbf{Q}_k\}_{k\in I_K} \quad I_K = \{1, \ldots, K\},$$

from the density p_Q. Each $\mathbf{Q} \in \{\mathbf{Q}_k\}_{k\in I_K}$ is multivariate with the number of dimensions $D > 1$. Generating realizations from a given density p_Q is often non-trivial, at least when D is large. A very common assumption made in uncertainty quantification is that each dimension in Q consists of stochastically independent components. Stochastic independence allows for a joint sampling scheme to be reduced to a series of univariate samplings, drastically reducing the complexity of generating a sample Q.

Unfortunately, the assumption of independence does not always hold in practice. We have examples from many research fields where stochastic dependence must be assumed, including modelling of climate [10], iron-ore minerals [11], finance [12], and ion channel densities in detailed neuroscience models [13]. There also exists examples where introducing dependent random variables is beneficial for the modelling process, even though the original input was stochastically independent [14]. In any cases, modelling of stochastically dependent variables are required to perform uncertainty quantification adequately. A strong feature of Chaospy is its support for stochastic dependence.

All random samples are in Chaospy generated using Rosenblatt transformations T_Q [15]. It allows for a random variable $\boldsymbol{U}$, generated uniformly on a unit hypercube $[0, 1]^D$, to be transformed into $\mathbf{Q}=T_Q^{-1}(\mathbf{U})$, which behaves as if it were drawn from the density p_Q. It is easy to generate pseudo-random samples from a uniform distribution, and the Rosenblatt transformation can then be used as a method for generating samples from arbitrary densities.

The Rosenblatt transformation can be derived as follows. Consider a probability decomposition, for example for a bivariate random variable $\boldsymbol{Q} = (Q_0, Q_1)$:

$$p_{Q_0,Q_1}(q_0,q_1)=p_{Q_0}(q_0)p_{Q_1|Q_0}(q_1|q_0),$$

were $pQ0$ is an marginal density function, and $p_{Q_1|Q_0}$ is a conditional density. For the multivariate case, the density decomposition will have the form

$$p_{\mathbf{Q}}(\boldsymbol{q}) = \prod_{d=0}^{D-1} p_{Q'_d}(q'_d) \tag{4}$$

where

$$Q_d{}'=Q_d|Q_0,\ldots,Q_{d-1} \quad q_d{}'=q_d|q_0,\ldots,q_{d-1} \quad (5)$$

denotes that Q_d and Q_d are dependent on all components with lower indices. A forward Rosenblatt transformation can then be defined as

$$(6)\mathbf{T}_Q(\mathbf{q})=(F_{Q0}{}'(q_0{}'),\ldots,F_{QD-1}{}'(q_{D-1}{}')),$$

where $F_{Qd}{}'$ is the cumulative distribution function:

$$F_{Q'_d}(q'_d) = \int_{-\infty}^{q_d} p_{Q'_d}(r \mid q_0, \ldots, q_{d-1})\mathrm{d}r$$

This transformation is bijective, so it is always possible to define the inverse Rosenblatt transformation $T_Q{}^{-1}$ in a similar fashion.

3.2. Numerical estimation of inverse Rosenblatt transformations

To implement the Rosenblatt transformation in practice, we need to identify the inverse transform $T_Q{}^{-1}$. Unfortunately, T_Q is often non-linear without a closed-form formula, making analytical calculations of the transformation's inverse difficult. In the scenario where we do not have a symbolic representation of the inverse transformation, a numerical scheme has to be employed. To the authors' knowledge, there are no standards for defining such a numerical scheme. The following paragraphs therefore describe our proposed method for calculating the inverse transformation numerically.

The problem of calculating the inverse transformation $T_Q{}^{-1}$ can, by decomposing the definition of the forward Rosenblatt transformation in (6), be reformulated as

$$\begin{aligned} &F_{Q'_d}^{-1}(u \mid q_0, \ldots, q_{d-1}) \\ &= \left\{r : F_{Q'_d}(r \mid q_0, \ldots, q_{d-1}) = u\right\} \qquad d = 0, \ldots, D-1 \end{aligned}$$

In other words, the challenge of calculating the inverse transformation can be reformulated as a series of one dimensional root-finding problems. In Chaospy, these roots are found by employing a Newton–Raphson scheme.

However, to ensure convergence, the scheme is coupled with a bisection method. The bisection method is applicable here since the problem is one-dimensional and the functions of interest are by definition monotone. When the Newton–Raphson method fails to converge at an increment, a bisection step gives the Newton–Raphson a new start location away from the previous location. This algorithm ensures fast and reliable convergence towards the root.

The Newton–Raphson-bisection hybrid method is implemented as follows. The initial values are the lower and upper bounds $[lo_0, up_0]$. If p_{Qd}' is unbound, the interval is selected such that it approximately covers the density. For example for a standard normal random variable, which is unbound, the interval [−7.5,7.5] will approximately cover the whole density with an error about 10^{-14}. The algorithm starts with a Newton–Raphson increment, using the initial value $r_0 = (up_0 - lo_0)u + lo_0$:

$$r_{k+1} = r_k - \frac{F_{Q_d'}(r_k \mid q_0, \ldots, q_{d-1}) - u}{p_{Q_d'}(r_k \mid q_0, \ldots, q_{d-1})}, \quad 8)$$

where the density p_{Qd}' can be approximated using finite differences. If the new value does not fall in the interval $[lo_k, up_k]$, this proposed value is rejected, and is instead replaced with a bisection increment:

$$r_{k+1} = \frac{up_k + lo_k}{2}. \quad (9)$$

In either case, the bounds are updated according to

$$(lo_{k+1}, up_{k+1}) = \begin{cases} (lo_k, r_{k+1}) & F_{Q_d'}(r_{k+1} \mid q_0, \ldots, q_{d-1}) > u \\ (r_{k+1}, up_k) & F_{Q_d'}(r_{k+1} \mid q_0, \ldots, q_{d-1}) < u \end{cases} \quad (10)$$

The algorithm repeats the steps in (8)–(10), until the residual $|F_{Q_d'}(r_k \mid q_0, \ldots, q_{d-1}) - u|$ is sufficiently small.

The described algorithm overcomes one of the challenges of implementing Rosenblatt transformations in practice: how to calculate the inverse transformation. Another challenge is how to construct a transformation in the first place. This is the topic of the next section.

3.3. Constructing distributions

The backbone of distributions in Chaospy is the Rosenblatt transformation T_Q. The method, as described in the previous section, assumes that p_Q is known to be able to perform the transformation and its inverse. In practice, however, we first need to construct p_Q, before the transformation can be used. This can be a challenging task, but in Chaospy a lot of effort has been put into constructing novel tools for making the process as flexible and painless as possible. In essence, users can create their own custom multivariate distributions using a new methodology as described next.

Following the definition in (6), each Rosenblatt transformation consists of a collection of conditional distributions. We express all conditionality through distribution parameters. For example, the location parameter of a normal distribution can be set to be uniformly distributed, say on [− 1, 1]. The following interactive Python code defines a normal variable with a normally distributed mean:

```
>>>uniform=cp.Uniform(lo= -1,up=1)
```

```
>>> normal=cp.Normal(mu=uniform,sigma=0.1
```

We now have two stochastic variables, uniform and normal, whose joint bivariate distribution can be constructed through the cp.J function:

```
>>>joint =cp.J (uniform,normal)
```

The software will, from this minimal formulation, try to sort out the dependency ordering and construct the full Rosenblatt transformation. The only requirement is that a decomposition as in (4) is in fact possible. The result is a fully functioning forward and inverse Rosenblatt transformation. The following code evaluates the forward transformation (the density) at (1, 0.9), the inverse transformation at (0.4, 0.6), and draws a random sample from the joint distribution:

```
>>> print joint.fwd([1, 0.9])
[ 1.          0.15865525]
>>> print joint.inv([0.4, 0.6])
[-0.2        -0.17466529]
>>> print joint.sample()
[-0.05992158 -0.07456064]
```

Distributions in higher dimensions are trivially obtained by including more arguments to the cp.J function.

As an alternative to the explicit formulation of dependency through distribution parameters, it is also possible to construct dependencies implicitly through arithmetic operators. For example, it is possible to recreate the example above using addition of stochastic variables instead of letting a distribution parameter be stochastic. More precisely, we have a uniform variable on [− 1, 1] and a normally distributed variable with location at $x = 0$. Adding the uniform variable to the normal variable creates a new normal variable with stochastic location:

```
>>> uniform = Uniform(lo=-1, up=1)
>>> normal0 = Normal(mu=0, scale=0.1)
>>> normal = normal0 + uniform
>>> joint = J(uniform, normal)
```

As before, the software automatically sorts the dependency ordering from the context. Here, since the uniform variable is present as first argument, the software recognises the second argument as a normal distribution, conditioned on the uniform distribution, and not the other way around.

Another favorable feature in Chaospy is that multiple transformations can be stacked on top of each other. For example, consider the example of a multivariate log-normal random variable Q with three dependent components. (Let us ignore for a moment the fact that Chaospy already offers such a distribution.) Trying to decompose this distribution is a very cumbersome task if performed manually. However, this process can be drastically simplified through variable transformations, for which Chaospy has strong support. A log-normal distribution, for example, can be expressed as

$$Q = e^{ZL+b},$$

where $\boldsymbol{Z}$ are standard normal variables, and L and $\boldsymbol{b}$ are predefined matrix and vector, respectively. To implement this particular transformation, we only have to write

```
>>> Z = cp.J(cp.Normal(0,1), cp.Normal(0,1), cp.Normal(0,1))
>>> Q = e**(Z*L + b)
```

The resulting distribution is fully functional multivariate log-normal, assuming L and b are properly defined.

One obvious prerequisite for using univariate distributions to create conditionals and multivariate distributions, is the availability of univariate distributions. Since the univariate distribution is the fundamental building block, Chaospy offers a large collection of 64 univariate distributions. They are all listed in Table 1. The list also shows that Dakota's support is limited to 11 distributions, and Turns has a collection of 26 distributions.

Table 1. List of supported continuous distributions in software. The titles 'D', 'T' and 'C' represents Dakota, Turns and Chaospy respectively. The elements 'y' and 'n' represent the answers 'yes' and 'no' indicating if the distribution is supported or not.

Distribution	D	T	C	Distribution	D	T	C
Alpha	n	n	y	Anglit	n	n	y
Arcsinus	n	n	y	Beta	y	y	y
Brandford	n	n	y	Burr	n	y	y
Cauchy	n	n	y	Chi	n	y	y
Chi-square	n	y	y	Double Gamma	n	n	y
Double Weibull	n	n	y	Epanechnikov	n	y	y
Erlang	n	n	y	Exponential	y	y	y
Exponential power	n	n	y	Exponential Weibull	n	n	y
Birnbaum–Sanders	n	n	y	Fisher–Snedecor	n	y	y
Fisk/log-logistic	n	n	y	Folded Cauchy	n	n	y
Folded normal	n	n	y	Frechet	y	n	y
Gamma	y	y	y	Gen. exponential	n	n	y
Gen. extreme value	n	n	y	Gen. gamma	n	n	y
Gen. half-logistic	n	n	y	Gilbrat	n	n	y
Truncated Gumbel	n	n	y	Gumbel	y	y	y
Hypergeometric secant	n	n	y	Inverse-normal	n	y	n
Kumaraswamy	n	n	y	Laplace	n	y	y
Levy	n	n	y	Log-gamma	n	n	y
Log-laplace	n	n	y	Log-normal	y	y	y
Log-uniform	y	y	y	Logistic	n	y	y
Lomax	n	n	y	Maxwell	n	n	y
Mielke's beta-kappa	n	n	y	Nakagami	n	n	y
Non-central chi-squared	n	y	y	Non-central Student-T	n	y	y
Non-central F	n	n	y	Normal	y	y	y

Pareto (first kind)	n	n	y	Power log-normal	n	n	y
Power normal	n	n	y	Raised cosine	n	n	y
Rayleigh	n	y	y	Reciprocal	n	n	y
Rice	n	y	n	Right-skewed Gumbel	n	n	y
Student-*T*	n	y	y	Trapezoidal	n	y	n
Triangle	y	y	y	Truncated exponential	n	n	y
Truncated normal	n	y	y	Tukey-Lambda	n	n	y
Uniform	y	y	y	Wald	n	n	y
Weibull	y	y	y	Wigner	n	n	y
Wrapped Cauchy	n	n	y	Zipf–Mandelbrot	n	y	n

The Chaospy software supports in addition custom distributions through the function cp.constructor. To illustrate its use, consider the simple example of a uniform random variable on the interval [*lo*, *up*]. The minimal input to create such a distribution is

```
>>> Uniform = cp.constructor(
...     cdf=lambda self,x,lo,up: (x-lo)/(up-lo),
...     bnd=lambda self,x,lo,up: (lo,up) )
>>> uniform = Uniform(lo=-1, up=1)
```

Here, the two provided arguments are a cumulative distribution function (cdf), and a boundary interval function (bnd), respectively. The cp.constructor function also takes several optional arguments to provide extra functionality. For example, the inverse of the cumulative distribution function –the point percentile function – can be provided through the ppf keyword. If this function is not provided, the software will automatically approximate it using the method described in Section 3.2.

3.4. Copulas

Dakota and Turns do not support the Rosenblatt transformation applied to multivariate distributions with dependencies. Instead, the two packages model dependencies using copulas [16]. A copula consists of stochastically independent multivariate distributions made dependent using a parameterized function g. Since the Rosenblatt transformation is general purpose, it is possible to construct any copula directly. However, this can quickly become a very cumbersome task since each copula must be decomposed individually for each combination of independent distributions and parameterization of g. To simplify the user's efforts, Chaospy has dedicated constructors that can reformulate a copula coupling into a Rosenblatt transformation. This is done following the work of Lee [17] and approximated using finite differences. The implementation is based of the software toolbox RoseDist [18]. In practice, this approach allow copulas to be defined in a Rosenblatt transformation setting. For example, to construct a bivariate normal distribution with a Clayton copula in Chaospy, we do the following:

```
>>> joint=cp.j(cp.Normal (0,1),cp.Normal (0,1))
>>> clayton=cp.clayton(joint,theta=2)
```

A list of supported copulas are listed in Table 2. It shows that Turns supports 7 methods, Chaospy 6, while Dakota offers 1 method.

Table 2. The list of supported copulas in the various software packages.

Supported copulas	Dakota	Turns	Chaospy
Ali–Mikhail–Haq	No	Yes	Yes
Clayton	No	Yes	Yes
Farlie–Gumbel–Morgenstein	No	Yes	No
Frank	No	Yes	Yes
Gumbel	No	Yes	Yes
Joe	No	No	Yes
Minimum	No	Yes	No
Normal/Nataf	Yes	Yes	Yes

3.5. Variance reduction techniques

As noted in the beginning of Section 3, by generating samples $\{\mathbf{Q}_k\}_{k\in I_K}$ and evaluating the response function f, it is possible to draw inference upon Y without knowledge about p_Y, through Monte Carlo simulation. Unfortunately, the number of samples K to achieve reasonable accuracy can often be very high, so if f is assumed to be computationally expensive, the number of samples needed frequently make Monte Carlo simulation infeasible for practical applications. As a way to mitigate this problem, it is possible to modify $\{\mathbf{Q}_k\}_{k\in I_K}$ from traditional pseudo-random samples, so that the accuracy increases. Schemes that select non-traditional samples for $\{\mathbf{Q}_k\}_{k \in I_K}$ to increase accuracy are known as *variance reduction techniques*. A list of such techniques are presented in Table 3, and it shows that Dakota, Turns and Chaospy support 4, 7, and 7 variance reduction techniques, respectively.

Table 3. The different sampling schemes available.

	Dakota	Turns	Chaospy
Quasi-Monte Carlo scheme			
Faure sequence [20]	No	Yes	No
Halton sequence [21]	Yes	Yes	Yes
Hammersley sequence [22]	Yes	Yes	Yes
Haselgrove sequence [23]	No	Yes	No
Korobov latice [24]	No	No	Yes
Niederreiter sequence [25]	No	Yes	No
Sobol sequence [26]	No	Yes	Yes
Other methods			
Antithetic variables [1]	No	No	Yes
Importance sampling [1]	Yes	Yes	Yes
Latin Hypercube sampling [27]	Yes	Limited	Yes

One of the more popular variance reduction technique is the *quasi-Monte Carlo scheme*[1]. The method consists of selecting the samples $\{\mathbf{Q}_k\}_{k \in I_K}$ to be a low-discrepancy sequence instead of pseudo-random samples. The idea is that samples placed with a given distance from each other increase the coverage over the sample space, requiring fewer samples to reach a given accuracy. For example, if standard Monte Carlo requires 10^6 samples for a given accuracy, quasi-Monte Carlo can often get away with only 10^3. Note that this would break some of the statistical properties of the samples [19].

Most of the theory on quasi-Monte Carlo methods focuses on generating samples on the unit hypercube $[0, 1]^N$. The option to generate samples directly on to other distributions exists, but is often very limited. To the authors' knowledge, the only viable method for including most quasi-Monte

Carlo methods into the vast majority of non-standard probability distributions, is through the Rosenblatt transformation. Since Chaospy is built around the Rosenblatt transformation, it has the novel feature of supporting quasi-Monte Carlo methods for all probability distributions. Turns and Dakota only support Rosenblatt transformations for independent variables and the Normal copula.

Sometimes the quasi-Monte Carlo method is infeasible because the forward model is too computationally costly. The next section describes polynomial chaos expansions, which often require far fewer samples than the quasi-Monte Carlo method for the same amount of accuracy.

4. POLYNOMIAL CHAOS EXPANSIONS

Polynomial chaos expansions represent a collection of methods that can be considered a subset of polynomial approximation methods, but particularly designed for uncertainty quantification. A general polynomial approximation can be defined as

$$\hat{f}(\boldsymbol{x}, t, \boldsymbol{Q}) = \sum_{n \in I_N} c_n(\boldsymbol{x}, t)\boldsymbol{\Phi}_n(\boldsymbol{Q}) \qquad I_N = \{0, \ldots, N\} \quad , \tag{11}$$

where $\{c_n\}_{n \in I_N}$ are coefficients (often known as Fourier coefficients) and $\{\boldsymbol{\Phi}_n\}_{n \in I_N}$ are polynomials. If $\hat{f}$ is a good approximation of f, it is possible to either infer statistical properties of $\hat{f}$ analytically or through cheap numerical computations where $\hat{f}$ is used as a surrogate for f.

A polynomial chaos expansion is defined as a polynomial approximation, as in (11), where the polynomials $\{\boldsymbol{\Phi}_n\}_{n \in I_N}$ are orthogonal on a custom weighted function space L_Q:

$$\begin{aligned} \langle \boldsymbol{\Phi}_n, \boldsymbol{\Phi}_m \rangle &= \mathbb{E}\boldsymbol{\Phi}_n(\boldsymbol{Q})\boldsymbol{\Phi}_m(\boldsymbol{Q}) \\ &= \int \cdots \int \boldsymbol{\Phi}_n(\boldsymbol{q})\boldsymbol{\Phi}_m(\boldsymbol{q})p_{\boldsymbol{Q}}(\boldsymbol{q})d\boldsymbol{q} = 0 \quad n \neq m \end{aligned} \quad .(12)$$

As a side note, it is worth noting that in parallel with polynomial chaos expansions, there also exists an alternative collocation method based on multivariate Lagrange polynomials [28]. This method is supported by Dakota and Chaospy, but not Turns.

To generate a polynomial chaos expansion, we must first calculate the polynomials $\{\boldsymbol{\Phi}_n\}_{n\in I_N}$ such that the orthogonality property in (12) is satisfied. This will be the topic of Section 4.1. In Section 4.2 we show how to estimate the coefficients $\{c_n\}_{n\in I_N}$. Last, in Section 4.7, tools used to quantify uncertainty in polynomial chaos expansions will be discussed.

4.1. Orthogonal polynomials construction

From (12) it follows that the orthogonality property is not in general transferable between distributions, since a new set of polynomials has to be constructed for each p_Q. The easiest approach to construct orthogonal polynomials is to identify the probability density p_Q in the so-called Askey-Wilson scheme [29]. The polynomials can then be picked from a list, or be built from analytical components. The continuous distributions supported in the scheme include the standard normal, gamma, beta, and uniform distributions respectively through the Hermite, Laguerre, Jacobi, and Legendre polynomial expansion. All the three mentioned software toolboxes support these expansions.

Moving beyond the standard collection of the Askey-Wilson scheme, it is possible to create custom orthogonal polynomials, both analytically and numerically. Unfortunately, most methods involving finite precision arithmetics are ill-posed, making a numerical approach quite a challenge [30]. This section explores the various approaches for constructing polynomial expansions. A full list of methods is found in Table 4. It shows that Dakota, Turns and Chaospy support 4, 3 and 5 orthogonalisation methods, respectively.

Table 4. Methods for generating expansions of orthogonal polynomials.

Orthogonalization Method	Dakota	Turns	Chaospy
Askey–Wilson scheme [29]	Yes	Yes	Yes
Bertran recursion [31]	No	No	Yes
Cholesky decomposition [14]	No	No	Yes
Discretized Stieltjes [32]	Yes	No	Yes
Modified Chebyshev [32]	Yes	Yes	No
Modified Gram–Schmidt [32]	Yes	Yes	Yes

Looking beyond an analytical approach, the most popular method for constructing orthogonal polynomials is the discretized Stieltjes procedure [33]. As far as the authors know, it is the only truly numerically stable method for orthogonal polynomial construction. It is based upon one-dimensional recursion coefficients that are estimated using numerical integration. Unfortunately, the method is only applicable in the multivariate case if the components of p_Q are stochastically independent.

Generalized polynomial chaos expansions. One approach to model densities with stochastically dependent components numerically, is to reformulate the uncertainty problem as a set of independent components through generalised polynomial chaos expansion [34]. As described in detail in Section 3.1, a Rosenblatt transformation allows for the mapping between any domain and the unit hypercube $[0, 1]^D$. With a double transformation we can reformulate the response function f as

$$f(\boldsymbol{x}, t, \boldsymbol{Q}) = f(\boldsymbol{x}, t, T_{\boldsymbol{Q}}^{-1}(T_{\boldsymbol{R}}(\boldsymbol{R}))) \approx \hat{f}(\boldsymbol{x}, t, \boldsymbol{R}) = \sum_{n \in I_N} c_n(\boldsymbol{x}, t)\boldsymbol{\Phi}_n(\boldsymbol{R})$$

where $\boldsymbol{R}$ is any random variable drawn from $p_{\boldsymbol{R}}$, which for simplicity is chosen to consists of independent components. Also, $\{\boldsymbol{\Phi}_n\}_{n \in I_N}$ is constructed to be orthogonal with respect to L_R, not $L_{\boldsymbol{Q}}$. In any case, $\boldsymbol{R}$ is either selected from the Askey-Wilson scheme, or calculated using the discretized Stieltjes procedure. We remark that the accuracy of the approximation deteriorate if the transformation composition $T_Q^{-1} \circ T_R$ is not smooth [34]. Dakota, Turns, and Chaospy all support generalized polynomial chaos expansions for independent stochastic variables and the Normal/Nataf copula listed in Table 2. Since Chaospy has the Rosenblatt transformation underlying the computational framework, generalized polynomial chaos expansions are in fact available for all densities.

The direct multivariate approach. Given that both the density p_Q has stochastically dependent components, and the transformation composition $T_Q^{-1} \circ T_{\mathbf{R}}$ is not smooth, it is still possible to generate orthogonal polynomials numerically. As noted above, most methods are numerically unstable, and the accuracy in the orthogonality can deteriorate with polynomial order, but the methods can still be useful [14]. In Table 4, only Chaospy's implementation of Bertran's recursion method [31], Cholesky decomposition [35] and modified Gram-Schmidt orthogonalization [32] support construction of orthogonal polynomials for multivariate dependent densities directly.

Custom polynomial expansions. In the most extreme cases, an automated numerical method is insufficient. Instead, a polynomial expansion has to be constructed manually. User-defined expansions can be created conveniently, as demonstrated in the next example involving a second-order Hermite polynomial expansion, orthogonal with respect to the normal density [29]:

$$\left\{\Phi_n\right\}_{n \in I_6} = \left\{1, Q_0, Q_1, Q_0^2 - 1, Q_0 Q_1, Q_1^2 - 1\right\}$$

The relevant Chaospy code for creating this polynomial expansion looks like

```
>>> q0, q1 = cp.variable(2)
>>> phi = cp.Poly([1, q0, q1, q0**2-1, q0*q1, q1**2-1])
>>> print phi
[1, q0, q1, q0^2-1, q0q1, -1+q1^2]
```

Chaospy contains a collection of tools to manipulate and create polynomials, see Table 5.

Table 5. List of tools for creating and manipulating polynomials.

Function	Description
all	Test all coefficients for non-zero
any	Test any coefficients for non-zero
around	Round to a given decimal
asfloat	Set coefficients type as float
asint	Set coefficient type as int
basis	Create monomial basis
cumprod	Cumulative product
cumsum	Cumulative sum
cutoff	Truncate polynomial order
decompose	Convert from series to sequence
diag	Construct or extract diagonal
differential	Differential operator
dot	Dot-product
flatten	Flatten an array

gradient	Gradient (or Jacobian) operator
hessian	Hessian operator
inner	Inner product
mean	Average
order	Extract polynomial order
outer	Outer product
prod	Product
repeat	Repeat polynomials
reshape	Reshape axes
roll	Roll polynomials
rollaxis	Roll axis
rolldim	Roll the dimension
std	Empirical standard deviation
substitute	Variable substitution
sum	Sum along an axis
swapaxes	Interchange two axes
swapdim	Swap the dimensions
trace	Sum along the diagonal
transpose	Transpose the coefficients
tril	Extract lower triangle of coefficients
tricu	Extract cross-diagonal upper triangle
var	Empirical variance
variable	Simple polynomial constructor

One thing worth noting is that polynomial chaos expansions suffers from the curse of dimensionality: The number of terms grows exponentially with the number of dimensions [36]. As a result, Chaospy does not support neither high dimensional nor infinite dimensional problems (random fields). One approach to address such problems with polynomial chaos expansion is to first reduce the number of dimension through techniques like Karhunen–Loeve expansions [37]. If software implementations of such methods can be provided, the user can easily extend Chaospy to high and infinite dimensional problems.

Chaospy includes operators such as the expectation operator E. This is a helpful tool to ensure that the constructed polynomials are orthogonal, as defined in (12). To verify that two elements in phi are indeed orthogonal under the standard bivariate normal distribution, one writes

```
>>> dist = cp.J(cp.Normal(0,1), cp.Normal(0,1))
>>> print cp.E(phi[3]*phi[5], dist)
0.0
```

More details of operators used to perform uncertainty analysis are given in Section 4.7.

4.2. Calculating coefficients

There are several methodologies for estimating the coefficients $\{c_n\}_{n \in I_N}$, typically categorized either as non-intrusive or intrusive, where non-intrusive means that the computational procedures only requires evaluation of f (i.e., software for f can be reused as a black box). Intrusive methods need to incorporate information about the underlying forward model in the computation of the coefficients. In case of forward models based on differential equations, one performs a Galerkin formulation for the coefficients in probability space, leading effectively to a D-dimensional differential equation problem in this space [38]. Back et al. [39] demonstrated that the computational cost of such an intrusive Galerkin method in some cases was higher than some non-intrusive methods. None of the three toolboxes discussed in this paper have support for intrusive methods.

Within the realm of non-intrusive methods, there are in principle two viable methodologies available: pseudo-spectral projection [40] and the point collocation method [41]. The former applies a numerical integration scheme to estimate Fourier coefficients, while the latter solves a linear system arising from a statistical regression formulation. Dakota and Chaospy support both methodologies, while Turns only supports point collocation. We shall now discuss the practical, generic implementation of these two methods in Chaospy.

4.3. Integration methods

The pseudo-spectral projection method is based on a standard least squares minimization in the weighted function space L_Q. Since the polynomials are orthogonal in this space, the associated linear system is diagonal, which allows a closed-form expression for the Fourier coefficients. The expression involves high-dimensional integrals in L_Q. Numerical integration is then required,

$$
\begin{aligned}
c_n &= \frac{\mathbb{E}Y\boldsymbol{\Phi}_n}{\mathbb{E}\boldsymbol{\Phi}_n^2} = \frac{1}{\mathbb{E}\boldsymbol{\Phi}_n^2}\int\cdots\int p_Q(\boldsymbol{q})f(\boldsymbol{x},t,\boldsymbol{q})\boldsymbol{\Phi}_n(\boldsymbol{q})\mathrm{d}\boldsymbol{q} \\
&\approx \frac{1}{\mathbb{E}\boldsymbol{\Phi}_n^2}\sum_{k\in I_K} w_k p_Q(\boldsymbol{q}_k)f(\boldsymbol{x},t,\boldsymbol{q}_k)\boldsymbol{\Phi}_n(\boldsymbol{q}_k) \quad I_K = \{0,\ldots,K-1\}
\end{aligned}
\tag{13}
$$

where w_k are weights and q_k nodes in a quadrature scheme. Note that f is only evaluated for the nodes q_k, and these evaluations can be made once. Thereafter, one can experiment with the polynomial order since any c_n depends on the same evaluations of f.

Table 6 shows the various quadrature schemes offered by Dakota and Chaospy (recall that Turns does not support pseudo-spectral projection). All techniques for generating nodes and weights in Chaospy are available through the cp.generate_quadrature function. Suppose we want to generate optimal Gaussian quadrature nodes for the normal distribution. We then write

Table 6. Various numerical integration strategies implemented in the three software toolboxes.

Node and weight generators	Dakota	Turns	Chaospy
Clenshaw-Curtis quadrature [42]	Yes	No	Yes
Cubature rules [43]	Yes	No	No
Gauss-Legendre quadrature [44]	Yes	No	Yes
Gauss-Patterson quadrature [45]	Yes	No	Yes
Genz-Keister quadrature [46]	Yes	No	Yes
Leja quadrature [47]	No	No	Yes
Monte Carlo integration [1]	Yes	No	Yes
Optimal Gaussian quadrature [44]	Yes	No	Yes

```
>>> nodes, weights = cp.generate_quadrature(
...             3, cp.Normal(0, 1), rule="Gaussian")
>>> print nodes
[[-2.33441422 -0.74196378  0.74196378  2.33441422]]
>>> print weights
[ 0.04587585  0.45412415  0.45412415  0.04587585]
```

Most quadrature schemes are designed for univariate problems. To extend a univariate scheme to the multivariate case, integration rules along each axis can be combined using a tensor product. Unfortunately, such a product suffers from the curse of dimensionality and becomes a very costly integration procedure for large *D*. In higher-dimensional problems one can replace the full tensor product by a Smolyak sparse grid [48]. The method works by taking multiple lower order tensor product rules and joining them together. If the rule is nested, i.e., the same samples found at a low order are also included at higher order, the number of evaluations can be further reduced. Another feature is to add anisotropy such that some dimensions are sampled more than others [49]. In addition to the tensor product rules, there are a few native multivariate cubature rules that allow for low order multivariate integration [43]. Both Dakota and Chaospy also support the Smolyak sparse grid and anisotropy.

Chaospy has support for construction of custom integration rules defined by the user. The cp.rule_generator function can be used to join a list of univariate rules using tensor grid or Smolyak sparse grid. For example, consider the trapezoid rule:

```
>>> def trapezoid(n):
...     X = np.linspace(0, 1, n+1)
...     W = np.ones(n+1)/n
...     W[0] *= 0.5; W[-1] *= 0.5
...     return X, W
...
>>> nodes, weights = trapezoid(2)
>>> print nodes
[ 0.   0.5  1. ]
>>> print weights
[ 0.25  0.5   0.25]
```

The cp.rule_generator function takes positional arguments, each representing a univariate rule. To generate a rule for the multivariate case, with the same one-dimensional rule along two axes, we do the following:

```
>>> mvtrapezoid = cp.rule_generator(trapezoid, trapezoid)
>>> nodes, weights = mvtrapezoid(2, sparse=True)
>>> print nodes
[[ 0.   0.5  1.   0.   0.   1. ]
 [ 0.   0.   0.   0.5  1.   1. ]]
>>> print weights
[ 0.     0.25   0.125  0.25   0.125  0.25 ]
```

Software for constructing and executing a general-purpose integration scheme is useful for several computational components in uncertainty quantification. For example, in Section 4.1when constructing orthogonal polynomials using raw statistical moments, or calculating discretized Stieltjes' recurrence coefficients, numerical integration is relevant. Like the ppffunction noted in Section 3.3, the moments and recurrence coefficients can be added directly into each distribution. However, when these are not available, Chaospy will automatically estimate missing information by quadrature rules, using the cp.generate_quadrature function described above.

To compute the Fourier coefficients and the polynomial chaos expansion, we use the cp.fit_quadrature function. It takes four arguments: the set of orthogonal polynomials, quadrature nodes, quadrature weights, and the user's function for evaluating the forward model (to be executed at the quadrature nodes). Note that in the case of the discretized Stieltjes method discussed in Section 4.1, the nominator $\mathbb{E}\boldsymbol{\Phi}_n^2$ in (13) can be calculated more accurately using recurrence coefficients [32]. Special numerical features like this can be added by including optional arguments in cp.fit_quadrature.

4.4. Point collocation

The other non-intrusive approach to estimate the coefficients $\{\mathbf{Q}_k\}_{k\in I_K}$ is the point collocation method. One way of formulating the method is to require the polynomial expansion to equal the model evaluations at a set of collocation nodes $\{\mathbf{Q}_k\}_{k\in I_K}$, resulting in an over-determined set of linear equations for the Fourier coefficients:

$$\begin{bmatrix} \Phi_0(\boldsymbol{q}_0) & \cdots & \Phi_N(\boldsymbol{q}_0) \\ \vdots & & \vdots \\ \Phi_0(\boldsymbol{q}_{K-1}) & \cdots & \Phi_N(\boldsymbol{q}_{K-1}) \end{bmatrix} \begin{bmatrix} c_0 \\ \vdots \\ c_N \end{bmatrix} = \begin{bmatrix} f(\boldsymbol{q}_0) \\ \vdots \\ f(\boldsymbol{q}_{K-1}) \end{bmatrix} \tag{14}$$

Unlike pseudo spectral projection, the locations of the collocation nodes are not required to follow any integration rule. Hosder [41] showed that the solution using Hammersley samples from quasi-Monte Carlo samples resulted in more stable results than using conventional pseudo-random samples. In other words, well placed collocation nodes might increase the accuracy. In Chaospy these collocation nodes can be selected from integration rules or from pseudo-random samples from Monte Carlo simulation, as discussed in Section 3.5. In addition, the software accepts user defined strategies for choosing the sampling points. Turns also allows for user-defined points, while Dakota has its predefined strategies.

The obvious way to solve the over-determined system in (14) is to use least squares minimization, which resembles the standard statistical linear regression approach of fitting a polynomial to a set of data points. However, from a numerical point of view, this might not be the best strategy. If the numerical stability of the solution is low, it might be prudent to use Tikhonov regularization [50], or if the problem is so large that the number of coefficients is very high, it might be useful to force some of the coefficients to be zero through least angle regression [51]. Being able to run and compare alternative methods is important in many problems to see if numerical stability is a potential problem. Table 7 lists the regression methods offered by Dakota, Turns, and Chaospy.

Table 7. List of supported regression methods for estimating Fourier coefficients.

Regression schemes	Dakota	Turns	Chaospy
Basis pursuit [52]	Yes	No	No
Bayesian auto. relevance determination [53]	No	No	Yes
Bayesian ridge [54]	No	No	Yes
Elastic net [55]	Yes	No	Yes
Forward stagewise [56]	No	Yes	No
Least absolute shrinkage &Selection[51]	Yes	Yes	Yes
Least angle &Shrinkage with AIC/BIC [57]	No	No	Yes
Least squares minimization	Yes	Yes	Yes
Orthogonal matching pursuit [58]	Yes	No	Yes
Singular value decomposition	No	Yes	No
Tikhonov regularization [50]	No	No	Yes

Generating a polynomial chaos expansion using linear regression is done using Chaospy's cp.fit_regression function. It takes the same arguments as cp.fit_quadrature, except that quadrature weights are omitted, and optional arguments define the rule used to optimize (14).

4.5. Model evaluations

Irrespectively of the method used to estimate the coefficients c_k, the user is left with the job to evaluate the forward model (response function) *f*, which is normally by far the most computing-intensive part in uncertainty quantification. Chaospy does not impose any restriction on the simulation code used to compute the forward model. The only requirement is that the user can provide an array of values of *f* at the quadrature or collocation nodes. Chaospy users will usually wrap any complex simulation code for *f* in a Python function f(q), where q is a node in probability space (i.e., q contains values of the uncertain parameters in the problem). For example, for pseudo-spectral projection, samples of *f* can be created as

```
Samples=[f(node) for node in nodes.T]
```

or perhaps done in parallel if *f* is time consuming to evaluate:

```
import multiprocessingas mp

pool=mp.Pool(mp.cpu_count( ))

samples =pool.map(f,nodes.T)
```

The evaluation of all the *f* values can also be done in parallel with MPI in a distributed way on a cluster using the Python module like mpi4py. Both Dakota and Turns support parallel evaluation of *f* values, but the feature is embeded into the code, potentially limiting the customization options of the parallelization.

4.6. Extension of polynomial expansions

There is much literature that extends on the theory of polynomial chaos expansion [36]. For example, Isukapalli showed that the accuracy of a polynomial expansion could be increased by using partial derivatives of the model response [59]. This theory is only directly supported by Dakota. In Turns and Chaospy the support is indirect by allowing the user to add the feature manually.

To be able to incorporate partial derivatives of the response, the partial derivative of the polynomial expansion must be available as well. In both Turns and Chaospy, the derivative of a polynomial can be generated easily. This derivative can then be added to the expansion, allowing us to incorporate Isukapalli's theory in practice. This is just an example on how manipulation of the polynomial expansions and model approximations can overcome the lack of support for a particular feature from the literature.

To be able to support many current and possible future extensions of polynomial chaos, a large collection of tools for manipulating polynomials must be available. In Dakota, no such tools exist from a user perspective. In Turns, there is support for some arithmetic operators in addition to the derivative. In Chaospy, however, the polynomial generated for the model response is of the same type as the polynomials generated in Sections 4.1 and 4.2, and the rich set of manipulations of polynomials is then available for $\hat{f}$ as well.

Beyond the analytical tools for statistical analysis of $\hat{f}$, either from the

toolbox or custom ones by the user, there are many statistical metrics that cannot easily be expressed as simple closed-form formulas. Such metrics include confidence intervals, sensitivity indices, p-values in hypothesis testing, to mention a few. In those scenarios, it makes sense to perform a secondary uncertainty analysis through Monte Carlo simulation. Evaluating the approximation $\hat{f}$ is normally computationally much cheaper than evaluating the full forward model *f*, thus allowing a large number of Monte Carlo samples within a cheap computational budget. This type of secondary simulations are done automatically in the background in Dakota and Turns, while Chaospy does not feature automated tools for secondary Monte Carlo simulation. Instead, Chaospy allows for simple and computationally cheap generation of pseudo-random samples, as described in Section 3.5, such that the user can easily put together a tailored Monte Carlo simulation to meet the needs at hand. Within a few lines of Python code, the samples can be analyzed with the standard Numpy and the Scipy libraries [60] or with more specialized statistical libraries like Pandas [9], Scikit-learn [61], Scikit-statsmodel [62], and Python's interface to the rich R environment for statistical computing. For example, for the specific $\hat{f}$ function illustrated above, the following code computes a 90 percent confidence interval, based on 10^5 pseudo-random samples and Numpy's functionality for finding percentiles in discrete data:

```
>>> q_samples = cp.Normal(0,1).sample(10**5)
>>> samples = f_approx(*q_samples)
>>> p05, p95 = np.percentile(samples, [5, 95], axis=-1)
>>> print p05[:3]
[ 1.          1.0000004  1.0000016]
>>> print p95[:3]
[ 1.          1.00038886  1.00155544]
```

Since the type of statistical analysis of $\hat{f}$ often strongly depends on the physical problem at hand, we believe that the ability to quickly compose custom solutions by putting together basic building blocks is very useful in uncertainty quantification. This is yet another example of the need for a package with a strong focus on easy customization.

4.7. Descriptive tools

The last step in uncertainty quantification based on polynomial chaos expansions is to quantify the uncertainty. In polynomial chaos expansion this is done by using the uncertainty in the model approximation f_approx as a substiute for the uncertainty in the model *f*. For the most popular statistical metrics, like mean, variance, correlation, a polynomial chaos expansion allows for analytical analysis, which is easy to calculate and has high accuracy. This property is reflected in all the three toolboxes. To calculate the expected value, variance and correlation of a simple (here univariate) polynomial approximation f_approx, with a normally distributed ξ_0 variable, we can with Chaospy write

```
>>> f_approx = fit_quadrature(orth, nodes, weights, samples)
>>> print f_approx
[q0, q0^2, q0^3]
>>> dist = Normal(0,1)
>>> print E(f_approx, dist)
[ 0.  1.  0.]
>>> print Var(f_approx, dist)
[  1.   2.  15.]
>>> print Corr(f_approx, dist)
[[ 1.          0.          0.77459667]
 [ 0.          1.          0.        ]
 [ 0.77459667  0.          1.        ]]
```

A list of supported analytical metrics is listed in Table 8.

Table 8. List of common statistical operators that can be used for analytical evaluation of polynomials.

Method	Dakota	Turns	Chaospy
Covariance/correlation	Yes	Yes	Yes
Expected value	Yes	Yes	Yes
Conditional expectation	No	No	Yes
Kurtosis	Yes	Yes	Yes
Sensitivity index	Yes	Yes	Yes
Skewness	Yes	Yes	Yes
Variance	Yes	Yes	Yes

5. CONCLUSION AND FURTHER WORK

Until now there have only been a few real software alternatives for implementing non-intrusive polynomial chaos expansions. Two of the more popular implementations, Dakota and Turns, are both high-quality software that can be applied to a large array of problems. The present paper has introduced a new alternative: Chaospy. Its aim is to be an experimental foundry for scientists. Besides featuring a vast library of state-of-the-art tools, Chaospy allows for a high degree of customization in a user-friendly way. Within a few lines of high-level Python code, the user can play around with custom distributions, custom polynomials, custom integration schemes, custom sampling schemes, and custom statistical analysis of the result. Throughout the text we have compared the built-in functionality of the three packages, and Chaospy do very well in this comparison, which is summarized in Table 9. But the primary advantage of the package is the strong emphasis on offering well-designed software building blocks, with a high abstraction level, that can easily be combined to create tailored uncertainty quantification algorithms for new problems.

Table 9. A summary of the various features in Dakota, Turns and Chaospy.

Feature	Dakota	Turns	Chaospy
Distributions	11	26	64
Copulas	1	7	6
Sampling schemes	4	7.5	7
Orthogonal polynomial schemes	4	3	5
Numerical integration strategies	7	0	7
Regression methods	5	4	8
Analytical metrics	6	6	7

Although the primary aim of the software is to construct polynomial chaos expansions, the software is also a state-of-the-art toolbox for performing Monte Carlo simulation, either directly on the forward model or in combination with polynomial chaos expansions. Variance reduction techniques are included to speed up the convergence, and because Chaospy is based on Rosenblatt transformations, efficient quasi-Monte Carlo sampling is available for any distribution. Another novel feature of Chaospy is the ability to handle *stochastically dependent* input variables through a new mathematical technique.

ACKNOWLEDGEMENTS

The work is supported by funding from Statoil ASA through the Simula School of Research and Innovation, and by a Center of Excellence grant from the Research Council of Norwaythrough the Center for Biomedical Computing.

Thanks also go to Vinzenz Gregor Eck, Stuart Clark, Karoline Hagane, Samwell Tarly, and a random distribution of unnamed bug fixers for their contributions.

REFERENCES

1. R.Y. Rubinstein, D.P. Kroese**SIMULATION AND THE MONTE CARLO METHOD** (2nd ed.), Wiley-Interscience (2007)
2. D. Xiu**NUMERICAL METHODS FOR STOCHASTIC COMPUTATIONS: A SPECTRAL METHOD APPROACH** Princeton University Press (2010)
3. M.S. Eldred, A.A. Giunta, B.G. van Bloemen Waanders, S.F. Wojtkiewicz, W.E. Hart, M.P. Alleva**DAKOTA, A MULTILEVEL PARALLEL OBJECT-ORIENTED FRAMEWORK FOR DESIGN OPTIMIZATION, PARAMETER ESTIMATION, UNCERTAINTY QUANTIFICATION, AND SENSITIVITY ANALYSIS: VERSION 4.1 REFERENCE MANUAL** Sandia National Laboratories, Albuquerque, NM (2007)
4. G. Andrianov, S. Burriel, S. Cambier, A. Dutfoy, I. Dutka-Malen, E. De Rocquigny, B. Sudret, P. Benjamin, R. Lebrun, F. Mangeant**OPEN TURNS, AN OPEN SOURCE INITIATIVE TO TREAT UNCERTAINTIES, RISKS' N STATISTICS IN A STRUCTURED INDUSTRIAL APPROACH** Procedings ESREL'2007 Safety and Reliability Conference, Stavanger, Norway (2007)
5. B.J. Debusschere, H.N. Najm, P.P. Pébay, O.M. Knio, R.G. Ghanem, O.P. Le Maître**NUMERICAL CHALLENGES IN THE USE OF POLYNOMIAL CHAOS REPRESENTATIONS FOR STOCHASTIC PROCESSES** SIAM J. Sci. Comput., 26 (2004), p. 2
6. P.R. Conrad, Y.M. Marzouk**ADAPTIVE SMOLYAK PSEUDOSPECTRAL APPROX IMATIONS** SIAM J. Sci. Comput., 35 (2013), p. 6
7. B. Efron**BOOTSTRAP METHODS: ANOTHER LOOK AT THE JACKKNIFE** Ann. Stat. (1979)
8. J. Feinberg, H.P. Langtangen**CHAOSPY SOFTWARE PACKAGE FOR UNCERTAINTY QUANTIFICATION** (2014) https://github.com/hplgit/chaospy
9. W. McKinney**PYTHON FOR DATA ANALYSIS: DATA WRANGLING WITH PANDAS, NUMPY, AND IPYTHON** O'Reilly Media (2012)
10. P. Laux, G. Jackel, R.M. Tingem, H. Kunstmann**IMPACT OF CLIMATE CHANGE ON AGRICULTURAL PRODUCTIVITY UNDER RAINFED CONDITIONS IN CAMEROON – A METHOD TO IMPROVE ATTAINABLE CROP YIELDS BY PLANTING DATE ADAPTATIONS** Agric. For. Meteorol., 150 (9) (2010)
11. R.C. Boardman, J.E. Vanna**A REVIEW OF THE APPLICATION OF COPULAS TO IMPROVE MODELLING OF NON-BIGAUSSIAN BIVARIATE RELATIONSHIPS**

(WITH AN EXAMPLE USING GEOLOGICAL DATA) International Congress on Modelling and Simulation (2011)

12. J. Dobric, F. Schmid**A GOODNESS OF FIT TEST FOR COPULAS BASED ON ROSENBLATT'S TRANSFORMATION** Comput. Stat. Data Anal., 51 (9) (2007)
13. P. Achard, E. De Schutter**COMPLEX PARAMETER LANDSCAPE FOR A COMPLEX NEURON MODEL** PLoS Comput. Biol., 2 (7) (2006)
14. J. Feinberg, H.P. Langtangen, Multivariate Polynomial Chaos with Dependent Variables Unpublished Journal Article. http://bit.ly/1Bkp72S.
15. M. Rosenblatt**REMARKS ON A MULTIVARIATE TRANSFORMATION** Ann. Math. Stat., 23 (3) (1952)
16. R.B. Nelsen**AN INTRODUCTION TO COPULAS** Springer (1999)
17. A.J. Lee**GENERATING RANDOM BINARY DEVIATES HAVING FIXED MARGINAL DISTRIBUTIONS AND SPECIFIED DEGREES OF ASSOCIATION** Am. Stat., 47 (3) (1993)
18. J. Feinberg, S. Clark**ROSEDIST. GENERALIZED TOOL FOR SIMULATING WITH NON-STANDARD PROBABILITY DISTRIBUTIONS** J. Piantadosi, R.S. Anderssen, J. Boland (Eds.), International Congress on Modelling and Simulation, Adelaide, Australia (2013), pp. 367-372http://www. mssanz. org.au/modsim 2013/ A7/feinberg.pdf
19. Tezuka**UNIFORM RANDOM NUMBERS: THEORY AND PRACTICE** Kluwer Academic Publishers, Boston (1995)
20. Galanti, A. Jung**LOW-DISCREPANCY SEQUENCES: MONTE CARLO SIMULATION OF OPTION PRICES** J. Deriv., 5 (1) (1997)
21. J.H. Halton**ON THE EFFICIENCY OF CERTAIN QUASI-RANDOM SEQUENCES OF POINTS IN EVALUATING MULTI-DIMENSIONAL INTEGRALS** Numer. Math., 2 (1) (1960)
22. J.M. Hammersley**MONTE CARLO METHODS FOR SOLVING MULTIVARIABLE PROBLEMS** Ann. N. Y. Acad. Sci., 86 (3) (1960)
23. C.B. Haselgrove**A METHOD FOR NUMERICAL INTEGRATION** Math. Comput. (1961)
24. N.M. Korobov**THE APPROXIMATE CALCULATION OF MULTIPLE INTEGRALS USING NUMBER THEORETIC METHODS** Dokl. Acad. Nauk SSSR (1957)
25. H. Niederreiter**POINT SETS AND SEQUENCES WITH SMALL DISCREPANCY** Mon. Math., 104 (4) (1987)
26. I.M. Sobol**ON THE DISTRIBUTION OF POINTS IN A CUBE AND THE APPROXIMATE EVALUATION OF INTEGRALS**USSR Comput. Math. Math. Phys., 7 (4) (1967)
27. M.D. McKay, R.J. Beckman, W.J. Conover**COMPARISON OF THREE METHODS FOR SELECTING VALUES OF INPUT VARIABLES IN THE ANALYSIS OF OUTPUT FROM A COMPUTER CODE** Technometrics, 21 (2) (1979)
28. D. Xiu, J.S. Hesthaven**HIGH-ORDER COLLOCATION METHODS FOR DIFFERENTIAL EQUATIONS WITH RANDOM INPUTS** SIAM J. Sci. Comput. (2005)
29. R. Askey, J.A. Wilson**SOME BASIC HYPERGEOMETRIC ORTHOGONAL POLYNOMIALS THAT GENERALIZE JACOBI POLYNOMIALS** Am. Math. Soc. (1985)
30. W. Gautschi**CONSTRUCTION OF GAUSS-CHRISTOFFEL QUADRATURE FORMUL AS** Math. Comput., 22 (102) (1968), 10.2307/2004654
31. M. Bertran**NOTE ON ORTHOGONAL POLYNOMIALS IN V-VARIABLES** SIAM J. Math. Anal., 6 (2) (1975), 10.1137/0506025
32. W. Gautschi**ORTHOGONAL POLYNOMIALS: COMPUTATION AND APPROXIMATION** Oxford University Press, USA (2004)
33. T.J. Stieltjes**QUELQUES RECHERCHES SUR LA THÉORIE DES QUADRATURES DITES MÉCANIQUES** Ann. Sci. École Norm. Sup., 3 (1) (1884)
34. D. Xiu, D. Lucor, C.H. Su, G.E. Karniadakis**STOCHASTIC MODELING OF FLOW-STRUCTURE INTERACTIONS USING GENERALIZED POLYNOMIAL CHAOS** J. Fluids Eng. (2002)
35. J. Feinberg, H.P. Langtangen**UNCERTAINTY QUANTIFICATION OF DIFFUSION IN LAYERED MEDIA BY A NEW METHOD BASED ON POLYNOMIAL CHAOS**

EXPANSIONH.I. Andersson, B. Skallerud (Eds.), Seventh National Conference on Computational Mechanics MekIT'13, Akademika Publishing, Norwegian University of Science and Technology (2013)

36. D. Xiu**FAST NUMERICAL METHODS FOR STOCHASTIC COMPUTATIONS: A REVIEW** Commun. Comput. Phys., 5 (2–4) (2009)
37. S. Sakamoto, R. Ghanem**POLYNOMIAL CHAOS DECOMPOSITION FOR THE SIMULATION OF NON-GAUSSIAN NONSTATIONARY STOCHASTIC PROCESSES** J. Eng. Mech., 128 (2) (2002)
38. R. Ghanem, P.D. Spanos**STOCHASTIC FINITE ELEMENTS: A SPECTRAL APPROACH** Courier Dover Publications (2003)
39. J. Back, F. Nobile, L. Tamellini, R. Tempone**STOCHASTIC SPECTRAL GALERKIN AND COLLOCATION METHODS FOR PDES WITH RANDOM COEFFICIENTS: A NUMERICAL COMPARISON** Spectral and High Order Methods for Partial Differential Equations, Springer (2011), pp. 43-62
40. D. Gottlieb, S.A. Orszag**NUMERICAL ANALYSIS OF SPECTRAL METHODS: THEORY AND APPLICATIONS** Society for Industrial and Applied Mathematics (1977)
41. S. Hosder, R.W. Walters, M. Balch**EFFICIENT SAMPLING FOR NON-INTRUSIVE POLYNOMIAL CHAOS APPLICATIONS WITH MULTIPLE UNCERTAIN INPUT VARIABLES** Proceedings of the 48th Structures, Structural Dynamics, and Materials Conference, Vol. 125, Honolulu, HI (2007)
42. C.W. Clenshaw, A.R. Curtis**A METHOD FOR NUMERICAL INTEGRATION ON AN AUTOMATIC COMPUTER** Numer. Math., 2 (1) (1960)
43. A.H. Stroud**APPROXIMATE CALCULATION OF MULTIPLE INTEGRALS, VOL. 431** Prentice-Hall Englewood Cliffs, NJ (1971)
44. G.H. Golub, J.H. Welsch**CALCULATION OF GAUSS QUADRATURE RULES** Math. Comput. (1967)
45. T. Patterson**THE OPTIMUM ADDITION OF POINTS TO QUADRATURE FORMULAE** Math. Comput., 22 (104) (1968)
46. Genz, B. Keister**FULLY SYMMETRIC INTERPOLATORY RULES FOR MULTIPLE INTEGRALS OVER INFINITE REGIONS** J. Comput. Appl. Math. (1996)
47. Naraya, J. Jakeman, Adaptive Leja Sparse Grid Construction for Stochastic Collocation and High-dimensional Approximation, 2014. arXiv e-print https://cfwebprod.sandia.gov/cfdocs/CompResearch/docs/1404.5663v1.pdf.
48. S.A. Smolyak**QUADRATURE AND INTERPOLATION FORMULAS FOR TENSOR PRODUCTS OF CERTAIN CLASSES OF FUNCTIONS** Dok. Akad. Nauk SSSR, 4 (1963), p. 123
49. J. Burkardt**THE "COMBINING COEFFICIENT" FOR ANISOTROPIC SPARSE GRIDS** Tech. rep., Virginia Tech., 125 (2009)
50. R.M. Rifkin, R.A. Lippert**NOTES ON REGULARIZED LEAST SQUARES** J. Linear Algebra (2009)
51. Efron, T. Hastie, I. Johnstone, R. Tibshirani**LEAST ANGLE REGRESSION** Ann. Stat., 32 (2) (2004)
52. S.S. Chen, D.L. Donoho, M.A. Saunders**ATOMIC DECOMPOSITION BY BASIS PURSUIT** SIAM J. Sci. Comput., 20 (1) (1998)
53. D.P. Wipf, S.S. Nagarajan**A NEW VIEW OF AUTOMATIC RELEVANCE DETERMINATION** Advances in Neural Information Processing Systems (2007), pp. 1625-1632
54. D.J.C. MacKay**BAYESIAN INTERPOLATION** Neural Comput., 4 (3) (1992)
55. H. Zou, T. Hastie**REGULARIZATION AND VARIABLE SELECTION VIA THE ELASTIC NET** J. R. Stat. Soc.: Ser. B (Stat. Methodol.), 67 (2) (2005)
56. T. Hastie, J. Taylor, R. Tibshirani, G. Walther**FORWARD STAGEWISE REGRESSION AND THE MONOTONE LASSO** Electr. J. Stat. (2007)
57. H. Zou, T. Hastie, R. Tibshirani**ON THE "DEGREES OF FREEDOM" OF THE LASSO** Ann. Stat., 35 (5) (2007)
58. S.G. Mallat, Z. Zhang**MATCHING PURSUITS WITH TIME-FREQUENCY DICTIONARIES** IEEE Trans. Signal Process., 41 (12) (1993)

59. S.S. Isukapalli**UNCERTAINTY ANALYSIS OF TRANSPORT-TRANSFORMATION MODEL** The State University of New Jersey, Rutgers (1999) (Ph.D. thesis)
60. E. Jones, T. Oliphant, P. Peterson**SCIPY: OPEN SOURCE SCIENTIFIC TOOLS FOR PYTHON, AIP PUBLISHING** (2001) http://scipy.org/
61. F. Pedregosa, G. Varoquaux, A. Gramfort, V. Michel, B. Thirion, O. Grisel, M. Blondel, P . Prettenhofer, R.Weiss, V. Dubourg**SCIKIT-LEARN: MACHINE LEARNING IN PYTHON** J. Mach. Learn. Res. (2011)
62. S. Seabold, J. Perktold**STATSMODELS: ECONOMETRIC AND STATISTICAL MODELING WITH PYTHON** Proceedings of the 9th Python in Science Conference (2010), pp. 57-61

CHAPTER 10

Numerical Simulation of Compressible Reactive Flows

Bing Wang*, Zhao Xin Ren and Wei Wei

School of Aerospace Engineering, Tsinghua University, Beijing, China

ABSTRACT

Numerical simulation has been widely employed to investigate the compressible flows since it is difficult to carry out the experimental measurements, especially in the reactive flows. The shock-wave capturing scheme will be necessary for resolving the compressible flows, and moreover the careful treatments of chemical reaction should be considered for proceeding numerical simulation stable and fast. Recently, the present authors have tried to establish a high-resolution numerical solver for computing the compressible reactive flows. This chapter presents the numerical procedures acquired in this solver for computing the fluxes using weighted essentially non-oscillatory (WENO) scheme, dealing with chemical stiffness problems, and tracing droplets and their interaction with the compressible fluids. As examples, the Taylor-Green vortex (TGV) problem considering the passive scalar mixing, the spatially developing reactive mixing layer flows taken gas-phase hydrogen, and liquid *n*-decane as fuel are simulated, respectively. Many important characteristics of flow, flame, and ignition are analyzed under the supersonic condition.

Keywords: compressible reactive flows, spray combustion, numerical simulation, WENO, Lagrangian trajectory model

1. INTRODUCTION

Numerical simulation has become one of important approaches in academia and industry to research-complicated reactive flows in recent years, though it is still a fast developing subject. For combustion in low-speed flows, usually the low-Mach-number assumption is used, which assumes that the change of temperature is unrelated to the hydrodynamic pressure [1,2]. While for transonic or supersonic reactive flows, such as in the scramjet combustor, the compressible effects could not be ignored and the set of fully compressible Navier-Stokes equations should be solved.

There are three numerical simulation methods for compressible reactive flows via solving Navier-Stokes equations, including Reynolds-averaged Navier-Stokes (RANS) equations, large eddy simulation (LES), and direct numerical simulation (DNS). RANS simulates all scales of fluctuations using turbulent models. However, there is still lack of a general model suitable for all flow types, since RANS ignores the details and impacts of turbulent fluctuations. LES divides the flow scales into large and small scales via a low-pass filter. The large-scale flows are solved using the governing equations, and thus LES is able to give the temporal and spatial evolution of large-scale eddies. The small-scale flows are simulated using a sub-grid-scale (SGS) model, but there is also no general SGS model. DNS solves the Navier-Stokes equations for all scales of flows without introducing any models and assumptions, and thus it can accurately predict the full-scale flow characteristics. However, its main limitation in practice is the high cost of calculation.

In the early days, RANS or LES was used to simulate the compressible reactive flows considering the computing cost. Dauptain et al. [3] simulated the supersonic hydrogen flame using the large eddy simulation, and their numerical results agreed well with the experiment measurements. LES was also used to simulate the supersonic flow and combustion in a model scramjet combustor [4], and the velocity and temperature at different cross-sections were compared with the experimental data. Edwards et al. used the hybrid method of LES/RANS to simulate the supersonic reacting wall-jet flow [5]. There are also other examples using RANS or LES [6–9]. Recently, some researchers have used DNS to simulate the compressible reactive flows. For example, O'Brien et al. [10] simulated the time-developing, H_2/air-reacting mixing layers and investigated the dynamics of backscatter of kinetic energy using DNS; Diegelmann et al. [11] did the direct simulations of reactive shock-bubble interaction with detailed chemistry under varying initial pressure.

One of the main difficulties in the simulation of compressible flows is how to distinguish the discontinuities which may widely exist in supersonic flows. Jiang and Shu [12] proposed the weighted essentially non-oscillatory (WENO) scheme to capture shock waves with almost no oscillations. But the WENO scheme is too dissipative in regions with large shear rates. To overcome this drawback, the hybrid scheme was hence developed, which switches the WENO scheme and the linear (high-order) scheme according to a discontinuity detector. Pirozzoli [13] proposed a hybrid compact-WENO scheme to obtain high resolutions for simulation of shock-turbulence interaction. Furthermore, Ren et al. [14] proposed a Roe type, characteristic-wise compact-WENO scheme. Considering overcoming the complexity and program parallelization of compact schemes, Hu et al. [15] proposed a simple WENO scheme, which switches between the WENO scheme and its optimal linear scheme. A detailed review of numerical methods used for high-speed flows can be seen in [16].

The spray combustion can occur in the scenario of compressible flows [17–24]. Therefore, the methods to treat the dispersed droplets in the flows will be concerned by different approaches, such as Eulerian and Lagrangian methods. The Eulerian one needs complicated modelling of two-phase correlation terms, which is yet to be done. Even though the high computation cost will be taken by the Lagrangian method to track each particle released to the flow fields, it is the physically clear and mathematically simple method in simulations of two-phase reactive flows. This chapter will also consider the spray combustion in mixing layer flows as a computation example.

This chapter tries to contribute the numerical simulation in the direction of compressible reactive flows. It is organized as follows. In section 2, the governing equations are presented; the methods to calculate the multispecies physical parameters and to deal with the chemical stiffness are also presented. In section 3, the discretization schemes and time integration methods are discussed, including the discretization of inviscid/viscous terms. In section 4, three examples of numerical simulations of compressible flows are presented. In section 5, the conclusions are drawn finally.

2. GOVERNING EQUATIONS AND MODELS OF GAS PROPERTIES

2.1. GOVERNING EQUATIONS

Since this chapter will extend the simulations to the compressible two-phase turbulent flows considering the dispersed droplets' evaporation and gas-phase combustion, the governing equations will describe the Lagrangian transport of droplets through a continuous, compressible and chemically reacting carrier gas flow.

The equations for the gas-phase including mass, momentum, and energy exchange between the gas and the dispersed phase read as

$$\frac{\partial \rho}{\partial t}+\frac{\partial}{\partial x_j}\left(\rho u_j\right)=S_m \tag{1}$$

$$\frac{\partial}{\partial t}\left(\rho u_i\right)+\frac{\partial}{\partial x_j}\left(\rho u_i u_j+P\delta_{ij}-\tau_{ij}\right)=S_F \tag{2}$$

$$\frac{\partial}{\partial t}\left(\rho e_t\right)+\frac{\partial}{\partial x_j}\left(\left(\rho e_t+P\right)u_j-\lambda\frac{\partial T}{\partial x_j}-u_i\tau_{ij}\right)=S_Q \tag{3}$$

$$\frac{\partial}{\partial t}\left(\rho Y_k\right)+\frac{\partial}{\partial x_j}\left(\rho Y_k\left(u_j+V_j^c\right)-\rho D_k\frac{\partial Y_k}{\partial x_j}\right)=S_{\mathrm{combustion},k}+S_{Y_k} \tag{4}$$

Here, ρ is the gas-phase density, u_i is the velocity, T is the temperature, P is the pressure, Y_k is the mass fraction of the kth species, and δ_{ij} is the Kronecker delta function. The right-hand side terms of the above equations S_m, S_F, and S_Q describe the two-phase couplings of mass, momentum, and energy, respectively. $S_{\mathrm{combustion},k}$ is the source term due to combustion. τ_{ij} is the Newtonian viscous stress tensor

$$\tau_{ij}=\mu\left(\frac{\partial u_i}{\partial x_j}+\frac{\partial u_j}{\partial x_i}-\frac{2}{3}\frac{\partial u_k}{\partial x_k}\delta_{ij}\right) \tag{5}$$

A correction velocity V^c_j is added to the convection velocity in the species equations to ensure global mass conservation

$$V_j^c=\sum_{k=1}^{N_s}D_k\frac{W_k}{W}\frac{\partial X_k}{\partial x_i} \tag{6}$$

Here, N_S is the total number of species. W_k and X_k are the molecule weight and the mole fraction of the kth species, respectively. e_t is the total energy, that is, kinetic energy plus internal (containing chemical) energy, which is defined as follows:

$$e_t=\sum_{k=1}^{N_s}Y_k\left(\int_{T_{\mathrm{ref}}}^{T}c_{p,k}\mathrm{d}T+h_{f,k}^0\right)-\frac{P}{\rho}+\frac{u_iu_i}{2} \tag{7}$$

where $c_{p,k}$ and $h^0_{f,k}$ are the specific heat capacity at constant pressure and specific chemical formation enthalpy at the reference temperature, T_{ref}, of the kth species, respectively. The values of *cp,k* are represented by a fifth-order polynomial, and the coefficients can be found in [25]

$$\frac{c_{p,k}W_k}{R} = a_{1,k} + a_{2,k}T + a_{3,k}T^2 + a_{4,k}T^3 + a_{5,k}T^4 \tag{8}$$

The equation of state used for ideal multispecies gas is introduced to close the above equations as

$$P = RT\sum_{k=1}^{N_S}\frac{Y_k}{W_k} \tag{9}$$

where R is the universal gas constant.

2.2. Models Of Gas Properties

The kinetic theory for gases is used to compute the transport and thermodynamic properties of the mixture. The Lennard-Jones potentials are used to calculate the inter-molecular forces. The thermal conductivity of each species is determined by the modified Eucken model. The dynamic viscosity of each species and the binary diffusivity are computed with the Chapman-Enskog theory, and their formulas can be found in [26]

$$\mu_k = \frac{26.69\sqrt{W_k T}}{\sigma_k^2 \Omega_{v,k}} \tag{10}$$

$$\frac{\lambda_k W_k}{\mu_k C_{v,k}} = 1.32 + \frac{1.77}{\left(C_{p,k}/R\right) - 1} \tag{11}$$

$$\mathrm{D}_{AB} = \frac{0.00266\mathrm{T}^{3/2}}{\mathrm{P}\mathrm{W}_{AB}^{1/2}\sigma_{AB}^2\Omega_D} \tag{12}$$

The viscosity and thermal conductivity of the mixture are calculated by the expressions of Wake and Wassiljewa [26], respectively,

$$\mu_m = \sum_{i=1}^{n}\frac{X_i\mu_i}{\sum_{j=1}^{n}X_j\phi_{ij}} \tag{13}$$

where $\varphi_{ij} = \dfrac{(1 + (\mu_i / \mu_j)^{1/2}(W_j / W_i)^{1/4})^2}{(8(1 + W_i / W_j))^{1/2}}$.

$$\lambda = \sum_{i=1}^{n} \frac{X_i \lambda_i}{\sum_{j=1}^{n} X_j A_{ij}} \tag{14}$$

where $A_{ij} = \dfrac{(1 + (\lambda_i / \lambda_j)^{1/2}(W_j / W_i)^{1/4})^2}{(8(1 + W_i / W_j))^{1/2}}$.

2.3. Equations of Spray Droplets

The evaporating fuel droplets are solved individually under the Lagrangian framework. Before describing the Lagrangian model, several usual assumptions should be formulated. The spray of droplets is sparsely dispersed and every droplet is unaware of the existence of the others. The droplet rotations are neglected and an infinite heat conduction coefficient is assumed. Therefore, the inner temperature distribution of each droplet remains uniform. Because of the high-density ratio between the liquid and gas phases, only the drag force acting on the droplets is significant. The Lagrangian equations for the position (in the ith direction), $x_{d,i}$, velocity, $u_{d,i}$, temperature, T_d, and mass, m_d, of a single droplet are given by

$$\frac{du_{d,i}}{dt} = \frac{F_d}{m_d} = \left(\frac{f_1}{\tau_d}\right)(u_i - u_{d,i}) \tag{15}$$

$$\frac{\mathrm{d}u_{\mathrm{d},i}}{\mathrm{d}t} = \frac{F_\mathrm{d}}{m_\mathrm{d}} = \left(\frac{f_1}{\tau_\mathrm{d}}\right)(u_i - u_{\mathrm{d,i}}) \tag{16}$$

$$\frac{dT_d}{\mathrm{d}t} = \frac{Q_\mathrm{d} + \dot{m}_\mathrm{d} L_V}{m_\mathrm{d} c_\mathrm{L}} = \left(\frac{f_2}{\tau_\mathrm{d}}\right)\left(\frac{\mathrm{Nu}}{3\mathrm{Pr}}\right)\left(\frac{c_p}{c_\mathrm{L}}\right)(T - T_\mathrm{d}) + \left(\frac{\dot{m}_\mathrm{d}}{m_\mathrm{d}}\right)\frac{L_\mathrm{V}}{c_\mathrm{L}} \tag{17}$$

$$\frac{\mathrm{d}m_\mathrm{d}}{\mathrm{d}t} = \dot{m}_\mathrm{d} = -m_\mathrm{d}\left(\frac{1}{\tau_\mathrm{d}}\right)\left(\frac{\mathrm{Sh}}{3\mathrm{Sc}}\right)\mathrm{In}(1 + B_\mathrm{M}) \tag{18}$$

where cp is the specific heat of mixture gas, c_L is the specific heat of the liquid. The momentum response time, τ_d, is defined as

$$\tau_d = \frac{\rho_d d_d^2}{18\mu} \tag{19}$$

where d_d is the droplet diameter. The gas-phase Prandtl and Schmidt numbers in gaseous phase are given by

$$\mathrm{Pr} = \frac{\mu c_p}{\lambda}, \mathrm{Sc} = \frac{\mu}{\rho D} \tag{20}$$

respectively. The Nusselt and Sherwood numbers are

$$\mathrm{Nu} = 2 + 0.552\,\mathrm{Re}_d^{1/2}\,\mathrm{Pr}^{1/3} \tag{21}$$

$$\mathrm{Sh} = 2 + 0.552\,\mathrm{Re}_d^{1/2}\,\mathrm{Sc}^{1/3} \tag{22}$$

respectively. Here, the droplet Reynolds number based on the slip velocity is defined as

$$\mathrm{Re}_d = \frac{\rho \left| u_i - u_{d,i} \right| d_d}{\mu} \tag{23}$$

f_1 is an empirical correction to the Stokes drag law and is given as

$$f_1 = 1 + 0.15\,\mathrm{Re}_d^{0.687} \tag{24}$$

The corrections of heat transfer for an evaporating droplet are given as

$$f_2 = \frac{\beta}{e^{\beta} - 1} \tag{25}$$

where the nondimensional evaporation parameter is given by

$$\beta = -1.5\,\mathrm{Pr}\,\tau_d \left(\frac{\dot{m}_d}{m_d} \right) \tag{26}$$

The evaporation rate is controlled by the mass transfer number, B_M, and is given by

$$B_{\mathrm{M}} = \frac{Y_{\mathrm{sf}} - Y_{V}}{1 - Y_{\mathrm{sf}}} \tag{27}$$

Here, Y_{V} is the mass fraction of the vapor on the far-field condition for the droplets and Y_{sf} is the vapor surface mass fraction calculated directly from the surface molar fraction (χ_{sf}), which is obtained using the equilibrium assumption

$$Y_{\mathrm{sf}} = \frac{\chi_{\mathrm{sf}}}{\chi_{\mathrm{sf}} + (1 - \chi_{\mathrm{sf}}) W / W_{\mathrm{V}}} \tag{28}$$

$$\chi_{\mathrm{sf}} = \frac{P_{\mathrm{atm}}}{P} \exp\left(\frac{L_{\mathrm{V}}}{R / W_{\mathrm{V}}} \left(\frac{1}{T_{\mathrm{B,L}}} - \frac{1}{T_{\mathrm{d}}} \right) \right) \tag{29}$$

where W is the mean molecular weight of mixture gas, W_{V} is the molecular weight of the vapor, P_{atm} is the atmospheric pressure, R is the universal gas constant, and $T_{\mathrm{B,L}}$ is the liquid boiling temperature at P_{atm}. L_{V} is the latent heat of fuel vapor at liquid temperature, T_{d}, and is given as follows:

$$L_{\mathrm{V}} = L_{\mathrm{V},T_{\mathrm{B,L}}} \left(\frac{T_{\mathrm{C,L}} - T_{\mathrm{d}}}{T_{\mathrm{C,L}} - T_{\mathrm{B,L}}} \right) \tag{30}$$

where LV,TB,LLV,TB,L is the latent heat at $T_{\mathrm{B,L}}$ and $T_{\mathrm{C,L}}$ is the critical temperature.

2.4. Two-Way Coupling Models

The source terms S, due to interactions between continuous phase and dispersed phase, are expressed by summating total droplets existing in one computation cell, denoting as N_{c},

$$S_{m} = -\frac{1}{\Delta V} \sum_{N_{\mathrm{c}}} (\dot{m}_{\mathrm{d}}) \tag{31}$$

$$S_{F} = -\frac{1}{\Delta V} \sum_{N_{\mathrm{c}}} (F_{\mathrm{d}} + \dot{m}_{\mathrm{d}} u_{\mathrm{d},i}) \tag{32}$$

$$S_Q = -\frac{1}{\Delta V}\sum_{N_c}\left(Q_d + \dot{m}_d (\frac{u_{d,i}u_{d,i}}{2} + h_{V,sf}) \right) \tag{33}$$

$$S_{Y_k} = \begin{cases} -\frac{1}{\Delta V}\sum_{N}(\dot{m}_d) & \text{for fuel} \\ 0 & \text{for other species} \end{cases} \tag{34}$$

where m_d and m˙dm˙d denote the droplet mass and dm_d/dt, respectively, $u_{d,i}$ is the droplet velocity, ΔV is the volume of the computational grid for the gas-phase calculation, and $h_{V,sf}$ is the evaporated vapor enthalpy at the droplet surface. F_d and Q_d are the drag force and the convective heat transfer, respectively, described later. For the combustion reaction model, a one-step global reaction is adopted for the reaction of hydrocarbon-air mixtures. The source term, $S_{combustion,k}$, is expressed using the combustion reaction rate per unit volume, R_F, as follows:

$$S_{\text{combustion},k} = -\frac{n_k}{n_F}\frac{W_k}{W_F}R_F \tag{35}$$

where nk and n_F are the molar stoichiometric coefficients of the kth species and the fuel for the one-step global reaction (positive for the production side), respectively. Wk and W_F are the molecular weights of the kth species and fuel, respectively.

3. NUMERICAL DISCRETIZATION PROCEDURES

A finite difference methodology is used to solve the above governing equations. An explicit Runge-Kutta time-integration methodology is applied, obtaining a third-order time-accurate computation. For example, for a semi-discrete equation of dUj=Lj(U)dtdUj=Lj(U)dt, the iteration from n to iteration $n+1$ is performed as

$$\begin{aligned} U_j^{(1)} &= U_j^n + \Delta t L_j\left(U^n\right) \\ U_j^{(2)} &= \frac{3}{4}U_j^n + \frac{1}{4}U_j^{(1)} + \frac{1}{4}\Delta t L_j\left(U_j^{(1)}\right) \\ U_j^{n+1} &= \frac{1}{3}U_j^n + \frac{2}{3}U_j^{(2)} + \frac{2}{3}\Delta t L_j\left(U_j^{(2)}\right) \end{aligned} \tag{36}$$

where U_j is the conserved variable at the direction j.

The higher-resolution numerical scheme applied for the resolution of supersonic flow is essential to the reliability of simulation. The nonviscous flux $\hat{f}_{j+1/2}$, for the interface at $j+1/2$, is evaluated using a fifth-order hybrid Compact-Weighted Essentially Non-Oscillatory scheme [14] for the resolution of turbulent field and shock-capturing calculation in the supersonic flow. The numerical flux of the hybrid scheme is calculated as

$$\hat{f}_{j+1/2} = \sigma_{j+1/2}\hat{f}^{CU}_{j+1/2} + \left(1+\sigma_{j+1/2}\right)\hat{f}^{WENO}_{j+1/2} \tag{37}$$

which is constructed by hybridizing a fifth-order Compact Upwind (CU) scheme, $\hat{f}^{CU}_{j+1/2}$, and a fifth-order WENO one, $\hat{f}^{WENO}_{j+1/2}$, through a smoothness indicator r_c. $\sigma_{j+1/2}$ is the weight of the numerical flux,

$$\sigma_{j+1/2} = \min\left(1, \frac{r_{j+1/2}}{r_c}\right) \tag{38}$$

A sixth-order symmetric compact difference scheme is applied for the viscous diffusion terms,

$$\frac{1}{3}\hat{F}_{j-1} + \hat{F}_j + \frac{1}{3}\hat{F}_{j+1} = \frac{28\left(\hat{f}_{j+1} - \hat{f}_{j-1}\right) + \left(\hat{f}_{j+2} - \hat{f}_{j-2}\right)}{36\Delta x} \tag{39}$$

where $\hat{F}_j$ is the difference approximation for $\left(\frac{\partial \hat{f}}{\partial x}\right)_j$ at the node j.

Because of the stiffness of the equation due to the reactions, the time integration of the equations is performed by means of the point-implicit method [27]. For instance, U is supposed as the unknown variable at time step $n+1$, and the below equation is solved to obtain $U^{n+1}=U^n+\Delta U$

$$\left(I - \left(\frac{\partial S}{\partial U}\right)^n \Delta t\right)\Delta U = \left(-C^n + V^n + S^n\right)\Delta t \tag{40}$$

where C is the convection term, V is the viscous term, S is the source term, and I is the unity matrix.

In order to obtain the physical quantities of the fluid at a droplet position, the fluid velocities computed on the Eulerian mesh were interpolated at the droplet locations. A fourth-order Lagrangian interpolation was used for this purpose [28]. When the droplet is located at (x_d, y_d, z_d) in a cell whose coordinates are $(x_0, y_0, z_0)...(xl, yl, zl)$, the fluid velocity at the droplet position is obtained via the Lagrangian interpolation, which is expressed as

$$u(x_d, y_d, z_d) = \sum_{i=1-n/2}^{n/2} \sum_{j=1-n/2}^{n/2} \sum_{k=1-n/2}^{n/2} \times \left(\prod_{\substack{l=1-n/2 \\ (l \neq i)}}^{n/2} \frac{x_d - x_l}{x_i - x_l} \prod_{\substack{l=1-n/2 \\ (l \neq j)}}^{n/2} \frac{y_d - y_l}{y_i - y_l} \prod_{\substack{l=1-n/2 \\ (l \neq k)}}^{n/2} \frac{z_d - z_l}{z_i - z_l} \right) u_{i,j,k} \tag{41}$$

where the subscripts *i, j, k,* and *l* represents the cell indexes and $n = 4$.

In the Lagrangian droplet-tracking method, the droplet equation of motion was integrated with the third-order Adams-Bash-forth scheme in direction, *i*, for droplet position,

$$x_{d,i}^{n+1} = x_{d,i}^{n} + \Delta t \left(\frac{23}{12} u_{d,i}^{n} - \frac{16}{12} u_{d,i}^{n-1} + \frac{5}{12} u_{d,i}^{n-2} \right) \tag{42}$$

The equations of velocity and temperature of the droplet are solved using the same integration method, as Eq. (42).

4. NUMERICAL EXAMPLES

4.1. Taylor-Green Vortex (TGV) Case

Taylor-Green vortex problem is an idea test for a numerical scheme whether it has the ability to simulate the transition from laminar to turbulence, since it contains the stages of laminar, transition, and finally turbulence as time develops. The three-dimensional Taylor-Green vortex was first introduced by Taylor and Green, and its initial condition is given as follows:

$$\begin{aligned} u &= U_0 \sin x \cos y \sin z \\ v &= -U_0 \cos x \sin y \sin z \\ w &= 0 \\ p &= p_0 + \rho_0 U_0^2 (\cos 2z + 2)(\cos 2x + \cos 2y)/16 \end{aligned} \tag{43}$$

The computing domain is [0,2π]×[0,2π]×[0,2π][0,2π]×[0,2π]×[0,2π]. The parameters are set as p0=100p0=100, ρ0=1ρ0=1, U0=1U0=1, then the Mach number is about 0.08 and the flow is nearly incompressible. The boundary conditions in the three dimensions are all periodic. Three numerical schemes are used here, WENO [7], WENO-compact upwind [9] (denoted as WENO-CU), and WENO-linear upwind [10] (denoted as WENO-LU), respectively. Two sets of computation grid are taken as 64^3 and 32^3 for different flow Reynolds numbers.

Figure 1 shows the results of the total kinetic energy integrated in the entire flow field. For Re = 30,000, the kinetic energy decays very slowly at the beginning and it has almost no change in the first few seconds. However, when the time is at about 4 s, the sub-grid scales are produced and the kinetic energy begins to decay due to the numerical dissipation, which also can be taken as the implicit sub-grid dissipation. The results of Re number equalling 3000 are similar, since in both cases, the numerical viscosity is larger compared to the physical viscosity. For the results of Re = 300, there is only slight difference between the curve of WENO-LU and WENO-CU, since in this case the physical viscosity is dominated.

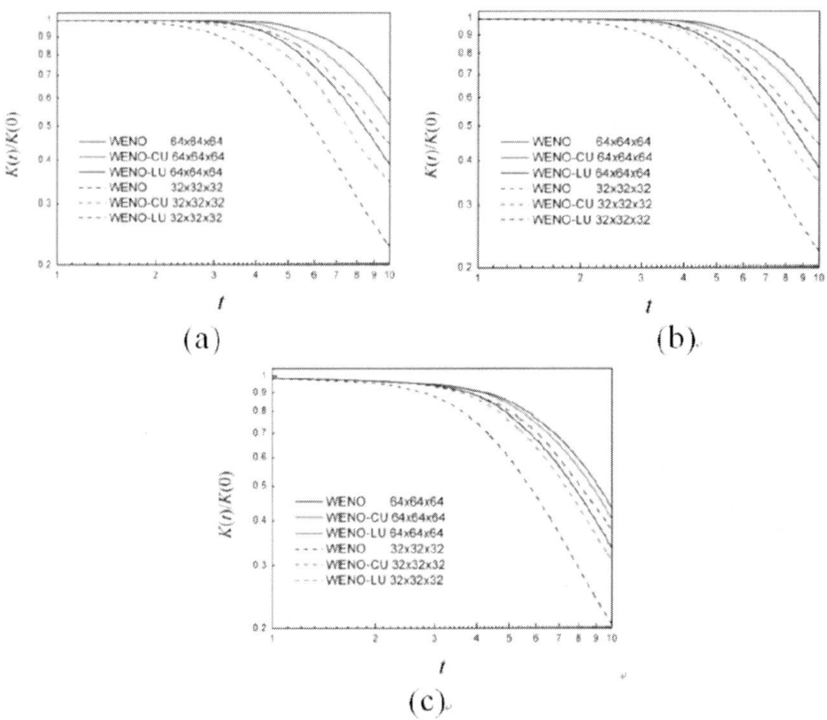

FIGURE 1. Comparison of the total kinetic energy. (a) Re = 30,000; (b) Re = 3000; (c) Re = 300.

Figure 2 shows the energy spectrum of the TGV with Re = 30,000. As the time increases, the slope of energy spectrum also changes. If the time is longer than 4, the turbulence develops and the kinetic energy builds up a Kolmogorov inertial range. The slope is then approximately decayed with the −5/3 law. It implies that it can predict the characteristics of turbulence.

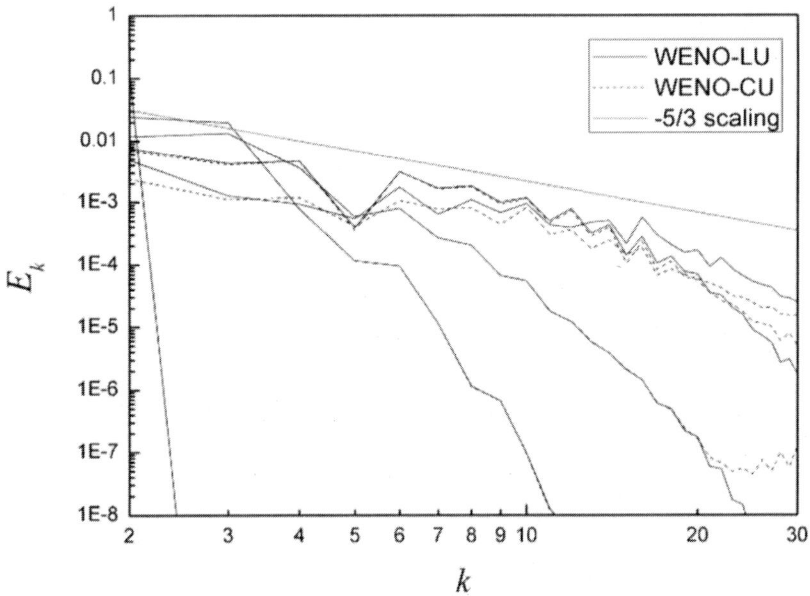

FIGURE 2. Energy spectrum of TGV with Re = 30,000.

Since the passive scalar transportation is very important in many fields, especially in the chemical-reacting process, we also solved the conserved passive scalar transport equation

$$\frac{\partial(\rho f)}{\partial t} + \nabla \cdot (\rho f \vec{U}) = \nabla \cdot (\rho D \nabla f) \tag{44}$$

Here, only the WENO-LU scheme is used since it has less dissipation and costs less computing time. Equation (44) presents the initial condition for f and it is a function of the initial Q distribution, where Q is the value of q-criteria. Different Schmidt numbers are taken in the simulations, 1, 0.1, and 0.01, corresponding to different diffusivity coefficients, respected $D=\mu/(\rho Sc)$

$$f(t=0)=\frac{1+erf(C_0Q)}{2}, C_0=100 \tag{45}$$

Here, erf is the error function. **Figure 3** shows the initial distribution of *f*. As we can see, in most regions, *f* equals 1 or 0 and there are large gradients of *f* initially.

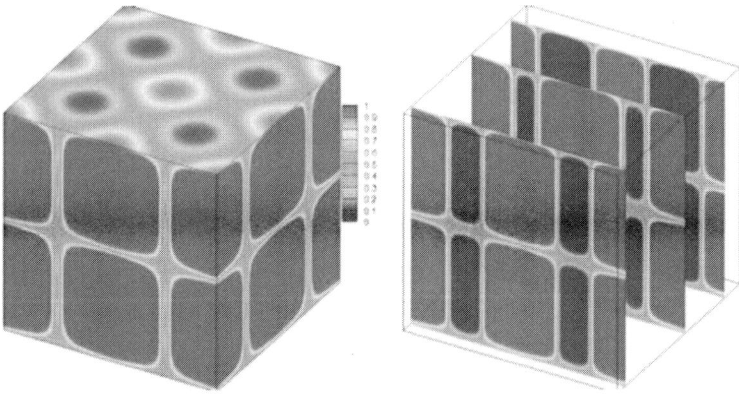

FIGURE 3. Initial distribution of passive scalar *f*.

Figure 4 shows the variance of the passive scalar for the Re = 30,000 flow in different Schmidt numbers and different grid levels. Similar to the kinetic energy decaying, the scalar variance decays very slowly in the first 4 s, and it can be concluded that the passive scalar mixing goes on with the laminar diffusion. However, the significant decaying can be observed due to the initial large gradients of passive scalar. For the time longer than four, the turbulent mixing occurs and the decaying of passive scalar variance is larger than that in the beginning stage.

Figure 5 shows the passive scalar contours for Re = 30,000 and Sc = 1.0 in the plane of $z = \pi$. During the stage of laminar diffusion, the interface between the high and low *f* region is clear while the interface is wrinkled in the stage of turbulent mixing, which enhances the mixing process.

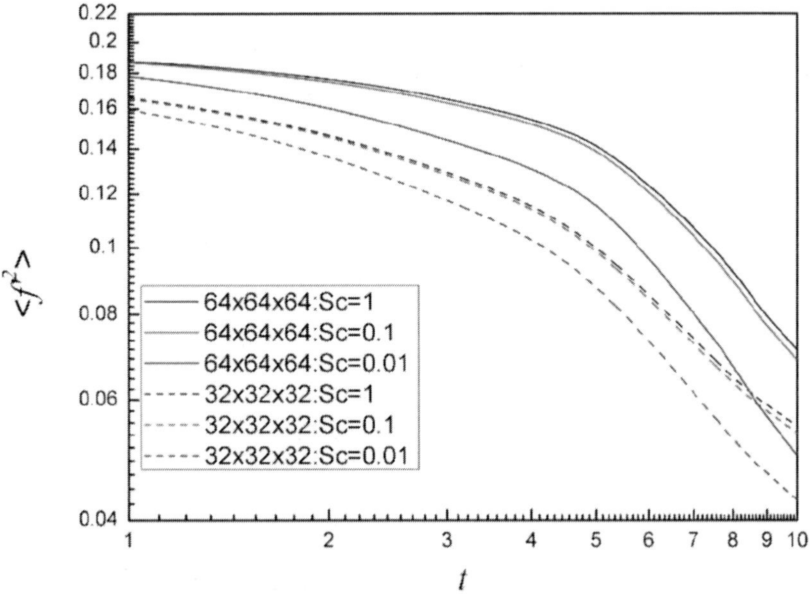

FIGURE 4. Variance curves of passive scalar.

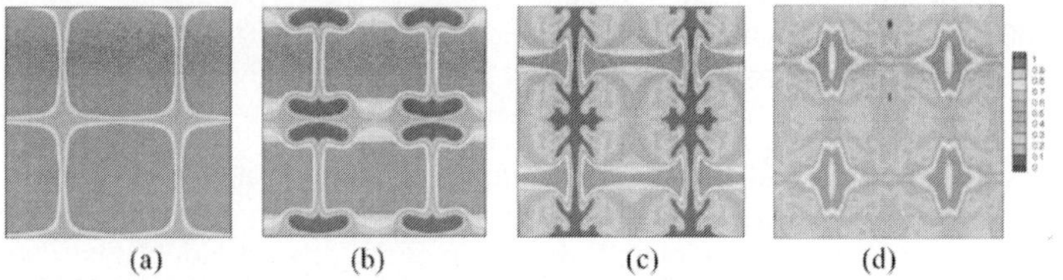

FIGURE 5. Passive scalar field in the plane of $z = \pi$. (a) $t = 0$ s; (b) $t = 3$ s; (c) $t = 6$ s; (d) $t = 10$ s.

4.2. Reactive H_2/Air Spatially Developing Mixing Layer

The schematic of the supersonic mixing layer is shown in **Figure 6**.

The inlet conditions are set as follows. The average velocity profile in the stream-wise direction is a hyperbolic tangent, and the average velocity in other directions is set to zero

$$\overline{U} = \frac{U_A + U_F}{2} + \frac{U_A - U_F}{2}\tanh\left(\frac{y - 0.5l_y}{2\delta_0}\right) \tag{46}$$

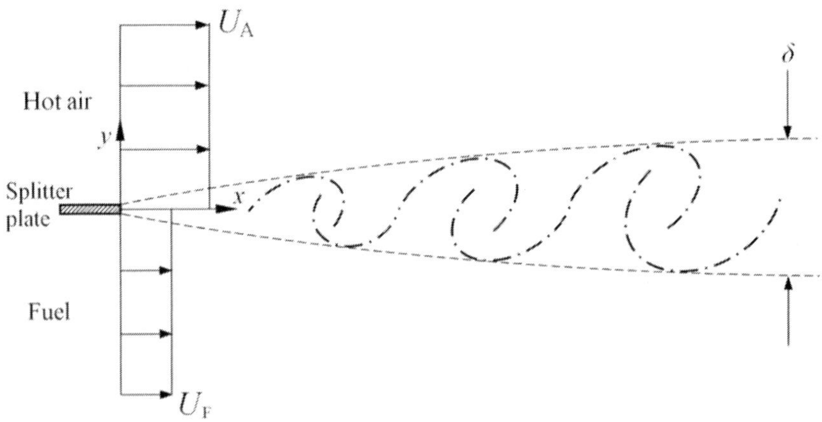

FIGURE 6. Schematic of computation model of spatially developing mixing layer.

where δ_0 is the inlet layer thickness given as 0.4 mm, and l_y is the length of computational domain in the y direction. U_A is the hot-air inflow velocity, 2000 m/s, U_F is the cold fuel inflow velocity, 1000 m/s. The temperatures of both streams are $T_A = 1600$ K and $T_F = 390$ K. The reactive system pressure is specified as $p_A = 1$ atm.

The stream-wise and transverse domain lengths are lx and ly, respectively. By setting $lx = 4ly$, the transverse domain size is large enough to have minimal influence on the main flow region of the droplet-laden shear layer. The computation grid in the two-dimensional case is taken as 512 × 128 equally spaced computational cells in the stream-wise and the transverse directions, respectively, after a series of grid independence tests.

In order to stimulate the flow instability, the inlet velocity perturbation is given as

$$v' = (U_A - U_F)G(y - 0.51_y)\, A\sin(2\pi f_o t + \xi) \tag{47}$$

where G is the Gauss function, A is the fluctuation amplitude, f_o is the maximum disturbance frequency, and ξ is the random phase. The nonreflecting conditions are used in the transverse directions. The outflow is treated by an extrapolation of each primitive variable. A 19-step detailed H_2/O_2 reaction mechanism is used, which can be referred in [29].

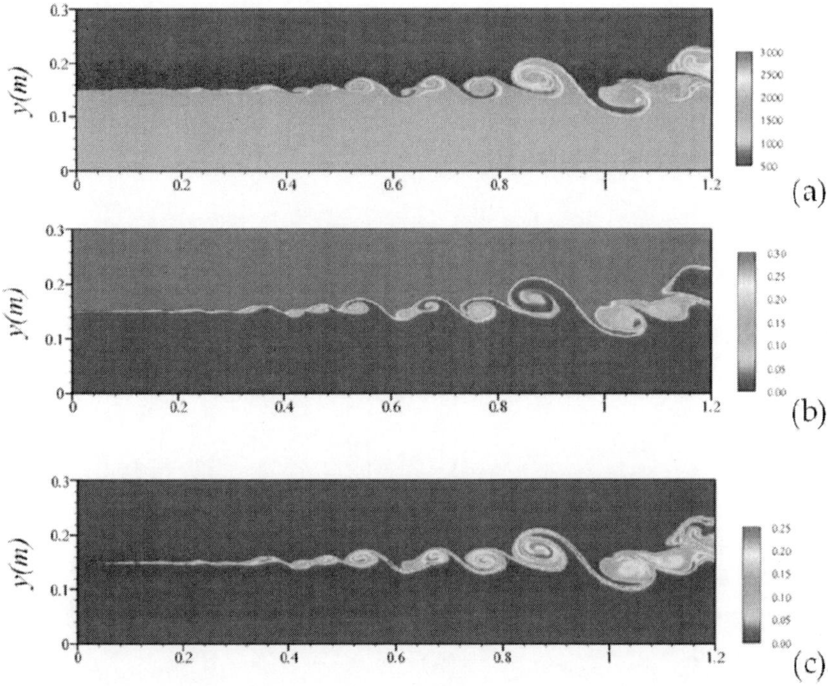

FIGURE 7. Instantaneous contour distribution of reactive flows. (a) temperature; (b) H_2; (c) H_2O.

Instantaneous distributions of temperature and mass fraction of H_2 and H_2O at t = 2 ms are shown in **Figure 7**. At about x = 0.4 m, the mixing layer loses stability, the large-scale vortices begin to form, and efficient mixing occurs inside the vortices. Due to the relatively high temperature of air, fast reaction occurs and a large amount of water concentrates inside the vortices.

Figure 8 shows the flame structures in the mixing layer flow. The flame structures were obtained by means of plotting all the combustion status in the whole computation domain within the time interval $2<t<2.5$ ms, that is, recording and plotting the temperature or species fraction at each calculation node in the observing time interval. Here, the mixture fraction is defined as

$$Z \equiv \frac{W_{\mathrm{H}}}{W_{\mathrm{H}} + W_{\mathrm{O}}} \tag{48}$$

where W_H and W_O are the mass fraction of element H and element O, respectively. For the stoichiometric ratio, $Z_{st} = 0.111$. It can be seen that the peaks of temperature and water are both near Z_{st}. The flame structure is similar with that in the flamelet model, since the non-premixed flame dominates in the current case.

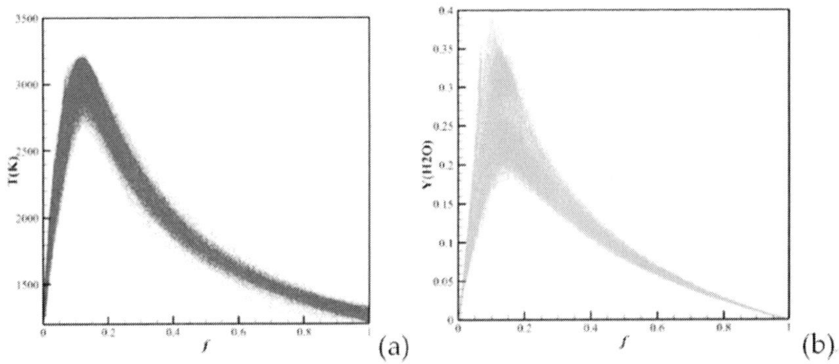

FIGURE 8. Flame structure. (a) temperature distribution and (b) water distribution.

4.3. Non-Premixed Combustion of *N*-Decane Droplets in Turbulent Mixing Layers

The considered flow configuration is that of a spatially developing shear layer, as depicted in **Figure 9**, which is formed between a stream of hot air moving at velocity U_A and a spray carried by cold air moving at velocity U_S. x and y refer to the stream-wise and transverse directions, respectively. The inflow parameters of the two streams are listed in **Table 1**.

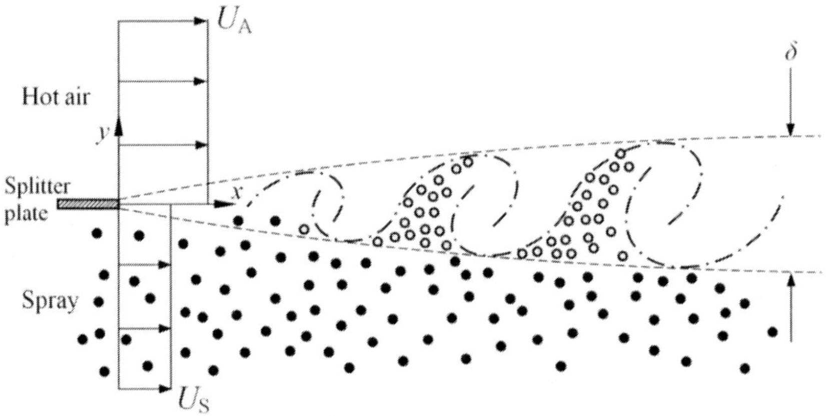

FIGURE 9. Schematic of a fuel spray in a two-dimensional turbulent shear layer.

TABLE 1. Inflow parameters.

Velocity ratio U_A/U_S	Temperature ratio T_A/T_S	Pressure ratio P_A/P_S	Convective Mach number $M_c = \frac{U_A - U_S}{a_A + a_S}$
2	4	1	0.4
U_S (m/s)	T_S (K)	P_S (MPa)	Inflow Reynolds number $Re = \frac{\rho \mid U_A - U_S \mid \delta_0}{\mu}$
458.3	363.3	0.1	2731.2
u_d (m/s)	T_d (K)	ρ_d (kg/m³)	Particle Stokes number (St_0) $St_0 = \frac{\tau_d}{\tau_f} = \left(\frac{\rho_d d_{d,0}^2}{18\mu_S}\right) / \left(\frac{\delta_0}{\mid U_A - U_S \mid}\right)$
458.3	298.2	642.0	10.5

$$k = A \cdot T^n \exp\left(-\frac{E}{RT}\right)\left(\frac{\rho \cdot Y_F}{W_F \times 10^6}\right)^a \left(\frac{\rho \cdot Y_{O_2}}{W_{O_2} \times 10^6}\right)^b \tag{50}$$

where A is the coefficient of frequency and E is the activation energy. In the reaction of n-decane, these parameters are given as follows: $A = 3.8 \times 10^{11}$, $E = 1.256 \times 10^5$ J/mol, $a = 0.25$, $b = 1.5$, and $n = 0$. Therefore, the combustion reaction rate per unit volume (g/cm³/s) is $R_F = W_V k \times 10^3$.

In all simulation cases, the fuel droplets are initially located in the cold air and cannot vaporize until they traverse into the hot air associated with the rolling-up process of vortices. The hot-air stream provides the heat needed for droplet vaporization. The fuel vapor diffuses into the air stream, and it begins to react with the local oxygen as it reaches the high-temperature regions.

Large eddies rotate and entrain the hot air, providing heat for the evaporation of fuel droplets that preferentially accumulate in the high-strain vortex-braid regions adjoining the hot air. The fuel vapor generated by droplet evaporation diffuses toward the hot air and mixes with the surrounding oxygen, forming reactive pockets. Auto-ignition occurs when the fuel mass fraction is high enough, as depicted in **Figure 10(a)**. Under the instantaneous condition, the time t_{ig} defines the moment that the incipient establishment of ignition kernels occurs. We normalize the gaseous temperature by $T^* = T/T_{in}$ and $T_{in} = (T_A + T_S)/2$. The droplet diameters are normalized by d^* d $= d_d/d_{d0}$ and d_{d0} is the initial droplet diameter.

Large eddies rotate and the reaction kernels between two convection vortices are strained at time $t = t_{ig} + 0.25t_L$, as shown in **Figure 10(b)**. Here, the time t_L is the large eddy turnover time and is defined as

$$t_L = \frac{\delta_L}{|U_A - U_S|} \tag{51}$$

where δ_L is the length of the large eddy. Compared with the reaction rates of the ignition kernels at time t_{ig}, the chemical heat release in the reaction kernels at time $t_{ig} + 0.25t_L$ results in a larger temperature increase and facilitates the self-acceleration of the chemical reaction rate. At time $t = t_{ig} + 0.5t_L$, the reaction kernels are separated into two parts due to the stretch effects, as shown by **Figure 10(c)**. One part of the reaction kernels still exists in the high-strain vortex-braid zone and suffers the shear strain continuously. Evaporating fuel droplets segregate in a thin layer along this reaction kernel and feed fuel vapors for the chemical reaction. The other separated part is entrained in the inner of the large cold eddy. It is difficult for the fuel droplets to penetrate into the high-vorticity core of the large eddy because of the large droplet inertia and fuel droplets surround far from this separated reaction kernel, forming a thick vaporization layer. However, the turbulence motion separates the vaporization layer and the flame kernel. Hence, the fuel vapor diffusion is difficult for feeding the chemical reaction in the flame kernel effectively. Here, we call it the weak-fueling mechanism due to the preferential concentration effects of droplets. If we track the separated reaction kernel in the inner of the cold eddy, it is found that the local chemical reaction rate of this flame kernel is decelerated, compared to that of the reaction kernel in the high-strain zone accompanied by the rich reactant supplement. **Figure 10(d)**depicts the extinction phenomena at time $t = t_{ig} + 0.75t_L$. Comparison with **Figure 10(c)** shows that the reaction kernels that have high chemical reaction

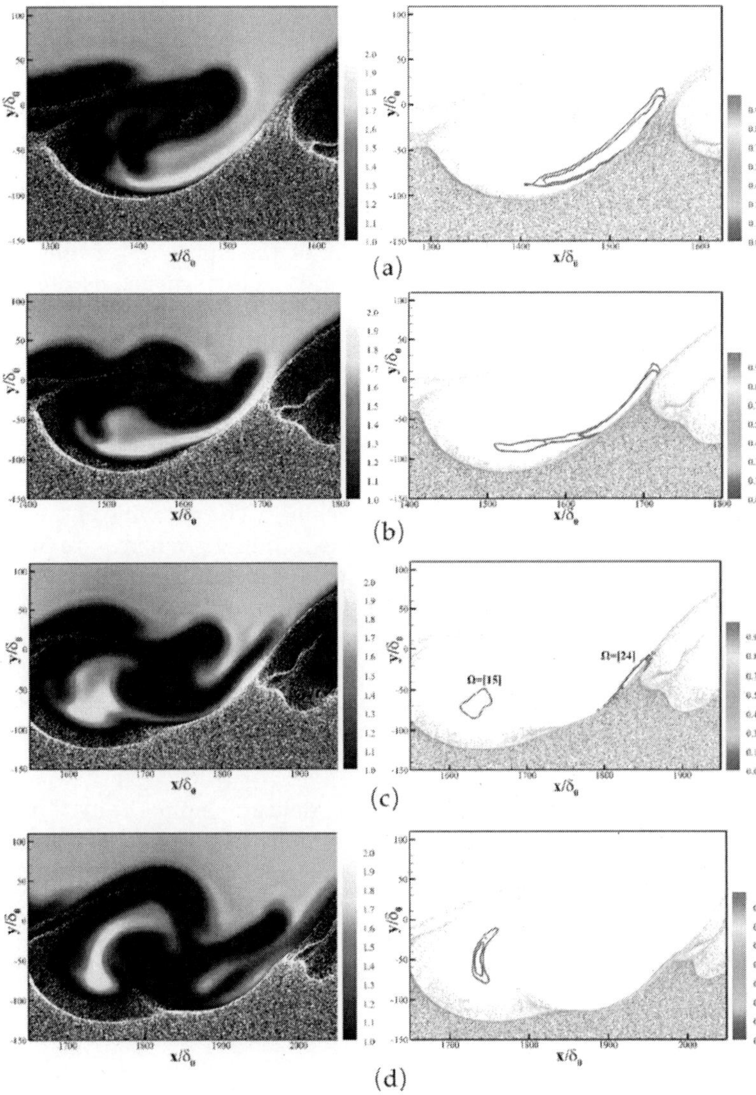

FIGURE 10. Flame kernels at (a) time $t = t_{ig}$; (b) time $t = t_{ig} + 0.25\ t_L$; (c) time $t = t_{ig} + 0.5\ t_L$; and (d) time $t = t_{ig} + 0.75\ t_L$. Instantaneous distributions of dimensionless gaseous temperature (left) and reaction rate (right), with red isolines given by $\Omega = [11, 15]$ ($kgm^{-3}s^{-1}$) in (a) and $\Omega = [15, 24]$ ($kgm^{-3}s^{-1}$) in (b)–(d). Here, the dots indicate fuel droplets in the figures, with colors corresponding to dimensionless droplet diameters in the right figures.

rates in the high-strain region at time $t = t_{ig} + 0.5t_L$disappear. The other reaction kernel entrained in large eddies is distorted due to the turbulence motions.

5. CONCLUSIONS

This chapter introduced the state of art of the numerical simulation on compressible flows. The high-order hybrid WENO scheme, as one of the important issues for simulating the supersonic flows, is required to compute the numerical fluxes, in which the WENO scheme is used near the discontinuities and linear scheme is used in the smooth regions of the flows. In the solver developed by the present authors, the point-implicit method is utilized to treat the chemical stiffness problem.

A TGV problem, defined to investigate the transports of passive scalar, is simulated by the present numerical procedures. The decaying of the kinetic energy and variance of passive scalar due to initial gradients are analyzed based on the numerical results. The spatially developing mixing layer flows are simulated, taking the hydrogen and liquid *n*-decane droplets as fuel and air as oxidizer in the scenario of supersonic inflows. The stable combustion of hydrogen/air and the flame structure are analyzed by the numerical results. The ignition considering the eddy turnover effects is carefully analyzed, since the large-scale structures and droplet dispersion can be captured by the present numerical methods.

This chapter tries to contribute the field of numerically analyzing compressible reactive flows. The new high-order schemes and numerical strategy will be developed in the future, so that the stabilities and robustness will be enhanced by the solver based on the Navier-Stokes equations. Thus, deeply understanding the flow physics of compressible reactive flows will be benefited from results obtained by numerical simulations.

REFERENCES

1. Tomboulides, AG, Lee, JCY and Orszag, SA. Numerical simulation of low Mach number reactive flows. Journal of Scientific Computing. 1997;12(2):139–167.
2. Desjardins O, Blanquart G, Balarac G, et al. High order conservative finite difference scheme for variable density low Mach number turbulent flows. Journal of Computational Physics. 2008;227(15):7125–7159.
3. Dauptain A, Cuenot B, Poinsot T. Large eddy simulation of a supersonic hydrogen-air diffusion flame. Complex Effects in Large Eddy Simulation. 2005;71:98.
4. Berglund M, Fureby C. LES of supersonic combustion in a scramjet engine model. Proceedings of the Combustion Institute. 2007;31(2):2497–2504.

5. Edwards J R, Boles J A, Baurle R A. Large-eddy/Reynolds-averaged Navier--Stokes simulation of a supersonic reacting wall jet. Combustion and Flame. 2012;159(3):1127–1138.
6. Selle L, Lartigue G, Poinsot T, et al. Compressible large eddy simulation of turbulent combustion in complex geometry on unstructured meshes. Combustion and Flame. 2004;137(4):489–505.
7. Lacaze G, Cuenot B, Poinsot T, et al. Large eddy simulation of laser ignition and compressible reacting flow in a rocket-like configuration . Combustion and Flame. 2009;156(6):1166–1180.
8. Boudier G, Gicquel L Y M and Poinsot T, et al. Comparison of LES, RANS and experiments in an aeronautical gas turbine combustion chamber. Proceedings of the Combustion Institute. 2007;31(2):3075–3082.
9. Saghafian A, Terrapon V E, Pitsch H. An efficient flamelet-based combustion model for compressible flows. Combustion and Flame. 2015;162(3):652–667.
10. O'Brien J, Urzay J, Ihme M, et al. Subgrid-scale backscatter in reacting and inert supersonic hydrogen–air turbulent mixing layers. Journal of Fluid Mechanics. 2014;743:554–584.
11. Diegelmann F, Tritschler V, Hickel S, et al. On the pressure dependence of ignition and mixing in two-dimensional reactive shock-bubble interaction. . Combustion and Flame. 2016;163:414–426.
12. Jiang G S, Shu C W. Efficient implementation of weighted ENO schemes. Journal of Computational Physics. 1996;126:202–228.
13. Pirozzoli S. Conservative hybrid compact-WENO schemes for shock-turbulence interaction. Journal of Computational Physics. 2002;178(1):81–117.
14. Ren Y X, Zhang H. A characteristic-wise hybrid compact-WENO scheme for solving hyperbolic conservation laws. Journal of Computational Physics. 2003;192(2):365--386.
15. Hu X Y, Wang B, Adams N A. An efficient low-dissipation hybrid weighted essentially non-oscillatory scheme. Journal of Computational Physics. 2015;301:415–424.
16. Pirozzoli S. Numerical methods for high-speed flows. Annual review of fluid mechanics. 2011;43(163–194)
17. Miller R S, Bellan J. Direct numerical simulation of a confined three-dimensional gas mixing layer with one evaporating hydrocarbon-droplet-laden stream. Journal of Fluid Mechanics. 1999;384:293–338.
18. Miller R S. Effects of nonreacting solid particle and liquid droplet loading on an exothermic reacting mixing layer. Physics of Fluids. 2001;13(11):3303–3320.
19. Reveillon J, Vervisch L. Analysis of weakly turbulent dilute-spray flames and spray combustion regimes. Journal of Fluid Mechanics. 2005;537:317–347. 4
20. Wang Y, Rutland C J. Direct numerical simulation of ignition in turbulent n-heptane liquid-fuel spray jets. Combustion and Flame. 2007;149(4):353–365.
21. Schroll P, Wandel A P, Cant R S, et al. Direct numerical simulations of autoignition in turbulent two-phase flows. Proceedings of the Combustion Institute. 2009;32(2):2275–2282.
22. Neophytou A, Mastorakos E, Cant R S. The internal structure of igniting turbulent sprays as revealed by complex chemistry DNS. Combustion and Flame. 2012;159(2):641–664.
23. Borghesi G, Mastorakos E, Cant R S. Complex chemistry DNS of n-heptane spray autoignition at high pressure and intermediate temperature conditions. Combustion and Flame. 2013;160(7):1254–1275.
24. Xia J, Zhao H, Megaritis A, et al. Inert-droplet and combustion effects on turbulence in a diluted diffusion flame. Combustion and Flame. 2013;160(2):366–383.
25. Burcat A, Ruscic B. Third millenium ideal gas and condensed phase thermochemical database for combustion with updates from active thermochemical tables. Argonne National Laboratory Argonne, IL; 2005.
26. Reid R C, Prausnitz J M, Poling B E. The properties of gases and liquids. McGraw-Hill, New York; 2001.

27. Eberhardt S, Imlay S. Diagonal implicit scheme for computing flows with finite rate chemistry. Journal of Thermophysics and Heat Transfer. 1992;6(2):208–216.
28. Wang Q, Squires K D, Wu X. Lagrangian statistics in turbulent channel flow. Atmospheric environment. 1995;29(18):2417–2427.
29. Nishioka M, Law C K. A numerical study of ignition in the supersonic hydrogen/air laminar mixing layer. Combustion and flame. 1997;108(1):199–219.
30. Westbrook C K, Dryer F L. Simplified reaction mechanisms for the oxidation of hydrocarbon fuels in flames. Combustion science and technology. 1981;27(1–2):31–43.

CHAPTER 11

Probabilistic, Statistical and Algorithmic Aspects of the Similarity of Texts and Application to Gospels Comparison

Soumaila Dembele[1,2], Gane Samb Lo[2,3]

[1]*LSTA, Université Pierre et Marie Curie, Paris, France*

[2]*LERSTAD, Université Gaston Berger de Saint-Louis, Saint-Louis, Sénégal*

[3]*Université des Sciences de Gestion de Bamako, Bamako, Mali*

ABSTRACT

The fundamental problem of similarity studies, in the frame of data-mining, is to examine and detect similar items in articles, papers, and books with huge sizes. In this paper, we are interested in the probabilistic, and the statistical and the algorithmic aspects in studies of texts. We will be using the approach of *k-shinglings*, a *k-shingling* being defined as a sequence of k consecutive characters that are extracted from a text ($k \geq 1$). The main stake in this field is to find accurate and quick algorithms to compute the similarity in short times. This will be achieved in using approxi- mation methods. The first approximation method is statistical and, is based on the theorem of Glivenko-Cantelli. The second is the banding technique. And the third concerns a modification of the algorithm proposed by Rajaraman et al. ([1]), denoted here as (RUM). The Jaccard index is the one being used in this paper. We finally illustrate these results of the paper on the four Gospels. The results are very conclusive.

Keywords: Similarity, Web Mining, Jaccard Similarity, RU Algorithm, Minhashing, Data Mining, Shingling, Bible's Gospels, Glivenko-Cantelli, Expected Similarity, Statistical Estimation

1. INTRODUCTION

In the modern context of open publication, particularly in Internet, similarity studies between classes of objects become crucial. For example, such studies can detect plagiarism of books, articles, and other works (see [2] [3] for example). Also they may reveal themselves as decision and management tools. An other illustration of the importance of such a knowledge concerns commercial firms. They may be interested in similarity patterns between clients from different sites or between clients who buy different articles. In the same order of ideas, movies renting companies may try to know the extent of similarity between clients subscribing for violence films and those renting action films for example.

As a probability concept, the notion of similarity is quite simple. However, in the context of Internet, the data may be huge. So that the main stake is the quick determination of some similarity index. The shorter the time of computation, the better the case. So similarity studies should rely on powerful algorithms that may give clear indications on similarities in seconds. The contextualization of the similarity, and forming the sets to be compared, and the similarity computations may take particular forms according to the domains of application.

The concept of similarity has been studied and is still studied by researchers from a variety of disciplines (mathematics, web-mining, information sciences, applied sciences, etc.) from several point of views (theoricical, algorithmic, high dimension reduction, computer calculations, etc.). The following references on top of [2] [3] already cited, to cite a few, may help the reader to have a broad idea on this concept : Chen et al. [4] for visual similarity, Cha [5] for the use of density functions in similarity detection, Gower et al. [6] and Zezula et al. [7] for the metric space approach, Strehl et al. [8] , Formica [9] in the context of information sciences, Bilenko et al. [10] and Theobold et al. [11] for focus similarity on on large-web collections.

In this paper, we will be focusing on similarity of texts. This leads us to consider the approach of shinglings, that we will define in Section 2.

The reader is referred to Rajaraman et al. ([1]) for a general introduction to similarity studies in the context of webmining. In their book, they provide methods of determination of approximated indices of similarity. Also, they propose an algorithm that we denote as RU (Rajaraman and Ullman).

However, this algorithm has not been yet investigated in the context of probability theory, up to our knowledge. Furthermore, an evaluation of the performances of such algorithms on usual texts may be of relevance to justify such methods.

First, we want to review these methods in a coherent probabilistic and statistical setting allowing to reach later—all the aspects of similarity in this field. Then we will describe the RU algorithm in details. We will point out its redundant sides, from which a modified algorithm-denoted RUM (for RU modified) will be proposed.

To evaluate the studied techniques, the four Gospels will be used with the ends of study of similarity. The techniques will be compared in terms of speed, request of time, request of computer science resources, and request of accuracy.

The obtained results constitute a plea for improving these techniques when dealing with larger sizes.

Regarding Gospels study, our results seem to be conclusive, that is the fourth canonical Gospels are significantly similar.

This paper is organized as follows. In the next section, we define the similarity of Jaccard and its metric and probabilistic approaches. Section 3 is concerned with the similarity of texts. In Section 4, we discuss about computation stakes of similarity. In Section 5, we present different methods to estimate the similarity index. Finally in Section 7, we deal with applications of the described methods to the similarity between the four Gospels. We conclude the paper by giving some perspectives.

2. SIMILARITY OF SETS

2.1. Definition

Let A and B be two sets. The Jaccard similarity of sets A and B, denoted $sim(A,B)$, is the ratio of the size of the intersection of A and B to the size of the union of A and B

$$sim(A,B) = \frac{\#(A \cap B)}{\#(A \cup B)}. \tag{1}$$

It is easy to see that for two identical sets, the similarity is 100% and for two totally disjoint sets, it is 0%.

2.2. Metric Approach

Let us consider a non-empty set S and its power set *p(s)*. Let us consider the application of dissimilarity

$$\forall (S_1, S_2) \in \mathcal{P}(\mathcal{S})^2, d(S_1, S_2) = 1 - sim(S_1, S_2).$$

We have this simple result.

Proposition 1. *The mapping d is metric.*

Proof. Proving this simple result is not so obvious one might think. Indeed, special techniques are required to demonstrate the triangle inequality. This is done, for example, in ([6]), page 15. Here, we just outline the other conditions for a metric:

1) First, let us show that $\forall (S_1, S_2) \in (\mathcal{P}(\mathcal{S}))^2, d(S_1, S_2) \geq 0$.

We have

$$\#(S_1 \cap S_2) \leq \#(S_1 \cup S_2).$$

Then

$$\frac{\#(S_1 \cap S_2)}{\#(S_1 \cup S_2)} \leq 1.$$

Next

$$1 - \frac{\#(S_1 \cap S_2)}{\#(S_1 \cup S_2)} \geq 0.$$

Therefore

$$d(S_1, S_2) \geq 0.$$

2) Let us show that $d(S_1, S_2) = 0 \Rightarrow S_1 = S_2$. From (2.1), we get

$$\#(S_1 \cap S_2) = \#(S_1 \cup S_2) \Rightarrow S_1 = S_2.$$

3) Let us remark that $d(S_1, S_2) = d(S_2, S_1)$, since we have $S_1 \cap S_2 = S_2 \cap S_1$, and $S_1 \cup S_2 = S_2 \cup S_1$, that is: the roles of S_1 and S_2 are symmetrical in what precedes.

So, studying of the similarity is equivalent to studying the distance of dissimilarity d between two sets. For more on the metric approach, see [7] .

2.3. Probabilistic Approach

Let us give a probabilistic approach of the similarity. For that, let us introduce the notion of the representation matrix. Let n be the size of the introduced set above.

Let us consider p subsets of S:$S_1,\cdots,S_p$. The representation matrix of $S_1,\cdots,S_p$ consists in this:

- We form a rectangular array of $p+1$ columns.
- We put $S,S_1,\cdots,S_p$ in the first row.
- We put in the column of S all the elements of S, that we might write from 1 to n in an arbitrary order.
- In the column of each S_i, we will put 1 or 0 on the row i depending on whether the ith element of S is in S_i or not. We then can see that for $h \neq k$, $(S_h \cup S_k)$ is the number of rows for which one of the columns of S_h or S_k has 1 on them and $(S_h \cup S_k)$ is the number of rows for which the two columns of S_h and S_k have 1 on them.

An illustration of the matrix representation is given in Table 1.

Let us denote $(S_{ih})_{1\le i\le n}$, the column of S_h. We obtain:

$$sim(S_h, S_k) = \frac{\#\{i, 1 \le i \le n, S_{ih} = S_{ik} = 1\}}{\#\{i, 1 \le i \le n, (S_{ih} + S_{ik} = 1) + (S_{ih} = S_{ik} = 1)\}}$$

This formula can be written also in the following form:

$$sim(S_h, S_k) = \frac{\#\{i, 1 \le i \le n, S_{ih} + S_{ik} = 2\}}{\#\{i, 1 \le i \le n, (S_{ih} + S_{ik} = 1) + (S_{ih} + S_{ik} = 2)\}}$$

In the next theorem, we will establish that the similarity is a conditional probability.

Theorem 1. *Let us randomly pick a row X among n rows. Let $S_{X,h}$ be the value of the row X for a column h,*

Table 1. Representation matrix.

Element	S_1	S_2	...	S_k	...	S_h	...	S_p
1	1	0	...	0	...	1	...	1
2	0	0	...	1	...	0	...	0
...	0	...	...	...	...	...	...	...
i	1	0	...	1	...	1	...	1
...	...	...	...	...	...	...	...	...
n	0	0	...	0	...	0	...	1

$1 \le h \le p$. Then the similarity between two sets S_h and S_k is the conditional probability of the event $(S_{X,h} = S_{X,k} = 1)$ with respect to the event $(S_{X,h} + S_{X,k} \ge 1)$, i.e.

$$sim(S_k, S_h) = \mathbb{P}\left[(S_{X,h} = S_{X,k} = 1)/(S_{X,h} + S_{X,k} \ge 1)\right]$$

Proof. We first observe that for the defined matrix below, the set of rows can be split into three classes, based on the columns S_k and S_h:

1) The rows X such as we have (1,1) on the two places for columns S_k and S_h.

2) The rows Y such as we have (1,0) or (0,1) on the two places for columns S_k and S_h.

3) The rows Z such as we have (0,0) on the two places for columns S_k and S_h.

Let us show that

$$sim(S_k, S_h) = \mathbb{P}\left[(S_{X,h} = S_{X,k} = 1)/(S_{X,h} + S_{X,k} \ge 1)\right].$$

Clearly, the similarity is the ratio of the number of rows X to the sum of the numbers of rows X and the number of rows Y. The rows Z are not involved in the similarity between S_h and S_k. Thus

$$sim(S_k, S_h) = \frac{\#\{i, 1 \le i \le n, S_{Xh} = 1, S_{Xk} = 1\}}{\#\{i, 1 \le i \le n, (S_{Xh} + S_{Xk} = 1) + (S_{Xh} = 1, S_{Xk} = 1)\}}.$$

Then, by dividing the numerator and the denominator by n, we will have

$$sim(S_k, S_h) = \frac{\dfrac{\#\{i, 1 \le i \le n, S_{Xh} = 1, S_{Xk} = 1\}}{n}}{\dfrac{\#\{i, 1 \le i \le n, (S_{Xh} + S_{Xk} = 1) + (S_{Xh} = 1, S_{Xk} = 1)\}}{n}}.$$

Hence, we get the result

$$sim(S_k, S_h) = \mathbb{P}\left[\left(S_{X,h} = S_{X,k} = 1\right) / \left(S_{X,h} + S_{X,k} \ge 1\right)\right].$$

This theorem will be the foundation of the statistical estimation of the similarity as a probability.

Important remark. When we consider the similarity of two subsets, say S_h and S_k and we use the global space as $(S_h \cup S_k)$, we may see that the similarity is, indeed, a probability. But when we simultaneously study the joint similarities of several subsets, say at least S_h, S_k and S_ℓ with the global set $S_h \cup S_k \cup S_\ell$, the simi- larity between two subsets is a conditional probability. Then, using the fact that the similarity is a probability to prove the triangle inequality is not justified, as claimed in [1] , page 76.

2.4. Expected Similarity

Here we shall use the language of the urns. Suppose that we have a reference set of size n that we consider as an urn U. We pick at random a subset X of size k and a subset Y of size m. If m and k have not the same value, the picking order of the sets does have an impact on our results. We then proceed at the beginning by picking at random the first subset, that will be picked all at once, next put it back in the urn U (reference set). Then we pick the other subset. Let us ask ourselves the question: what is the expected value of the similarity of Jaccard?

The answer at this question allows us later to appreciate the degree of similarity between the texts. We have the following result.

Proposition 2. *Let U be a set of size n. Let us randomly pick two subsets X and Y of U, of respective sizes m and k according to the scheme described above. We have*

$$\mathbb{P}(\text{Card}(X \cap Y) = j) = \begin{cases} \dfrac{1}{2}\left\{\dfrac{C_m^j C_{n-m}^{k-j}}{C_n^k C_n^m} + \dfrac{C_k^j C_{n-k}^{m-j}}{C_n^k C_n^m}\right\} & \text{if } 0 \le j \le \min(k, m) \\ 0 & \text{otherwise} \end{cases} \quad (2)$$

Further

$$\mathbb{E}\left(sim(X,Y)\right)=\sum_{j=0}^{\min(k,m)}\frac{j}{2(m+k-j)}\left\{\frac{C_m^j C_{n-m}^{k-j}}{C_n^k C_n^m}+\frac{C_k^j C_{n-k}^{m-j}}{C_n^k C_n^m}\right\}. \quad (3)$$

Proof. Let us use the scheme described above. Let us first pick the set X. We have $L=C_n^k$ possibilities. Let us denote the subsets that would take X by $X_1,\cdots,X_L$. The searched probability becomes

$$\begin{aligned}\mathbb{P}\left(\text{Card}(X\cap Y)=j\right)&=\sum_{s=1}^{L}\mathbb{P}\left(\left(\text{Card}(X\cap Y)=j\right)\cap X_s\right)\\&=\sum_{s=1}^{L}\mathbb{P}\left(\left(\text{Card}(X\cap Y)=j\right)/X_s\right)\mathbb{P}(X_s)\end{aligned}.$$

Once X_s is chosen and fixed, we get

$$\mathbb{P}\left(\left(\text{Card}(X\cap Y)=j\right)/X_s\right)=\frac{C_m^j C_{n-m}^{k-j}}{C_m^k}.$$

Since $\mathbb{P}(X_s)=1/C_n^k=1/L$, we conclude

$$\mathbb{P}\left(\text{Card}(X\cap Y)=j\right)=\sum_{s=1}^{L}\frac{C_m^j C_{n-m}^{k-j}}{C_m^k}(1/L)=\frac{C_m^j C_{n-m}^{k-j}}{C_n^k C_n^m}.$$

The result corresponding to picking up Y first, is obtained by symmetry of roles of k and n. We then get (2). The formula (3) comes out immediately since

$$sim(X,Y)=\frac{\#(X\cap Y)}{\#(X\cup Y)}=\frac{\#(X\cap Y)}{m+k-\#(X\cap Y)}. \quad (4)$$

3. SIMILARITY OF TEXTS

The similarity is an automatic tool to anticipate plagiarisms, abusive quotations, influences, etc. However the study of the similarity of texts relies for instance on the words and not on the meanings.

3.1. Forming of Sets for Comparison

If we want to compare two texts S_1 and S_2, we must transform them into *shinglings* sets. For $k>0$, a *k-shingling* is simply a word of k letters. For finding the *k-shinglings* of a string, we first consider the word of k letters beginning with the first letter, the word of k letters beginning with the second letter, the word of k letters beginning with the third, etc., until the word of k letters finishing by the last letter of the string. So, a string of n letters is transformed into (n-k+1) *k-shinglings.*

We observe a serious difficulty in the practice in using the notion of similarity defined on sets of *k-shinglings*. Indeed, when we consider the *k-shinglings* of a text, it is very probable that some *k-shinglings* will be repeated. Then the collection of *k-shinglings* cannot define a mathematical set (whose elements are supposed to be distinct).

But fortunately, a *k-shingling* is determined by its value and its rank. Suppose that a text has a length n. We can denote the *k-shinglings* by means of a vector t of n-k+1 dimensions so that$t(i)$ is the ith k-shingling. The *k-shinglings* set is defined by

$$\{(i,t(i)),i=1,\cdots,n-k+1\}.$$

With this definition, the *k-shinglings* are different and do form a *well-defined mathematical set.*

3.2. Interpretation of the Similarity of Texts

Does the similarity between two texts have necessarily another explanation other than randomness? To answer to this question, let us remark that in any language, a text is composed from an alphabet that is formed by a finite and even small number of characters. A text in English is a sequence of lowercase and uppercase letters of the alphabet, of numbers and of some signs such as punctuations, apostrophes, etc. This set doesn't exceed a hundred characters.

Suppose that the computed similarity between the two sets is p_0. From what point can we reasonably consider that there is a possible collision between the authors, either the two texts are based on similar sources, or one author has used the materials of the other? To answer this question, we have to know the part due to randomness. As a matter of fact, any text is written from a limited set of *k-shinglings*. Then each k-shingling is expected to occur many times and hence contributes to rise the similarity. Let us consider a set of size $n=m+\ell$ *k-shinglings* containing those of the two compared texts. If the two texts were randomly written, that is the same to saying that they were written by machines subjected to randomness, the expected similarity that we denote by P_R would be given by (3). So we can say that the two authors would have some kind relationship of mutual influence or that plagiarism is suspected, if P_0 is significantly greater than P_R.

It is therefore important to have an idea of the value of P_R for sizes of the order of those of studied texts. For example, with the Bible texts that we study, the texts sizes go approximately from 50.000 to 110.000. The values P_R for these sizes turn round 30%. This knowledge is important to interpret the results.

3.3. Implementation of the Algorithm for Computing the Similarity of Texts

Let A and B be two texts. For a fixed $k \geq 1$, let us consider the two *k-shinglings* sets

$$\left(\left(i,t_A\left(i\right)\right),i=1,\cdots,n_A-k+1\right)$$

and

$$\left(\left(j,t_B\left(j\right)\right),j=1,\cdots,n_B-k+1\right)$$

The determination of the similarity between the two texts is achieved through comparing each *k-shingling* of A with all *k-shinglings* of B. We will have two problems to solve.

Suppose that a *k-shingling* is represented many times in B. We have the risk that the same value of this *k-shingling* in A is used as many times when forming the intersection between of *k-shinglings* sets. This would result in a disaster.

To avoid that, we associate to each *k-shingling* $\left(i,t_A\left(i\right)\right)$ at most one *k-shingling* $\left(j,t_B\left(j\right)\right)$. Let us use the wedding language by considering the *k-shinglings* of A as husbands, and the *k-shinglings* of B as wives and, then, the association between a *k-shingling* of A to a k-shingling of B as a wedding. Our principle says that a *k-shingling* of A can marry at most one k-shingling of B. In the same way, a *k-shingling* of B can be married at most to one *k-shingling* of A. We are in a case of perfect symmetrical monogamy. How to put this in practice in a program?

It suffices to introduce the sentinel variables that identify if a *k-shingling* husband or a *k-shingling* wife has a wife or a husband at the moment of the comparison.

Let us introduce the vectors

$$\left(\text{test}_A(i), i=1,\cdots,n_A-k+1\right)$$

and

$$\left(\text{test}_B(j), j=1,\cdots,n_B-k+1\right)$$

We put $\text{test}_A(i)=1$ if k-shingling has already a wife, $\text{test}_A(i)=0$ otherwise. We define $\text{test}_B(j)$ in the same manner. We apply the following algorithm:

1. Set *sim*=0;

2. Repeat for $i=1$ to n_A-k+1;

2a. If $\text{test}_A(i)=1$: nothing to do

2b. Else

2b-1. Do for: $j=1$ to n_B-k+1;

2b-11. If $t_B(j)=1$: nothing to do;

2b-12. Else compare $t_A(i)$ to $t_B(j)$;

2b-13. If equality holds, increment *sim* and put $\text{test}_A(i)=1$, $\text{test}_B(j)=1$;

2b-14. Else go to the next *j*.

3. Report the similarity $\left(sim/\left(n_A+n_B-sim\right)\right)$.

4. COMPUTATION STAKES

The search of similarity faces many challenges in the Web context and at the local post of personal computer.

4.1. Limitation of the Random Access Memory (RAM)

When we want to compare two sources of texts, each leading to a large number of shinglings, say n_1 and n_2, using the direct method will load in memory the vectors t_A, t_B, test(A) and test(B). When n_1 and n_2 are very large with respect to the capacities of the machine, this approach becomes impossible. For example, for the values of n_1 and n_2 in order of 98,000,000, the declaration of vectors of that order leads to an overflow in Microsoft VB6R.

We are tempted to appeal to another method, that directly uses data from files. Here is how it works:

1) open the file of the text *A*;

2) read a row of the file *A*;

3) open the file *B*: read all these rows one by one and compare the *k-shinglings* of the file B with the *k-shinglings* at the current row of file *A*;

4) close the file *B*;

5) go to the next row of file *A*.

This method that we denote by the similarity by file does practically not use the RAM of the computer. We are then facing to two competing methods. Each of them has its qualities and its defects.

4.1.1. The Direct Method

It loads the vectors of *k-shinglings* in the RAM. It leads to quick calculations. However we have the risk to stuck the machine when the sizes of the files are huge.

4.1.2. The Method of Similarity by File

It spares the RAM of the machine and increases the computing speed. However it leads to considerable times of computations since, for example, the second file is opened as many time as the first contains rows. We spare the RAM but we lose time.

You have to notice that in the implementation of this method, we always have to carry the incomplete ends of each row at the next row.

Example 1. *Suppose that we compare the 5-shinglings of the first row of A and the first row of B. The last four letters of the row cannot form a 5-shinglings. We have to use them by adding them at the first place of the second row of A. These additional ends are denoted "boutavant 1"s in the procedures done in (4.1), when we implement the similarity by file method. We do the same thing for the rows of B that give the "boutavant 2" s.*

For example, in the work on the Gospel versions, where the numbers of k-shinglings are of the order of one hundred thousands, the method of similarity by file takes around thirty minutes and the direct method requires more or less ten minutes. We reduce the time of computation by three at the risk to block the RAM.

All what precedes advocates using approximated methods for computing similarity. Here, we are going to see three approaches but we only apply two of them in the study of the Gospel texts.

5. APPROXIMATED COMPUTATION OF SIMILARITY

5.1. Theorem of Glivenko-Cantelli

Since the similarity is a conditional probability in according to Theorem 1, we can deduce a law of Glivenko- Cantelli in the following way.

Theorem 2. *Let p be the similarity between two sets of total size n. Let us pick at random two subsets of respective sizes n_1 and n_2 so that $n=n_1+n_2$ and let us consider the random similarity P_n between these two subsets. Then P_n converges almost-surely to p with a speed of convergence in the order of $n^{-1/4}$ when n_1 and n_2 become very large.*

That is a direct consequence of the classical theorem of Glivenko-Cantelli. It then yields a useful tool. For example, for the similarity of Gospels for which the similarity is determined in more or less ten minutes, the random choice of subsets of size around ten thousand *k-shinglings* for each Gospel gives a computation time less than one minute, with an accuracy of 90%. To avoid the instability due to one random choice only, the average on ten random choices gives a better approximated similarity in more or less one minute. We will widely come back to this point in the applications.

5.2. Analysis of the Banding Technique

The banding technique is a supplementary technique based on the approximation of Theorem of Glivenko- Cantelli. Suppose that we divide the representation matrix, in *b* bands of *r* rows. The similarity can be computed first by considering the similarity between the different rows of one band then, between some bounds only. We do not use this approach here.

5.3. The RU Algorithm

It is based on the notion of minhashing to reduce documents of huge sizes into documents of small sizes called signatures. The computation of the similarity is done on their compressed versions, i.e. on their signatures. To better explain this notion, let us consider *p* subsets of a huge reference set and let us use the matrix be defined in **Table 2**.

The similarity between two sets is directly got as soon as this table is formed by using the formula (2.1) in a quick way. But the setting of this matrix takes time. This is serious drawback of the original RU algorithm that we will describe soon. For the moment, suppose that the table exists. On this basis, we are going to introduce the RU algorithm. By this algorithm, we do three things. First, we consider an arbitrary permutation of the rows. Then, we replace the columns transformations called minhashes by means of congruence functions. Then, a new table is formed to replace the original table. This new and shorter one, that we describe immediately below, is called signature matrix.

5.3.1. Minhashing Signature

Suppose that the elements of S are given in a certain order denoted from 1 to n. Let us consider p functions h_i $(i=1,\cdots,p)$ from $\{1,\cdots,n\}$ in itself in the following form:

$$h_i(x) = a_i x + b_i \bmod n \tag{5}$$

where a_i and b_i are given integers. We modify this function in the following way: $h_i(x)=n$ when the remainder of the euclidian division is zero. We then can transform the matrix as in **Table 3**.

The RU algorithm replaces this matrix by another smaller one called minhashing signature, represented in Table 4.

Table 2. Subsets representation matrix.

Element	S_1	S_2	...	S_s
1	1	0	...	0
2	0	0	...	1
...	0	...	...	...
i	1	0	...	1
...	...	...	...	...
n	0	0	...	0

Table 3. Completition of the representation matrix by minhashing columns.

Element	S_1	S_2	...	S_m	h_1	...	h_p
1	1	0	...	0	$h_1(1)$	...	$h_p(1)$
2	0	0	...	1	$h_1(2)$	...	$h_p(2)$
.	0	.	...	.	.	...	.
i	1	0	...	1	$h_1(i)$	...	$h_p(i)$
.	.	.	...	.	.	.	.
n	0	0	...	0	$h_1(n)$	...	$h_p(n)$

Table 4. Signature matrix.

minhashes	S_1	S_2	...	S_m
h_1	c_{11}	c_{12}	...	c_{1m}
h_2	c_{21}	c_{22}	...	c_{2m}
...	...	...	...	...
h_p	c_{p1}	c_{p2}	...	c_{pm}

To fill in the table above, Anand et al. ([1]), page 65, propose the algorithm below:

Algorithm of filling of the columns S_j:

1. *Set all the* c_{rj} equal to ∞.

2. For each column S_j, proceed like this.

2-a. For each element *i*, from 1 to *n*, compute $h_1(i), h_2(i), \cdots, h_p(i)$;

2-b. If i is not in S_j, then do nothing and go to *i*+1;

2-c. If *i* is in S_j, replace all the rows $(c_{rj})_{1 \leq r \leq p}$ by the minimum: $\min(c_{rj}, h_r(i))$;

2-d. Go to *i*+1.

3. Go to *j*+1.

4. End.

At the end of the procedure, each column will contain only integers between 1 and n. The computed simi- larity on this compressed table between S_i and S_j, denoted simRU (S_1,S_2), will be called RU approximated similarity . It is supposed to give an accurate approximation of the similarity.

However, we can simplify this algorithm in a very simple way, by saying this.

Criterion 1. *The transpose of the column* $(c_{rj})_{1 \leq r \leq p}$, *is the minimum of rows, when carried out coordinate by coordinate,* $(h_1(i), \cdots, h_p(i))$, *when i covers the elements of* S_j.

This simple remark allows to set up programs in a much easier way.

5.3.2. The RUM Algorithm or the Modified RU Algorithm

It is clear that by forming the matrix of Table 2, the similarity is automatically computed. Indeed, when we consider the columns S_i and S_j, we immediately see that the number of rows containing the unit number (1) on these two columns is the size of the intersection. Then the Jaccard similarity is already found and any further step is useless. The RU algorithm, on this basis, is not useful. Instead, forming this matrix is exactly applying the full method that requires comparison of each couple of shinglings of the two sets. This operation takes about thirty minutes for set of sizes one hundred thousands, for example. Based on this remark, we propose a modification for the implementation of the RU algorithm in that following way. Let us consider two sets S_1 and S_2 with respective sizes n_1 and n_2 to be compared. We proceed like that:

1) Form one set S by putting the elements of S_1 and then the elements of S_2 with the double elements. Let $n=n_1+n_2$;

2) Apply the RU algorithm at this collection by using Criterion 1.

We do not seek to find the intersections. Elements of the intersection are counted twice here. But it is clear that we still have a zero similarity index if the two sets S_1 and S_2 are disjoint, and a 100% index if the sets are identical.

The question is: how well the estimations of the similarity using RU or RUM algorithm are good approxi- mations of the true similarity index? In this paper, we give an empirical response based on the Gospels comparison by showing that the RUM approximation of the similarity of good while performing only in a few seconds in place of thirty minutes (1.800 seconds)!

The exact distribution of the RUM index is to be found depending on the laws of the stochastic laws of the coefficients a_i and b_i in (5). This paperis a preparation of theoritical study on the exact laws of similarity indices provided by the RU and the RUM random methods.

6. THE APPLICATIONS OF THE SIMILARITY OF THE BIBLE TEXTS

6.1. Textual Context of the Four Versions of the Gospels

Here, we are going to resume a few important points for the backgrounds of our Gospels analysis. In all this subsection, we refer to [12] [13] .

The Gospels (In Latin, Gospel means good news) are texts that relate the life and the teaching of Jesus of Nazareth, called Jesus Christ. Four Gospels were accepted as canonical by the Church: the Gospel according to Matthew, to Mark, to Luke and to John. The other unaccepted Gospels are qualified apocryphal ones. Numerous Gospels have been written in the first century in our era. Before to be consigned as written, the message of Christ was verbally transmitted. From tale stories, many texts were composed, among which the four Gospels that were retained in the Biblical canon. The canonical Gospels are anonymous. They were traditionally attributed to disciples of Jesus Christ. The Gospel according to Matthew and the Gospel according to John would have been from direct witnesses of the preaching of Jesus. Those of Mark and Luke are related to close disciples.

The first Gospel is the one attributed to Mark. It would have been written in about 70 years AD. In about 80 - 85, follows the Gospel according to Luke. The Gospel according to Matthew is dated between 80 and 90, and to finish, the one of John is dated in between 80 and 110. However, these uncertain dates vary according to the authors that propose chronologies of the evangelical texts. The original Gospels were written in Greek.

The Gospel according to Matthew, Mark and Luke are called synoptic. They tell the tale of Jesus in a relatively similar way. The Gospel according to John is written using another way of taling Jesus' life and mission (christology) qualified as Johannist. The first set of Gospel that has been written seems to be Mark's one. According to some researchers, the common parts between Matthew and Luke Gospels may depend on a more older text that was lost. This text is referred as the *Q* source.

The source *Q* or Document *Q* or simply *Q* (The letter is from the German word QUELLE, meaning source) is a hypothetical source, of whom some exegetes think it would be at the origin of common elements of Gospels of Matthew and Luke. Those elements are absent in Mark. It would be a collection of words of Jesus of Nazareth that some biblists attempted to reconstitute. This source is thought to date around of 50 AD.

The Gospels of Matthew and Luke are traditionally influenced by Mark's Gospel and the Old Testament. But though separately written, they have in common numerous extracts that don't come from the two first cited sources. This is why the biblists of XIXe century generally think that these facts suggest the existence of a second common source, called "document *Q*". Since the end of XIXe century, Logia (i.e. the speech in Greek) seems to have been an essentially collection of speeches of Jesus. With the hypothesis of the priority of the Gospel of Mark, the hypothesis of the existence of the document *Q* is part of what the biblists call the hypothesis of two sources.

This hypothesis of two sources is the most general solution that is accepted for the synoptic problem, that concerns the literary influences between the three canonical Gospels (Mark, Matthew, Luke), called Synoptic Gospels. These influences are sensitive by the similarities in the choice of words and the order of these words in the statement. The synoptic problem wonders about the origin and the nature of these relationships. From the hypothesis of two sources, not only Matthew and Luke learned all both on the Gospel according to Mark, independently one to other; but as we detect similarities between the Gospels of Matthew and Luke, that we cannot find in the Gospel of Mark, we have to suppose the existence of a second source.

Synoptic Gospels

The Gospels of Matthew, Mark, and Luke are considered synoptic Gospels on the basis of many similarities between them that are not shared by the Gospel of John. Synoptic means here that they can be seen or read together, indicating the many parallels that exist among the three.

The Gospel of John, on the contrary has been recognized, for a long time as distinct of first three Gospels so much by the originality of its themes, of its content, of the interval of time that it recovers, and of its narrative order and the style. Clément of Alexandria summarized the single character of the Gospel of John by saying: *John came last, and was conscious that the terrestrial facts had been already exposed in the first Gospel. He composed a spiritual Gospel.*

Indeed, the Gospel of John, presents a very different picture of Jesus and his ministry from the synoptics. In differentiating history from invention, some historians interpret the Gospel accounts skeptically but generally regard the synoptic Gospels as including significant amounts of historically reliable information about Jesus. The common parts of the Gospels of Matthew and of Luke depend on an antiquarian document but lost called source *Q* according to some researchers.

The synoptic Gospels effectively have many parallels between them: thus around 80% of verses of Mark may be found in Matthew and Luke Gospels. As the content is in three Gospels, one talks about of Triple Tradition. The passages of the Triple Tradition are essentially narrations but we can find in it some speeches of Christ.

But otherwise, we also find numerous identical passages between Matthew and Luke, but absent in the Gospel of Mark. Almost 25% of verses of the Gospel according to Matthew find an echo from Luke (but not from Mark). The common passages between Matthew and Luke are mentioned as the Double Tradition.

The four Gospels constitute the principle documentary concerning the life and the teachings of Christ. Each of them uses a particular perspective. But all of them use the same general scheme and convey the same philosophy. We stop here. For further details see [?]. We will attempt to explain the results in our own analysis of similarity below.

6.2. The General Setting

All the computations were done in the environment of VB6R. Once the four texts are chosen, we follow these steps. We first proceed to the editing files by dropping the words of less than three letters. Then we proceed to the computations of the similarity between the different Gospels.

Next, we find for each gospel, the number of the rows of files as well as the number of letters.

Table 5 gives the numbers of the rows, before and after editing, of each Gospel.

Now we are going to report the common numbers of *k-shinglings* with k=3 between the different Gospels and then compute the similarity between each couple of Gospels by the two exact methods.

The results are in Table 6 and Table 7.

Approximated Similarity

In this part, the computation of the similarity will be done by the direct method. Let us pick randomly 10,000 *k-shinglings* from first file and 10,000 *k-shinglings* from the second file. We remark that the time of com- putation of the similarity turns around 20 seconds. We get approximated values of similarities between the Gospels. Let us use the two methods of computation through a double approximation of the similarity i.e. approximation using the theorem of Glivenko-Cantelli and of the RUM algorithm. The results are given in **Table 8**.

This approach is simply extraordinary since we may use a very low number of hash functions (*pp*) and get good approximations. To guarantee the stability of the results, we report the average results got for *BB*=50 repetitions of the experience and the standard deviation of such a sequence of results in Tables 9-12.

Table 5. Total numbers of rows and letters of the synoptic Gospels.

	John	Luke	Mark	Matthew
Numbers of the rows	2534	3442	628	1319
Numbers of letters before editing	96.269	129.548	76.543	149.747
Numbers of letters after editing	69.316	94.766	55.555	108.722

Table 6. Computation durations of the exact Jaccard similarity beetwen two Gospels by using the direct methods.

	Luke	Mark	Matthew
John	Sim = 57.62% Kc = 59.981 Time = 755 s	Sim = 57.53% Kc = 45.600 Time = 816 s	Sim = 51% Kc = 60.134 Time = 510 s
Luke		Sim = 54.12% Kc = 52.782 Time = 640 s	Sim = 69.55% Kc = 83.468 Time = 1430 s
Mark			Sim = 48.74% Kc = 53.827 Time = 508 s

Table 7. Computation durations of the exact Jaccard similarity beetwen two Gospels by using the file method.

	Luke	Mark	Matthew
John	Sim = 57.62% Kc = 59.981 Time = 2312 s	Sim = 57.53% Kc = 4.5600 Time = 1376 s	Sim = 51% Kc = 60.134 Time = 3021 s
Luke		Sim =54.12% Kc = 52.782 Time = 3080 s	Sim = 69.55% Kc = 83.468 Time = 1457 s
Mark			Sim = 48.74% Kc = 53.827 Time = 1552 s

Table 8. Computation durations of the approximated Jaccard similarity beetwen two Gospels by using the glivenko-cantelli theorem.

	Luke	Mark	Matthew
John	Sim = 47.50% Time = 20 s	Sim = 46.46% Time = 34 s	Sim = 46.04% Time = 29 s
Luke		Sim = 50.79% Time = 19 s	Sim = 50.26% Time = 22 s
Mark			Sim = 52.28% Time = 27 s

6.3. Analysis of Results

6.3.1. Evaluation of Algorithms on the Similarity by the Direct Method

In this algorithm, we first form the *k-shinglings* sets for each text. Then we compute the similarity between them.

We remark that the time of the determination of the similarity between the different Gospels turns around ten minutes. The different similarity amounts are around 50%.

1) Algorithm on the similarity by the method by file

Here we remark that the times of the determination are much greater than those in the case of the similarity by the direct method. The time turns around 30 minutes. We naturally have the same similarities already given by the direct method.

2) Algorithm on the similarity by the theorem of Glivenko-Cantelli

We randomly pick a number *NG*=10000 *k-shinglings* from both files and next we compute the similarity as we did in the case of the direct method.

Table 9. Computation durations of the approximated Jaccard similarity beetwen two Gospels by using the RUM algorithm with *pp* = 5 minhashing functions.

	Luke	Mark	Matthew
John	Sim = 60% Ecart = 20.76 Time = 26 s	Sim = 58% Ecart = 22.1 Time = 25 s	Sim = 59.2% Ecart = 20.38 s Time = 22 s
Luke		Sim = 56.4% Ecart = 19.87 Time = 22 s	Sim = 63.2% Ecart = 22.03 Time = 22 s
Mark			Sim = 59.2% Ecart = 22.38 Time = 27 s

Table 10. Computation durations of the approximated Jaccard similarity beetwen two Gospels by using the RUM algorithm with *pp* = 10 minhashing functions.

	Luke	Mark	Matthew
John	Sim = 62% Ecart = 16.68 Time = 26 s	Sim = 61.6% Ecart = 15.91 Time = 25 s	Sim = 63.6% Ecart = 14.52 s Time = 22 s
Luke		Sim = 62% Ecart = 14.56 Time = 31 s	Sim = 58% Ecart = 13.41 Time = 27 s
Mark			Sim = 63.8% Ecart = 14.54 Time = 29 s

Table 11. Computation durations of the approximated Jaccard similarity beetwen two Gospels by using the RUM algorithm with *pp* = 15 minhashing functions.

	Luke	Mark	Matthew
John	Sim = 57% Ecart = 13.45 Time = 28 s	Sim = 59.33% Ecart = 11.33 Time = 28 s	Sim = 58.26% Ecart = 13.84 s Time = 30 s
Luke		Sim = 60.13% Ecart = 13.23 Time = 30 s	Sim = 58.26% Ecart = 13.51 Time = 28 s
Mark			Sim = 57.2% Ecart = 14.17 Time = 29 s

Table 12. Computation durations of the approximated Jaccard similarity beetwen two Gospels by using the RUM algorithm with pp = 20 minhashing functions.

	Luke	Mark	Matthew
John	Sim = 57.6% Ecart = 10.63 Time = 32 s	Sim = 60.8% Ecart = 10.11 Time = 31 s	Sim = 62.9% Ecart = 10.63 s Time = 31 s
Luk		Sim = 57.9% Ecart = 9.59 Time = 32 s	Sim = 63.6% Ecart = 9.22 Time = 31 s
Mark			Sim = 60.7% Ecart = 8.94 Time = 31 s

We remark a considerable reduction of the time of the determination of the similarity. The result is huge. The similarity indices are got in less a minute. The similarity also turns around 50%.

3) RUM Algorithm for the similarity computation

We randomly pick N_1 =10000 *k-shinglings* from of the first file and N_2 =10000 *k-shinglings* from the second file. We apply the RUM algorithm with a number of hashing *pp* taking the values 5, 10, 15, 20. To guarantee the stability of results, the RUM method is used fifty times (*BB* = 50) and the average similarity has been reported out in **Tables 9-12**.

Finally, we arrive at a tuning result: by using subsamples of the two sets and by using the approximation method via the RUM algorithm, we get an acceptable estimation of the similarity in a few number of seconds. But since the results may be biased, performing the process a certain number of times and reporting the average is better.

We may study the variability of the results. If we proceed BB=50 times with pp=20 hash functions, the different obtained values for the similarities present an empirical deviation of the order of 10%. This means that the reported value is accurate at 2%.

For the Gospels for example, we finally conclude that the true estimation of the similarity is in an interval centered at the approximated value given by the RUM method with magnitude 10%. This result, that is achieved only in seconds, is very significant for large sets.

We may also appreciate the power of this algorithm that allows estimation of the similarity of set around one hundred thousand (100.000) characters in only 6 seconds.

6.3.2. Comparison of Gospels

From the Tables 9-12, we notice that the Gospels of Luke and Matthew have the greatest similarity around 70%. From what we already said in Subsection 6.1, Luke and Matthew have used the Gospel of Mark and in addition, are based on unknown source Q. Likewise the similarity between the Gospel of John and the others might explained by the fact that the John Gospel is the last to be released in about year 100 or year 110 of our era. He might already be aware of the contents of the other three gospels.

We might hope to have a similarity around 90%. But many factors can influence on the outcomes. Actually, the Gospels are written by four different persons. Each of them may use his own words. Besides, we used translated versions. This latter fact can result in a significant decrease of the true similarity. An other point concerns the fact that a limited alphabet is used. This in turn is in favor of forming a structural part in the similarity. For example, for the considered sizes, this part is around 30%.

With the order of the sets sizes, we have the automatic and stochastic similarity of order of 30%. Since the similarities turn around 50% between the Gospels, we conclude that Gospels really have a significant similarity. By taking account the remarks that have been made above, we may expect that these similarities should be really much greater. This is in favor of the hypothesis of the existence of a common source that can be denamed as the source Q.

6.3.3. Recommendations and Perspectives

To conclude we recommend these following steps in assessing similarity:

1) Determine the automatic and stochastic part of the similarity, by simulation studies by using formula (4);

2) Form the sets of *k-shinglings* of the two studied sets;

3) Pick at random n_1 and n_2 *k-shinglings* for the two sets to study;

4) Apply the RUM algorithm;

5) Compare the finding similarity with the results of the point (1);

6) Conclude on a significant similarity if the reached similarity, is widely superior tothe stochastic similarity determined in (1). Otherwise the similarity is not accepted;

7) Apply the RUM algorithm a number of times before doing definitive conclusion.

6.3.4. Conclusions

In this paper, we described the main methods of determination of the similarity. We empirically estimated the incompressible stochastic similarity between two texts. We proposed a modification of the RU algorithm, named RUM, and we applied on subsamples of the studied texts. The combination of the Glivenko-Cantelli theorem and an empirical study of the RUM algorithm, led to the conclusion that the approximated similarity that was given by this procedure, was a good estimation of the true similarity. Since this approximated similarity was computed in seconds, the method showed remarkable performance. Hence it was recommended for the study of similarity for very large data sets.

We applied our methods to the four Gospels. The obtained results concerned the study of Gospels themselves as well as the evaluation of different methods of computation of the similarity. In conclusion, the Gospels had indices of similarity at least 50%.

In a coming paper, we would concentrate on the theoretical foundations of the RUM algorithm in the setting of Probability Theory and Statistics.

ACKNOWLEDGEMENTS

The reviewer drew our attention on profound sights and interesting facts about the Gospels writing history. For example, he pointed out that for each gospel, there are several authors, each of them providing a contribution. He also drew to our attention on other methods for texts analysis. As he concluded himself, this text is data analysis oriented. The Bible's Gospels has been used as a material on which the theory is illustrated. The ideas, the suggestions and the points of views he provided will certainly be extensively used in the final version of the PhD dissertation of the first author.

The first author thanks the *Programme de formation des formateurs des Universités de Bamako* which financed his stays in the LERSTAD of UGB while preparing his PhD dissertation. The authors acknowledge support from the Réseau EDP-Modélisation et Contrôle, of Western African Universities, that financed travel and accomodation of the second author while visiting USTTB in preparation of this work.

REFERENCES

1. Rajaraman, A. and Ullman, J.D. (2011) Mining of Massive Datasets. Cambridge University Press, Cambridge. http://dx.doi.org/10.1017/cbo9781139058452
2. Stein, B. and Eissen, S.M. (2006) Near Similarity Search and Plagiarism Analysis. In: Spiliopoulou, et al., Eds., From Data and Information Analysis to Knowledge Engineering Selected Papers from the 29th Annual Conference of the German Classification Society (GfKl) Magdeburg, Springer, Berlin Heidelberg, 430-437. http://dx.doi.org/10.1007/3-540-31314-1_52
3. Gionis, A., Indyk, P. and Motwani, R. (1999) Similarity Search in High Dimensions via Hashing. Proceedings of the 25th VLDB Conference, Edinburgh, Scotland, 1999, 518-529.
4. Chen, D.-Y., Tian, X.-P., Shen, Y.-T. and Ouhyoung, M. (2003) On Visual Similarity Based 3D Model Retrieval. Eurographics 2003, 22, 223-232.
5. Cha, S.H. (2007) Comprehensive Survey on Distance Similarity Measures between Probability Density Functions. International Journal of Mathematical Models and Methods in Applied Sciences, 4, 300-307.
6. Gower, J.C. and Legendre, P. (1986) Metric and Euclidean Properties of Dissimilarity Coefficients. Journal of Classification, 3, 5-48. http://dx.doi.org/10.1007/BF01896809
7. Zezula, P., Dohnal, V. and Amato, G. (2006) Similarity Search the Metric Space Approach. Advances in Database Systems, Springer Series, 32, 220 p.
8. Strehl, A., Ghosh, J. and Mooney, R. (2000) Impact of Similarity Measures on Web-Page Clustering. American Association for Artificial Intelligence, 78712-1084.
9. Formica A. (2005) Ontology-Based Concept Similarity in Formal Concept Analysis. Information Sciences, 176, 2624-2641.
10. Bilenko, M. and Mooney, R.J. (2003) Adaptive Duplicate Detection Using Learnable String Similarity Measures. Proceedings of the 9th ACM SIGKDD International Conference on Knowledge Discovery and Data Mining (KDD), Washington DC, 2003, 39-48, http://dx.doi.org/10.1145/956750.956759
11. Theobold, M., Siddharth, J. and Paepcke, A. SpotSigs: robust and efficient near duplicate detection in large web collections. Proceedings of the 31st Annual International ACM SIGIR Conference on Research and Development in Information Retrieval, Singapore, 2008, 563-570.

CHAPTER 12

Bounce Regime of Droplet Collisions: A Molecular Dynamics Study

Yi Ran Zhang[a], Xi Zhuo Jiang[b], Kai Hong Luo[a,b,*]

[a] *Center for Combustion Energy, Key Laboratory for Thermal Science and Power Engineering of Ministry of Education, Tsinghu*

[b] *University, Beijing 100084,ChinabDepartment of Mechanical Engineering, University College London, Torrington Place, London WC1E 7JE, United Kingdom*

ABSTRACT

Droplet collisions have complex dynamics, which can lead to many different regimes of outcomes. The head-on collision and bounce back regime has been observed in previous experiments but numerical simulations using macro- or mesoscale approaches have difficulties reproducing the phenomena, because the interfacial regions are not well resolved. Previous molecular dynamics (MD) simulations have not reproduced the bounce regime either but have reported the coalescence and/or shattering regimes. To scrutinize the dynamics and mechanisms of binary collisions especially the interfacial regions, head-on collision processes of two identical nano-droplets with various impact velocities both in vacuum and in an ambient of nitrogen gas are investigated by MD simulations. With the right combination of the impact velocity and ambient pressure, the head-on collision and bounce back phenomenon is successfully reproduced. The bounce phenomena are mainly attributed to the "cushion effect" of the in-between nitrogen molecules and evaporated water molecules from the two nano-droplets. The analysis has verified and also extended the current gas film theory for the bounce regime through including the effects of evaporated water molecules (vapour). Some similarities and some dissimilarities between nanoscale and macro-/meso-/microscale droplet collisions have been observed. The study provides unprecedented insight into the interfacial regions between two colliding droplets.

Keywords: Multiphase flow Binary droplet collision Nanoscale Molecular dynamics Droplet bounce

1. INTRODUCTION

Droplet collisions are encountered in both natural and industrial processes, such as, in the formation of clouds and rain drops [6], in the operation of nuclear reactors and in the process of spraying [5]. In recent decades, efforts have been made to unveil the fundamental mechanisms associated with coalescence and bounce of two identical droplets. Experimental studies have identified that under different atmospheric conditions most of head-on liquid-gas droplet collisions can be described using three distinct regimes: coalescence, bounce and reflexive separation [1,2,9,30]. Other studies using experimental and numerical methods such as the level set, volume of fluid, immersed boundary method, lattice Boltzmann method have already presented various phenomena of droplet collisions [4,7,14,17–19]. Even so, details of the collision process are difficult to obtain through experiments and continuum simulation methods, especially when the approaching droplets are within a distance comparable to the molecular mean free path. It should be noted too that the lattice Boltzmann method has been successful in reproducing all the collisions regimes except the head-on collision and bouncing regime [18]. Thus, computer simulations considering atomic effects become the only method to understand droplet dynamics when the separation distance comes within nanoscale. Yet up to the present time, the published simulations have only reproduced coalescence and shattering regimes [8,11,12,16,22,26,31] but not the bounce regime. Such an absence is intriguing as it suggests the possibility that the bounce regime may not exist in nano-droplet collisions while it was observed in micro- and macro-droplet collisions. Our study has, however, successfully observed the bounce behaviours with two nanoscale water droplets, thereby completing the regime map of the nano-droplet collision phenomena.

In this research, molecular dynamics (MD) simulations of droplet collisions in vacuum and in ambient gas have been conducted to reproduce the bounce behaviours of nano-droplets. Our research will demonstrate that bounce frequently occurs in nanoscale droplet collisions. Such occurrences are influenced by parameters including pressure and impact velocity. Moreover, a "cushion effect" will be proposed to reveal the impact of the ambient gas and vaporized water molecules on droplet behaviours during the bounce process.

2. MOLECULAR DYNAMICS SIMULATIONS

Fig. 1 illustrates the initial setup of the simulation system with two nano-droplets submerged in the nitrogen gas.

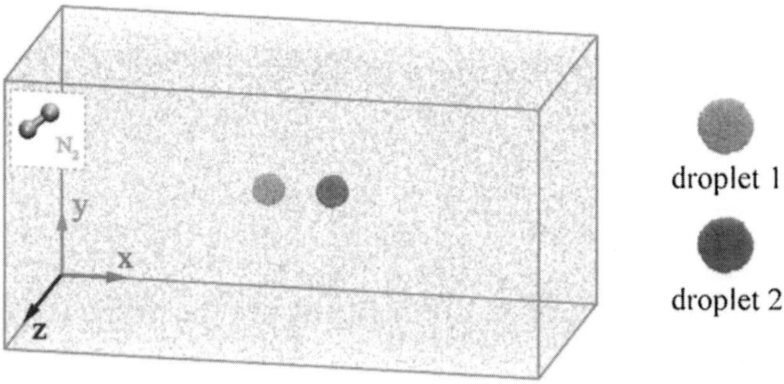

Figure. 1. The initial setup of the simulation system. Binary nano-droplets of 10 nanometers in diameter are submerged in nitrogen gas molecules and approach each other at equal speed along the *x* direction.

The dimension of the simulation box is 800 Å × 400 Å × 400 Å, and the origin of the coordinates was set at the left bottom back corner. Two identical droplets with diameters of 10 nm were placed along the *x* direction at the central part of the box with *x* coordinates being 500 Å and 1500 Å, respectively. The TIP3 P water model [10,15] was chosen for water molecules because it offers a sufficient description of intermolecular forces in water in MD simulations. Nitrogen molecules were treated as Lennard-Jones particles. Intermolecular interactions between water and nitrogen molecules included both Coulombic and Lennard-Jones (L-J) 12–6 potentials [3,24,27], as shown in Eq. (1).

$$U_{ij} = \frac{q_i q_j}{r_{ij}} + 4\varepsilon_{ij}\left[\left(\frac{\sigma_{ij}}{r_{ij}}\right)^{12} - \left(\frac{\sigma_{ij}}{r_{ij}}\right)^{6}\right]$$

where q represents the charge of particle, and r is the distance between any two atoms, and σ and ε represent the zero energy separation distance and the minimum energy, respectively. The subscripts i and j are the atom indexes. The Lorentz-Bertthelot mixing rule was employed for calculating the potential parameters between water and nitrogen molecules. L-J parameters used in this research are summarized in Table 1. A cut-off distance of 16 Å was adopted. Periodic boundary conditions were applied to all three directions. All the simulations were carried out using the LAMMPS platform [21].

Table 1. Values of potential parameters.

atom	σ (Å)	ε (kcal/mol)
O—O	3.188	0.102
H—H	0.000	0.000
N—N	0.072	3.320

Equilibrium simulations of droplets and ambient gases were conducted before the collision simulations. For droplets, the equilibrium was undertaken in vacuum at the temperature of 300 K, and the canonical ensemble (NVT) was employed. After equilibrium of 2 ns, a single droplet can be generated by removing all the vaporized molecules around. The second droplet was obtained by duplicating the first one. Meanwhile, ambient nitrogen molecules were also equilibrated at 300 K in a parallel equilibrium simulation.

After equilibrium, the two droplets were first placed in the required positions, and then equilibrated ambient gas molecules were added to fill the remaining space of the box. A further simulation was conducted for 2 ns to equilibrate the final system consisting of both droplets in ambient gas molecules. In the formal collision simulations, each droplet was assigned the same impact velocity along the *x* direction.

3. RESULTS AND DISCUSSION

3.1. Reproduction of bounce regime by MD simulations

Droplet collisions were first simulated in vacuum with impact velocities varying in a broad range from 5 to 500 m/s. Results have shown that the binary droplet collisions in vacuum follow either coalescence regime or shattering regime regardless of how the impact velocities vary, which is consistent with previous studies [13,20]. However, a bounce regime could never be found under the vacuum condition.

Binary collisions were then simulated with the existence of the ambient gas. With the impact velocity 8 m/s ($v = 8$ m/s) and the ambient pressure

2.7 atm ($p = 2.7$ atm), we found that the binary collision fell within the bounce regime (Fig. 2b and c). The whole process consists of three key stages. At the very beginning ($t = 0$–2.8 ns), the two droplets approach each other with very small deformation, whilst a small number of water molecules escape from the two droplets. The second stage ($t = 2.8$–4.2 ns) is reached when the distance between the two droplets reaches the minimum, where the droplets appear to be stationary. At this stage, the inter-droplet distance remains almost constant, whilst water molecules continue to escape from both droplets. Moreover, some of the water molecules escaping from one droplet are absorbed by the other droplet. The last stage ($t = 4.2$–5.6 ns) – the bouncing stage – occurs when an increase in the inter-droplet distance is clearly observed. Interestingly, the droplets are found to separate from each other along routes that deviate from the trajectories of the approaching droplets. Moreover, the separation velocity of the two droplets is detected to be smaller than the impact speed ($v = 8$ m/s). The deviating routes are attributed to the interactions between the water and nitrogen molecules, which are not entirely deterministic. The smaller separation velocity (as compared with the impact velocity) is due to the momentum loss through the escaping molecules as well as interactions with the ambient gas molecules.

3.2. Explanation for bounce regime

As for the cause for bouncing only in the presence of ambient gases, the theory of "gas film" is a popular explanation for droplet dynamics at microscale. Qian and Law [25] stated that the ambient gases form a gas film between the two droplets as the inter-droplet distance becomes tiny in the approaching stage. As the gap between the droplets decreases, the gas film pressure increases dramatically, leading the droplets to bounce back. Our results appear to support this physical picture, but at nanoscale. Furthermore, we have extended the "gas film" theory, which will be shown in detail in the following discussion.

The first step is to scrutinize the composition of the gas film. To distinguish whether a water molecule belongs to the parent droplet or has evaporated, the surface of the droplet should be defined. However, at the molecular scale, a nano-scale droplet has no well-defined surface as its macro-scale counterpart usually does. Thus, a criterion has been proposed to definite the surface. For two random molecules in the simulation box, if their distance (r_{ij}) is smaller than a critical value (R_{cl}), the two molecules are regarded in the same droplet. The droplet is defined as a set of molecules, each of which is within the R_{cl} from one or more molecules in the droplet Powles [23] had applied this method and suggested $R_{cl} = 1.9\sigma$. Any water molecule can be regarded exactly as an evaporated molecule or an intra-molecule of the droplet with this method.

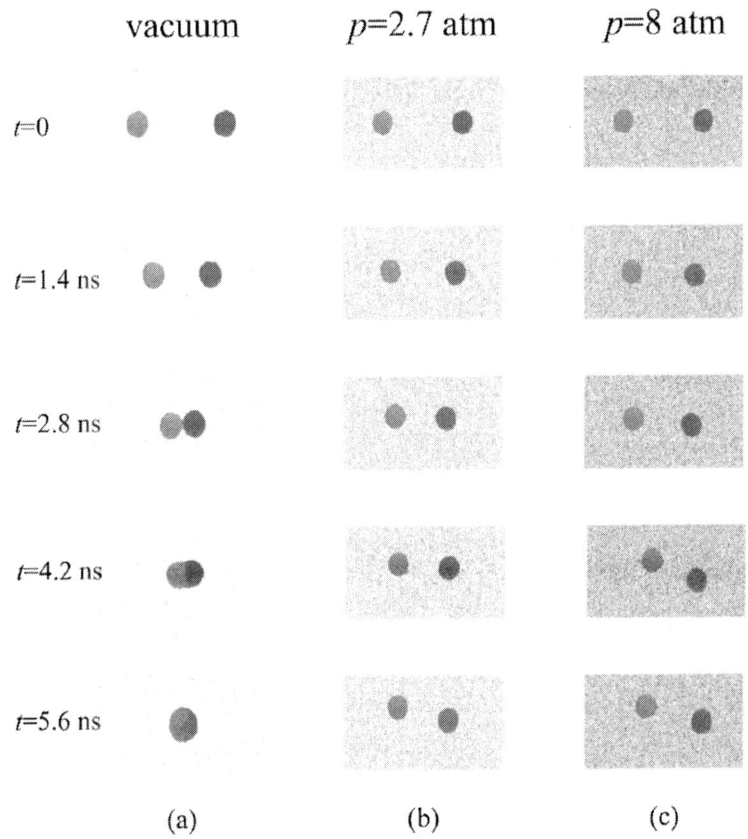

Figure. 2. Different collision regime for three different ambient gas pressure when v = 8 m/s.

Using the artificial surface method, the instantaneous numbers of nitrogen and evaporated water molecules in the gas film between the two droplets have been counted and shown in Fig. 3. The gap which is the rectangular region between the two droplets can be defined by the position of molecules in the surface in all the directions. For example, at time = 0 ns, the two droplets with diameters of 10 nm were placed along the *x* direction at the central part of the box with *x*coordinates being 200 Å and 600 Å. The 300 Å × 100 Å × 100 Å rectangular region between the two droplets is where the 225 nitrogen molecules in. In Fig. 3, when the two droplets are approaching each other (t = 0–2.8 ns), the number of nitrogen molecules between the two droplets decreases, which means that the nitrogen molecules are squeezed out of the space between the droplets. Meanwhile, an increasing number of evaporated water molecules occupies the in-between space. Interjacent molecules between the droplets therefore constitute both the nitrogen molecules and the evaporated water molecules. In the second stage (from 2.8 ns onwards), the two droplets have slowed down significantly, and the number of interjacent nitrogen molecules stops decreasing, but more water molecules continue to evaporate to fill the in-between space. As evaporating water molecules further reduce the momentum of the droplets, the droplets are not able to overcome the gas film and consequently come to an almost standstill (t = 2.8–4.2 ns). The process resembles droplets hitting a cushion, which can be called the "cushion effect" of the interjacent molecules. Afterwards, momentum exchange between the two droplets through the interjacent gases leads to the bounce back of the two droplets, which is aided by the continuing evaporation of water molecules from the droplet surfaces.

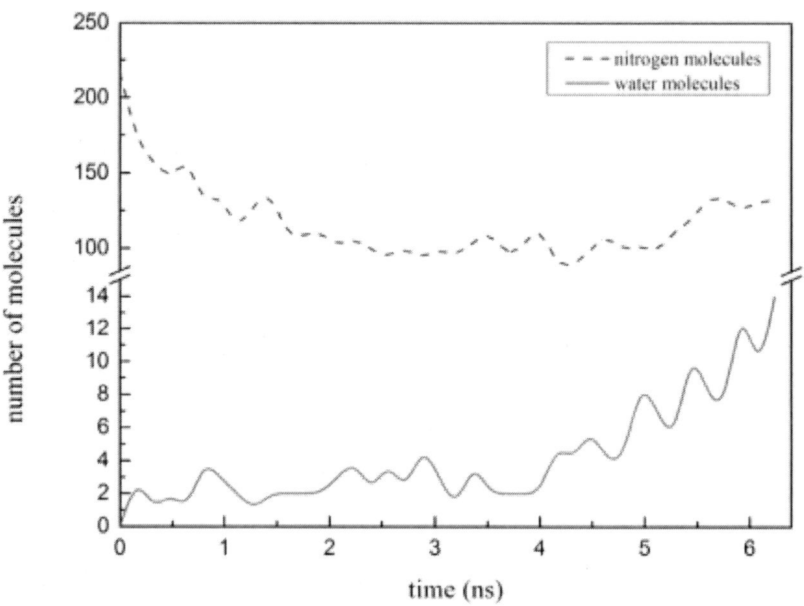

Figure. 3. The evolution of numbers of nitrogen and evaporated water molecules between the two droplets when v = 8 m/s and p = 2.7 atm.

3.3. Influencing parameters of bounce regime

In our research, ambient pressure and impact velocity of droplets are two key parameters which may influence the droplet behaviours. To scrutinize their effects on the bounce regime, simulations with different ambient pressure and impact velocity have been conducted.

3.3.1. Ambient pressure

As mentioned previously, coalescence happens (Fig. 2a) when two droplets collide in vacuum which is an extreme case of decreasing ambient pressure.

When the ambient pressure increases to 8 atm, the bounce regime occurs as expected. Compared with the lower pressure case (p = 2.7 atm) in which the minimum distance between the two droplet frontiers is 141 Å, the higher pressure causes a larger droplet frontier distance (155 Å), as shown in Fig. 4. The number of water molecules in each droplet at pressures of 2.7 atm and 8 atm are also recorded and shown in Fig. 5. Obviously, a higher pressure inhibits the evaporation of water molecules. In fact, higher pressure contributes to a more compact layout of molecules and shorter inter-molecules distances, and thus stronger atomic or molecular interactions. In the meantime, when pressure is higher, nitrogen molecules are more densely distributed and more nitrogen molecules stay in the in-between space. The gas film would feel more "stiff" less "elastic". Consequently, droplets would find it more difficult to approach each other and the minimum distance between the two droplet frontiers increases.

3.3.2. Impact velocity

When the impact velocity increases to 15 m/s, coalescence, instead of bounce, of the two droplets can be observed (Fig. 6a). Analogous to the vacuum low-impact-velocity collision simulation (Fig. 2a) and experimental observation [28,29], the two droplets rapidly approach each other before they finally merge into a large droplet (t = 1.6 ns). During the approaching process, the two droplets dramatically decrease the in-between space for accommodating nitrogen molecules and evaporated water molecules. Thus, "cushion effect" of the gas film is not strong enough and the two droplets combine into a large one. These results indicate that the bounce regime occurs as a balance between the competing factors of impact velocity and pressure, leading to different levels of the "cushion effect".

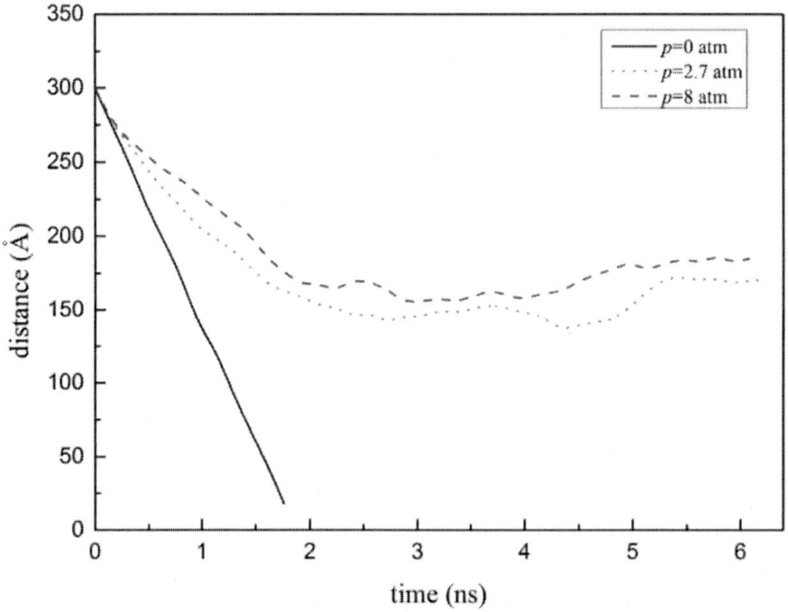

Figure. 4. Evolutions of two droplet frontier distances when $v = 8$ m/s and $p = 0$, 2.7 atm and 8 atm.

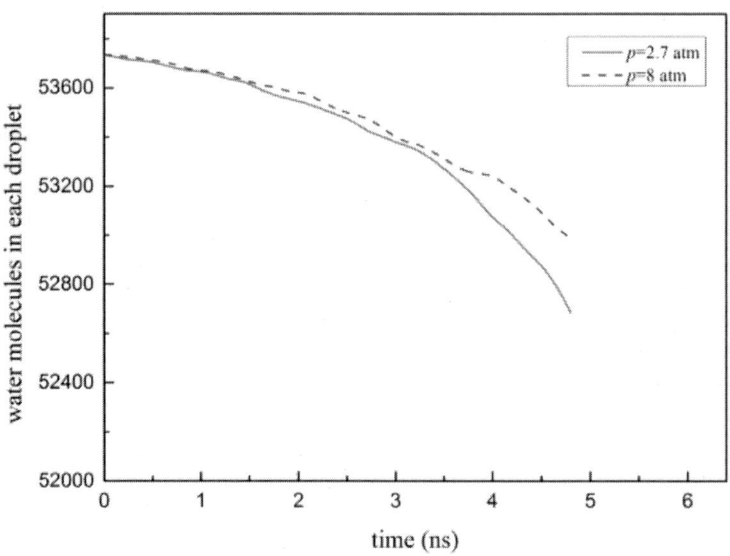

Figure. 5. Evolutions of water molecules at $v = 8$ m/s with pressures of 2.7 atm and 8 atm.

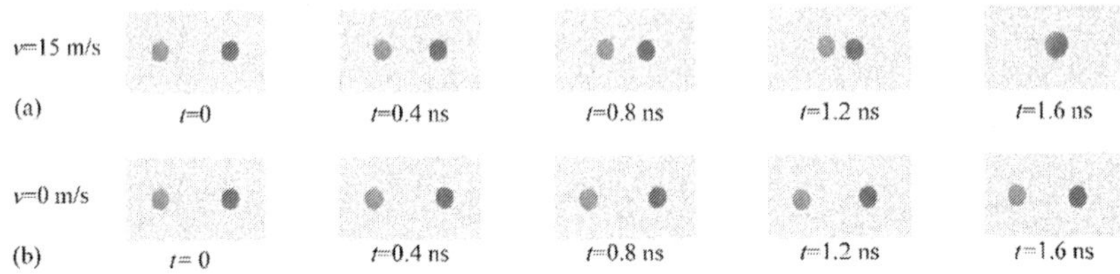

Figure. 6. Droplet behaviours at v = 15 m/s and v = 0 with p = 2.7 atm.

Meanwhile, another case in which two droplets in an ambient of nitrogen gas are placed at the original positions without impact velocities (that is, v = 0) has been simulated (Fig. 6b). Several cases have been considered for ambient pressures p = 0, 1 atm, 2.7 atm and 8 atm, respectively (with the corresponding nitrogen densities of 0, 1.138 kg/m^3, 3.074 kg/m^3, and 9.116 kg/m^3). The two droplets oscillate in all three directions with a maximum offset of 12 Å. At this separation distance, the van der Waals attractive forces are not effective while there are no bulk velocities to move the droplets in any specific directions. Instead, the oscillating positions of the nano-droplets reflect the random interactions between the in-between nitrogen molecules and the water vapour. These results have unveiled different behaviours of nano droplets from their macro or micro counterparts which tend to remain stationary without an impact velocity.

Both the higher ambient pressure and the lower impact velocity have been seen to promote bounce back, which can be attributed to the stronger interactions between interjacent molecules and two moving droplets. In vacuum, interjacent molecules only comprise water molecules evaporated from the droplets. The number of evaporated water molecules is too limited to push the droplets back, which allows further approaching, contacting and coalescence of the droplets. This kind of coalescence in vacuum at low impact velocities has been corroborated by previous studies [28,29]. When nitrogen molecules are introduced to provide an ambient pressure, interjacent molecules include both nitrogen and water molecules. The interactions between interjacent molecules and moving or stationary droplets become large enough to prevent coalescence, thereby enabling bounce to occur. When the impact velocity exceeds a certain threshold, however, the large momentum of the two moving droplets would overcome the "cushion effect" of the interjacent molecules, get into contact and coalesce.

3.4. Nano-scale collision regimes

The above zero impact velocity simulation indicates that nanoscale collisions have different behaviours from their macro- or micro-counterparts. The macroscale collision regimes successively contain stationary ($v = 0$), coalescence, bounce, coalescence and separation as the impact velocity increases [27]. The nanoscale collision regimes include approaching ($v = 0$), bounce, coalescence, shattering (in our follow-on paper) and possibly other regimes as the impact velocity increases. More detailed analysis of these issues are beyond the present paper and will be examined in future studies.

4. CONCLUSIONS

MD simulations of droplet collisions in vacuum and in ambient gas have been conducted to scrutinize the dynamics and mechanisms of binary collisions at molecular scale. The head-on collision and bounce phenomenon is successfully reproduced for the first time. The bounce phenomena are mainly attributed to the "cushion effect" of the in-between nitrogen molecules and evaporating water molecules from the two nano-droplets. The study has verified and extended the current gas film theory for the bounce regime by including the additional effects of evaporating water droplets. Moreover, the MD results show that low impact velocity and high ambient pressure promote bouncing. The behaviours of nanoscale droplet collisions have both similarities and dissimilarities to macro-/micro-droplet collisions. Finally, a set of regimes for nano-scale droplet collisions has been identified, but further details such as regime boundaries require extensive further investigations, which will be pursued in our future studies.

ACKNOWLEDGMENTS

The authors gratefully acknowledge the support from the Major Project of the National Science Foundation of China (Grant No. 91441120) and the Center for Combustion Energy at Tsinghua University. The simulations were performed on ARCHER funded under the EPSRC Grant No. EP/J016381/2.

REFERENCES

1. J.R. Adam, N.R. Lindblad, C.D. Hendrick **Collision coalescence and disruption of water droplets** J. Appl. Phys., 39 (1968), pp. 5173-5180, 10.1063/1.1655940
2. N. Ashgriz, J.Y. Poo **Coalescence and separation in binary collisions of liquid-drops** J. Fluid Mech., 221 (1990), pp. 183-204, 10.1017/S0022112090003536
3. J.V.L. Beckers, C.P. Lowe, S.W. De Leeuw **An iterative PPPM method for simulating Coulombic systems on distributed memory parallel computers** Mol. Simul., 20 (1998), pp. 369-383, 10.1080/08927029808022044
4. L. Dong, Z. Peng, C.K. Law, G. Yincheng **Collision dynamics and mixing of unequal-size droplets** Int. J. Heat Mass Transfer, 57 (2013), pp. 421-428, 10.1016 /j.ijheatmasstransfer.2012.10.023
5. G.M. Faeth **Current status of droplet and liquid combustion** Prog. Energy Combust. Sci., 3 (1977), pp. 191-224, 10.1016/0360-1285(77)90012-0
6. G. Falkovich, A. Fouxon, M.G. Stepanov**Acceleration of rain initiation by cloud turbulence** Nature, 419 (2002), pp. 151-154, 10.1038/nature00983
7. C. Focke, M. Kuschel, M. Sommerfeld, D. Bothe **Collision between high and low viscosity droplets: direct numerical simulations and experiments** Int. J. Multiph. Flow, 56 (2013), pp. 81-92, 10.1016/j.ijmultiphaseflow.2013.05.008
8. T. Hawa, M.R. Zachariah **Coalescence kinetics of unequal sized nanoparticles** J. Aerosol. Sci., 37 (2006), pp. 1-15, 10.1016/j.jaerosci.2005.02.007
9. Y.J. Jiang, A. Umemura, C.K. Law **An experimental investigation on the collision behavior of hydrocarbon droplets** J. Fluid Mech., 234 (1992), pp. 171-190, 10.1017/S0022112092000740
10. W.L. Jorgensen, J. Chandrasekhar, J.D. Madura, R.W. Impey, M.L. Klein**Comparison of simple potential functions for simulating liquid water** J. Chem. Phys., 79 (1983), pp. 926-935, 10.1063/1.445869
11. M. Kalweit, D. Drikakis **Collision dynamics of nanoscale Lennard-Jones clusters** Phys. Rev. B, 74 (2006), p. 235415, 10.1103/PhysRevB.74.235415
12. C.K. Kuan, K.L. Pan, W. Shyy **Study on high-Weber-number droplet collision by a parallel, adaptive interface-tracking method** J. Fluid Mech., 759 (2014), p. 30, 10.1017/jfm.2014.558
13. P. Kuo-Long, C.K. Law, Z. Biao **Experimental and mechanistic description of merging and bouncing in head-on binary droplet collision** J. Appl. Phys., 103 (2008), p. 064901, 10.1063/1.2841055
14. D. Lycett-Brown, K.H. Luo, R.H. Liu, P.M. Lv **Binary droplet collision simulations by a multiphase cascaded lattice Boltzmann method** Phys. Fluids, 26 (2014), 10.1063/1.4866146
15. A.D. MacKerell, D. Bashford, M. Bellott, R.L. Dunbrack, J.D. Evanseck, M.J. Field, S.Fischer, J. Gao, H. Guo, S. Ha, D. Joseph McCarthy, L. Kuchnir, K. Kuczera, F.T.K. Lau, C. Mattos, S. Michnick, T. Ngo, D.T. Nguyen, B. Prodhom, W.E. Reiher, B. Roux, M.Schlenkrich, J.C. Smith, R. Stote, J. Straub, M. Watanabe, J. Wiorkiewicz-Kuczera, D. Yin, M. Karplus **All-atom empirical potential for molecular modeling and dynamics studies of proteins** J. Phys. Chem. B, 102 (1998), pp. 3586-3616, 10.1021/jp973084f
16. L. Ming-Liang, J. Shin-Pon, Y. Sheng-Hui **Coalescence behavior of water nanoclusters: temperature and size** J. Phys. Chem. C, 111 (2007), pp. 6927-6932, 10.1021/jp066246t
17. E. Monaco, G. Brenner, K.H. Luo **Numerical simulation of the collision of two microdroplets with a pseudopotential multiple-relaxation-time lattice Boltzmann model** Microfluid. Nanofluid., 16 (2014), pp. 329-346, 10.1007/ s10404-013-1202-0
18. N. Nikolopoulos, G. Bergeles **The effect of gas and liquid properties and droplet size ratio on the central collision between two unequal-size droplets in the**

reflexive regime Int. J. Heat Mass Transfer, 54 (2011), pp. 678-691, 10.1016/j.ijheatmasstransfer.2010.09.002

19. N. Nikolopoulos, G. Strotos, K.S. Nikas, G. Bergeles **The effect of Weber number on the central binary collision outcome between unequal-sized droplets** Int. J. Heat Mass Transfer, 55 (2012), pp. 2137-2150, 10.1016/ j.ijheatm asstransfer. 2011.12.017
20. K.L. Pan, P.C. Chou, Y.J. Tseng **Binary droplet collision at high Weber number** Phys. Rev. E, 80 (2009), p. 036301, 10.1103/PhysRevE.80.036301
21. S.J. Plimpton **Fast parallel algorithms for short-range molecular dynamics** J. Comp. Phys., 117 (1995), pp. 1-19, 10.1006/jcph.1995.1039
22. J.C. Pothier, L.J. Lewis **Molecular-dynamics study of the viscous to inertial crossover in nanodroplet coalescence** Phys. Rev. B, 85 (2012), p. 115447, 10.1103/PhysRevB.85.115447
23. J.G. Powles, R.F. Fowler, W.A.B. Evans **A new method for computing surface-tension using a drop of liquid** Chem. Phys. Lett., 96 (1983), pp. 289-292, 10.1016/0009-2614(83)80674-5
24. S.L. Price, A.J. Stone, M. Alderton **Explicit formulas for the electrostatic energy, forces and torques between a pair of molecules of arbitrary symmetry** Mol. Phys., 52 (1984), pp. 987-1001, 10.1080/00268978400101721
25. J. Qian, C.K. Law **Regimes of coalescence and separation in droplet collision** J. Fluid Mech., 331 (1997), pp. 59-80, 10.1017/S0022112096003722
26. M. Svanberg, L. Ming, N. Markovic, J.B.C. Pettersson**Collision dynamics of large water clusters** J. Chem. Phys., 108 (1998), pp. 5888-5897, 10.1063/1.475999
27. L. Verlet **Computer experiments on classical fluids. I. Thermodynamical properties of Lennard-Jones molecules** Phys. Rev., 159 (1967), p. 98, 10. 1103/PhysRev.159.98
28. K. Willis, M. Orme **Binary droplet collisions in a vacuum environment: an experimental investigation of the role of viscosity** Exp. Fluids, 34 (2003), pp. 28-41, 10.1007/s00348-002-0526-4
29. K.D. Willis, M.E. Orme **Experiments on the dynamics of droplet collisions in a vacuum** Exp. Fluids, 29 (2000), pp. 347-358
30. A.L. Yarin **Drop impact dynamics: splashing, spreading, receding, bouncing** Annual Review of Fluid Mechanics, Annual Reviews, Palo Alto (2006), pp. 159-192
31. L.Y. Zhao, P. Choi **Molecular dynamics simulation of the coalescence of nanometer-sized water droplets in *n*-heptane** J. Chem. Phys., 120 (2004), pp. 1935-1942, 10.1063/1.1635804

CHAPTER 13

Mathematical Modeling for Nanofluids Simulation: A Review of the Latest Works

Mohammad Reza Safaei[1*], AminHossein Jahanbin[2], Ali Kianifar[3], Samira Gharehkhani[1], Akeel Shebeeb Kherbeet[4], Marjan Goodarzi[5] and Mahidzal Dahari[6]

[1] *Department of Mechanical Engineering, Faculty of Engineering, University of Malaya, Kuala Lumpur, Malaysia*

[2] *Department of Industrial Engineering (DIN), University of Bologna, Bologna, Italy*

[3] *Department of Mechanical Engineering, Engineering Faculty, Ferdowsi University of Mashhad, Mashhad, Iran*

[4] *Department of Mechanical Engineering, KBU International College, Petaling Jaya, Selangor, Malaysia*

[5] *Young Researchers and Elite Club, Mashhad Branch, Islamic Azad University, Mashhad, Iran*

[6] *Department of Electrical Engineering, Faculty of Engineering, University of Malaya, Kuala Lumpur, Malaysia*

ABSTRACT

Exploiting nanofluids in thermal systems is growing day by day. Nanofluids having ultrafine solid particles promise new working fluids for application in energy devices. Many studies have been conducted on thermophysical properties as well as heat and fluid flow characteristics of nanofluids in various systems to discover their advantages compared to conventional working fluids. The main aim of this study is to present the latest developments and progress in the mathematical modeling of nanofluids flow. For this purpose, a comprehensive review of different nanofluid computational fluid dynamics (CFD) approaches is carried out. This study provides detailed information about the commonly used formulations as well as techniques for mathematical modeling of nanofluids. In addition, advantages and disadvantages of each method are rendered to find the most appropriate approach, which can give valid results.

Keywords: nanofluid, CFD, numerical simulation, mathematical modeling, single- and two-phase methods

1. INTRODUCTION

In general, the assessment of the thermal performance of a system through numerical simulations is much affordable compared to experimental studies with high expenses of material and equipment. The significance of a numerical study is highlighted when a nanofluid is utilized as the working fluid. High costs for the production of nanofluids and difficulties in preparing stable nanofluids are the main barriers to perform experiments with nanofluids. Therefore, numerical modeling of nanofluids, where a suitable approach is selected to simulate the flow, could be the best solution for problems involved with nanoparticle suspensions.

However, in spite of considerable developments in computing power and methods, literature review reveals that there is no comprehensive study to conclude the best technique for the modeling of nanofluids. In particular, due to the ultrafine size of nanoparticles, the governing terms in multiphase models are still not entirely identified. In the present work, latest studies on numerical simulations of nanofluid flow are reviewed with a particular focus on different multiphase schemes.

2. NUMERICAL METHODS FOR NANOFLUIDS' FLOW SIMULATION

Nanofluid computational fluid dynamic (CFD) modeling can be classified into two main groups: single-phase and two-phase models. However, there are few other models that may not be included in these categories, such as Lattice-Boltzmann method (LBM). Moreover, different numerical approaches have been employed to solve models mentioned above to predict thermal and hydraulic characteristics of nanofluids flow. Finite volume method (FVM) and finite element method (FEM) are two main approaches for solving the governing equations of nanofluid problems. However, finite difference method (FDM), control volume-based finite element method, and some novel numerical approaches such as homotopy analysis method (HAM) and smoothed particle hydrodynamics (SPH) methods have also been utilized in the previous studies. In this study, a comprehensive review of various numerical methods for the simulation of nanofluids is accomplished.

2.1. Single-Phase Approaches

Although suspension of a nanofluid is inherently a two-phase fluid, if some proper assumptions are made, it can be considered as a homogeneous liquid. Due to the existence of ultrafine nanoparticles, it is assumed that these particles can be easily dispersed in the host fluid. For this purpose, both the nanoparticles and base fluid are considered to be in thermal balance without any slip between their molecules. Therefore, under such assumptions, in many studies, nanofluids have been assumed as a single-phase fluid.

2.1.1. Conventional single-phase model

Mass, momentum, and energy equations, which are used for conventional liquids, could also be applied to single-phase flow with the above assumptions. In this case, only thermophysical properties of nanofluids should be determined. Therefore, the governing equations in the steady state can be expressed as below [1]:

$$\nabla.(\rho_{eff}\vec{u}) = 0 \tag{1}$$

Conservation of momentum:

$$\nabla.(\rho_{eff}\vec{u}\vec{u}) = -\nabla P + \nabla.(\mu_{eff}\nabla\vec{u}) - (\rho\beta)_{eff}(T - T_0)g \tag{2}$$

Conservation of energy:

$$\nabla.((\rho c_p)_{eff}\vec{u}T) = \nabla.(k_{eff}\nabla T) \tag{3}$$

There are numerous publications simulating nanofluids characteristics as a single-phase fluid. Mixed convection heat transfer in a T-shaped lid-driven cavity was examined numerically by Mojumder et al. [2]. A residual FEM model was applied for the numerical simulation. They validated their simulation code against data of Abu-Nada, Chamkha [3]. The results revealed that higher Grashof number causes rise in the heat transfer rate.

Turbulent nanofluid flow for different nanoparticles such as alumina, cupric oxide, and titania was investigated by Rostamani et al. [4] at various concentrations in a long horizontal duct under constant heat flux. They employed the control volume approach and temperature-dependent thermophysical properties. It was found that increasing the number of dispersed nanoparticles in base fluid increases the pumping power and heat transfer rate. In addition, the predicted Nusselt numbers in some cases demonstrated good agreements to the obtained results by Pak, Cho [5], and Maiga et al. [6] correlations.

Abu-Nada [7] considered the Rayleigh-Benard-free convection heat transfer in CuO-water nanofluids using FVM, where the effect of temperature-dependent properties was the primary objective. The obtained results by temperature-dependent models were also compared with classic models (Maxwell-Garnett (MG) and Brinkman). Results displayed that for $Ra > 10^3$, the effect of temperature on fluid flow and heat transfer was insignificant due to the presence of high viscosity, which was caused by the nanoparticles volume fraction.

Bouhalleb and Abbassi [8] employed Control Volume-Finite Element method to study the free convection in an inclined cavity. They solved the free convection problem using Boussinesq approximation and employing the SIMPLER algorithm for velocity-pressure coupling. The results showed that the variation of inclination angel highlights a hysteresis behavior of the nanofluid flow. Also, increasing the diameter of solid nanoparticles led to a strong decay in heat transfer. Furthermore, they concluded that the efficiency of heat transfer strongly depends on the diameter of nanoparticles, not its concentration in base fluid.

While a considerable number of recent studies on numerical simulation of nanofluids have been employed by FVM [9–11] and FEM [12–14] for their studies, in several studies Finite Difference Method has been used for different applications of nanofluids. For instance, Buddakkagari and Kumar [15] studied laminar-forced convection of a nanofluid over a vertical cone/plate. The non-dimensional governing equations were solved using FDM model, Crank-Nicolson type. The results illustrated that the Prandtl number affects the boundary layer dramatically. Furthermore, the momentum boundary layer was more affected by higher values of Lewis number. Finally, it was concluded that the boundary layer growth depends on Brownian motion and thermophoresis force.

However, applying the single-phase model for nanofluids has some limitations. For instance, the results obtained from this model were strongly dependent on adopted thermophysical properties and in some cases using the single-phase model may underestimate the Nusselt number, compared to models adopting temperature-dependent properties [16, 17].

The reason behind this could be due to various factors such as gravity, friction forces, Brownian motion, and solid/liquid interface. Ding and Wen [18] showed that the primary assumption of homogeneous fluid is not always acceptable. However, the review of the previous works illustrates that choosing the appropriate thermophysical property correlations in the single-phase method results in a reasonable estimation of nanofluids properties [19, 20]. Therefore, selecting suitable thermophysical properties such as variable properties, and considering the chaotic movement of nanoparticles (dispersion model) may compensate, to some extent, the limitations of single-phase model.

2.1.2. Thermal Dispersion Model

Brownian and gravity forces, the friction force between the base fluid and nanoparticles, sedimentation, and dispersion may coexist in a nanofluid flow. In fact, slip motion between liquid molecules and solid particles is not negligible, and the random movement of nanoparticles ameliorates the thermal dispersion in nanofluids, which reinforces heat transfer. For the first time, the thermal dispersion model was suggested by Xuan and Roetzel [21]. They assumed that nanoparticles move randomly, causing small chaos in velocity and temperature magnitudes (T' and $\vec{u}'$). The essential phase averages can be written as [21]:

$$T = \langle T \rangle^f + T' \tag{4}$$

$$\vec{u} = \langle \vec{u} \rangle^f + \vec{u}' \tag{5}$$

where

$$\langle T \rangle^f = \frac{1}{V_f} \int_{V_f} T \, dV \tag{6}$$

$$\langle \vec{u} \rangle^f = \frac{1}{V_f} \int_{V_f} \vec{u} \, dV \tag{7}$$

$$\frac{1}{V_f} \int_{V_f} T' dV = 0 \tag{8}$$

By assuming that the boundary layer between the liquid and solid phases is negligible, unsteady-state, energy equation can be written as

$$(\rho c_p)_{nf}\,[\frac{\partial\langle T\rangle^f}{\partial t}+\langle \vec{u}\rangle^f.\nabla\langle T\rangle^f]=\nabla.(k_{nf}\nabla\langle T\rangle^f)-(\rho c_p)_{nf}\nabla\langle \vec{u'}T'\rangle^f \tag{9}$$

The thermal dispersion generates a heat flux in the flow that is equal to

$$q_d=(\rho c_p)_{nf}\langle \vec{u'}T'\rangle^f=-k_d\;.\nabla\langle T\rangle^f \tag{10}$$

$\mathbf{k}_d$ is the thermal conductivity tensor due to dispersion. Now, Eq. (9) can be rewritten as

$$\frac{\partial\langle T\rangle^f}{\partial t}+\langle \vec{u}\rangle^f.\nabla\langle T\rangle^f=\nabla.[(\alpha_{nf}\mathbf{I}+\frac{\mathbf{k}_d}{(\rho c_p)_{nf}})\;.\nabla\langle T\rangle^f] \tag{11}$$

where α is the thermal diffusivity and **I** is the identity tensor. The following form may be applicable for the effective thermal conductivity of nanofluid:

$$k_{eff}=k_{nf}+k_d \tag{12}$$

where $\mathbf{k}_d$ refers to the dispersed thermal conductivity coefficient. Despite the fact that this term plays a key role in dispersion model, just a few correlations were proposed in the literature. Xuan and Roetzel [21] with reference to some publications in porous media [22, 23] proposed the following forms:

$$k_d=C^*(\rho c_p)_{nf}u\,R,\;\; k_d=C_d(\rho c_p)_{nf}u\,Rd_p\varphi \tag{13}$$

C_d is an unknown constant and R is referred to pipe radius. Khaled and Vafai [24] investigated heat transfer enhancement by considering dispersion effects. They used the following linearized model introduced by Xuan and Roetzel [21] correlations for dispersed thermal conductivity of nanofluids:

$$k_d=C^*(\rho c_p)_{nf}\,\varphi H\,u_m \tag{14}$$

where H is the half of duct height, u_m is the average bulk velocity, and C^* is an unknown constant. By employing control volume approach, Mojarrad et al. [25] analyzed the heat transfer of α-alumina-water nanofluid in the entrance region of a rounded pipe. They also compared their results with experimental data. The results showed that dispersion model provides reasonable outcomes in spite of its simplicity. Moreover, they suggested a new correlation for radial dispersed thermal conductivity:

$$k_d = C_d(\rho c_p)_{nf} \frac{R\varphi}{d_p}\left(\frac{\partial T}{\partial r}\right) \tag{15}$$

By using dispersion model, Zeinali Heris et al. [26, 27] investigated convective heat transfer of different nanofluids. They studied heat transfer augmentation due to nanofluids flow through the tubes with different cross-section geometries. The numerical results were validated against experimental data [28, 29] and good agreement was observed. The results showed that the Nusselt number would enhance with increasing particle loading and decreasing particle size.

Thermal behavior of nanofluids in a cavity was analyzed by Kumar et al. [30]. The dimensionless-governing equations were resolved via semi-explicit FVM solver. The Grashof number, volume concentration, and nanoparticles shape effects on heat transfer rate were assessed. The results showed that the dispersed thermal conductivity is more intensive in the vicinity of walls in comparison with pipe center. Also, it was found that dispersed thermal conductivity and hydraulic diameter of the particles are strongly dependent to each other.

Akbaridoust et al. [31] examined laminar steady-state nanofluid flow through helically shaped tubes, both numerically and experimentally. The governing equations in three-dimensional (3-D) form were solved by finite difference approach, using a FORTRAN code. Dispersion model was modified in order to be applicable for helical tubes. This modification resulted in minimized difference between numerical results and experimental data. The results showed that higher curvature ratios cause more heat transfer rates.

2.2. Two-Phase Approaches

Due to some factors such as Brownian force, Brownian diffusion, friction force, thermophoresis, and gravity, nanofluids may be considered two-phase fluids by nature. Therefore, the classic theory of solid-liquid mixture can be applied to nanofluids. In such models, nanoparticles and base fluid are considered as two separate phases with different temperatures and velocities. Although two-phase approaches may obtain realistic results, they have high computational cost in comparison to single-phase models. Two-phase approaches can be categorized into two general models: Lagrangian-Eulerian and Eulerian-Eulerian.

2.2.1. Lagrangian-Eulerian Model

In the Lagrangian-Eulerian or discrete phase model, the fluid phase is considered as a continuum by solving the N-S equations in time-averaged form, while the dispersed phase was solved by tracking the particles in the Lagrangian frame. Also, in this model, the interaction between fluid and particles presented as a source term in both momentum and energy equations.

Mathematical formulations of the Lagrangian-Eulerian in two-phase model can be written as follows [32]:

Conservation of mass:

$$\nabla.(\rho\vec{u}) = 0 \tag{16}$$

Conservation of momentum:

$$\nabla.\left(\rho\vec{u}\,\vec{u}\right) = -\nabla P + \nabla.\left(\mu\,\nabla\vec{u}\right) + S_m \tag{17}$$

Conservation of energy:

$$\nabla.\,(\rho c_p \vec{u} T) = \nabla.(k\nabla T) + S_e \tag{18}$$

where S_m and S_e are source/sink terms representing the exchange of momentum and energy between liquid and solid phases. The momentum and energy source/sink terms are defined as [33, 34]

$$s_m = \frac{1}{\delta V}\sum_{np}\vec{F} \tag{19}$$

$$s_e = \frac{1}{\delta V}\sum_{np} Nu_p \pi d_p k_p (T - T_p) \tag{20}$$

where Nu_p can be computed from the Ranz correlation [35]:

$$Nu_p = 2 + 0.6Re_p^{0.5} Pr^{0.333} \tag{21}$$

In Lagrangian reference frame, the particle motion and energy equations are as follows:

$$m_p \frac{d\vec{u}_p}{dt} = \vec{F}_g + \vec{F}_D + \vec{F}_L + \vec{F}_{Br} + \vec{F}_b \tag{22}$$

$$m_p c_p \frac{dT_p}{dt} = Nu_p \pi d_p k_f (T_f - T_p) \tag{23}$$

In Eq. (22), *F* stands for various forces including gravity, drag, Saffman's lift, Brownian, and buoyancy, respectively.

Turbulent nanofluid flow in helical tubes was investigated numerically and experimentally by Bahremand et al. [36]. The numerical simulation was performed by both single-phase model and Lagrangian-Eulerian approach in connection with renormalization group (RNG) k-ε model. ANSYS CFX software was used for solving the governing equations. The results indicated that nanofluids with a higher concentration exhibit a greater heat transfer coefficient and pressure drop. Also, it was found that the two-phase model yields more accurate results compared to single-phase model.

Alumina-water nanofluid flow and heat transfer in a long tube with uniform heating at the walls were investigated by Moraveji and Esmaeili [37]. The simulations were conducted in both single- and two-phase methods where the governing equations were solved by FVM. Both temperature-dependent and constant thermophysical properties were considered in the study. The results of the modeling revealed that the temperature-dependent properties are more sensitive to the Reynolds number variations and led to higher values of the Nusselt number. Comparison between single-phase and two-phase (discrete phase) models showed the maximum difference of 11% for the average heat transfer coefficient.

Tahir and Mital [38] studied the laminar-forced convection of Al_2O_3-water nanofluid in a tube numerically. They analyzed the impacts of the Reynolds number, particle diameter, and volume fraction of the particles in their study. A good agreement was achieved between the simulation and experimental data using discrete phase method. The results of the survey demonstrated that the heat transfer coefficient increased linearly with both the Reynolds number and volume fraction of nanoparticles. However, there was a non-linear parabolic decrease with increasing nanoparticle size. It was concluded that the Reynolds number and volume fraction have the maximum and the minimum effects on heat transfer coefficient, respectively.

A comprehensive simulation of turbulent-forced convection for Cu-water was carried out by Behroyan et al. [39].The Reynolds number of the flow was chosen between 10,000 and 25,000, where the volume fraction of copper nanoparticles was taken in the range of 0.0–2.0%. Two single-phase models (Newtonian and non-Newtonian) and three two-phase models were employed in this study. The ANSYS commercial CFD package was utilized to solve the governing equations. The obtained results showed that the Newtonian single-phase method as well as discrete phase method is in better agreement with experimental data, compared to other numerical approaches.

The present literature survey reveals that using Lagrangian-Eulerian approach for modeling the heat transfer of nanofluids is in early stages. Therefore, more studies are required to be conducted to determine the capability of the Lagrangian-Eulerian model, especially in turbulent regime [40]. Also, in some other studies such as Safaei et al. [32] and Xu et al. [41], it is emphasized that Lagrangian-Eulerian is a suitable model just for low concentration two-phase suspensions (φ <1%). Moreover, it was mentioned that this model needs extremely high computational time due to a large amount of calculation.

2.2.2. Eulerian-Eulerian Model

The other important branch of two-phase models is the Eulerian-Eulerian model. Since the Eulerian-Eulerian model is suitable for mixtures with a high amount of particles, applying this model to nanofluids consisting of an extremely large number of nanoparticles is recommended. The main models of Eulerian-Eulerian available in the literature are three models including Mixture, Eulerian, and VOF (volume of fluid).

2.2.2.1. VOLUME OF FLUID MODEL

In the VOF model, the volume fractions of all phases are obtained for the entire domain of study, by solving the continuity equation for the secondary phases. A single set of momentum equations is solved for all the phases to find the velocity components. The sum of all employed phases' volume fractions is equal to unity. Accordingly, the primary phase volume fraction magnitude is achieved. In addition, all the physical properties are calculated by using an average weighted of different phases according to their volume fraction on each control volume. Finally, a shared temperature is calculated from a single energy equation [42].

Mass conservation for VOF model can be expressed as

$$\nabla.(\varphi_z \rho_z \vec{u}_z) = 0 \tag{24}$$

Where $\sum_{z=1}^{n} \varphi_z = 1$ and all properties are computed such as $N = \sum_{z=1}^{n} \varphi_z N_z$

In this model, Eqs. (2) and (3) are used as momentum and energy equations.

According to literature, a few studies have been done on using the VOF model for the simulation of nanofluids. Akbari et al. [43] studied turbulent-forced convective heat transfer of Al_2O_3-water as well as Cu-water nanofluids inside a horizontal tube under uniform heat flux. The governing equations were solved implementing different numerical approaches, for example, single-phase, VOF, mixture, and Eulerian models, using FLUENT software. The results showed that the thermal field forecasting by multiphase models was different from the results of experimental data and single-phase approach. However, single-phase and two-phase models predicted almost same hydrodynamic results. It was concluded that unlike the results of previous studies [17, 44], two-phase models overestimate the thermal field. Under similar conditions, however, Hejazian et al. [45] found different results when investigating the turbulent convection of TiO_2-water nanofluid in a horizontal tube using FVM method. The results of this study showed that the mixture and VOF models are more appropriate to predict the heat transfer field, compared to single-phase model.

Turbulent heat transfer of nanofluids flow through a mini-channel heat sink was analyzed by Naphon and Nakharintr [46]. The k-ε turbulence model with single-phase, mixture, and VOF approaches was employed to analyze the heat transfer and flow characteristics. Also, some experiments conducted to verify the predicted results and reasonable agreements were achieved. It was concluded that the single-phase model cannot predict the Nusselt number with accuracy as good as mixture and VOF models because the impacts of Brownian motion and non-uniform distribution of nano-particles in the solution domain are not considered in the single-phase model. In addition, under similar conditions, VOF and mixture models present more appropriate results compared to the single-phase model.

Hanafizadeh et al. [47] carried out a study to compare single and two-phase approaches for Fe_3O_4-water nanofluid in both developing and fully developed regions in a circular tube under constant heat flux. The study was conducted for 0.5–2 vol. % and $300 \leq Re \leq 1200$. The results showed that higher values of both Reynolds number and volume fraction would augment the average heat transfer coefficient, while just increasing the number of dispersed nanoparticles does not have a considerable impact on heat transfer enhancement. Also, in the fully developed region, a higher number of dispersed nanoparticles in base fluid would reduce the error of studied numerical methods. On the other hand, in developing region of a tube and for low Reynolds numbers, increase in nanofluid volume fraction would decrease the accuracy of numerical methods, while this trend was reversed for moderate and high Reynolds numbers.

In total, since a limited number of studies have used this numerical approach, further studies are needed to evaluate the capability of the VOF model.

2.2.2.2. MIXTURE MODEL

The mixture model is one of the most popular methods for modeling of multiphase slurry flows. The main feature of this approach is that only one set of velocity elements is solved for the mixture momentum conservation equations. The velocities of dispersed phases are extracted from the algebraic formulations [48]. Moreover, since the primary phase affects the secondary phase through drag force and turbulence, the effect of secondary phase on the primary phase could be found through mean momentum reduction and turbulence. The basic assumptions of mixture model are as follows [49, 50]:

- All phases share a single pressure.
- The interaction between different dispersed phases is assumed to be negligible.
- Nanoparticles in the secondarily dispersed phase are spherical in shape, with a uniform size.
- The concentration of the secondarily dispersed phases is solved by a scalar equation, considering the correction made by phase slip.

The governing equations of the nanofluids' mixture model can be written as follows [51]:

Continuity:

$$\frac{\partial}{\partial t}(\rho_m) + \nabla.(\rho_m \vec{u}_m) = 0 \tag{25}$$

where

$$\vec{u}_m = \frac{\sum_{Z=1}^{n} \varphi_Z \rho_Z \vec{u}_Z}{\rho_m} = \vec{u}_Z \tag{26}$$

and

$$\rho_m = \sum_{Z=1}^{n} \varphi_Z \rho_Z \tag{27}$$

Momentum:

$$\frac{\partial}{\partial t}(\rho_m \vec{u}_m) + \nabla.(\rho_m \vec{u}_m \vec{u}_m) = -\nabla P_m + \nabla.[\mu_m(\nabla \vec{u}_m + \nabla \vec{u}_m^{\,T})] + \rho_m \beta_m (T - T_0) g \tag{28}$$

where

$$\mu_m = \sum_{Z=1}^{n} \varphi_Z \mu_Z \tag{29}$$

Energy:

$$\frac{\partial}{\partial t}\rho_m h_m + \nabla \cdot (\rho_m h_m \vec{u}_m) + \nabla \cdot (P \vec{u}_m) = \nabla.(k_m \nabla T) \tag{30}$$

where

$$\rho_m h_m = \sum_{Z=1}^{n} (\varphi_Z \rho_Z h_Z) \tag{31}$$

and

$$k_m = \sum_{Z=1}^{n} \varphi_Z (k_Z) \tag{32}$$

As a pioneer, Behzadmehr et al. [44] employed single-phase approach as well as two-phase mixture model to study the turbulent heat transfer of copper-water nanofluid inside a circular tube. The results of their study revealed that the obtained results from mixture model are much closer to experimental data, compared to the results of single-phase model. In a similar study, Bianco et al. [52] analyzed the steady-state turbulent convection heat transfer of Al_2O_3-water nanofluid in a circular tube under constant heat flux. FLUENT commercial software was used to solve the governing equations. The results showed that single-phase and mixture models give approximately the same results at low concentrations (i.e., φ = 1%), while for higher concentrations, the difference between two models is considerable.

Shariat et al. simulated alumina-water nanofluid in an elliptic tube [53]. The impacts of nanoparticles mean diameter and buoyancy force on the nanofluid flow behaviors were investigated in that study. The three-dimensional equations of the mixture model were solved by using FVM. The results showed that at a specified value of Reynolds and Richardson numbers, an increase in nanoparticles size diminishes the Nusselt number while it does not have a remarkable effect on the friction factor. A non-linear relation between the nanoparticles size and heat transfer characteristics of nanofluid was also observed.

Laminar-free convection heat transfer of alumina-water nanofluid inside a cavity was studied by Corcione et al. [54]. The governing equations were solved by a CFD code based on a two-phase mixture model. Temperature-dependent effective properties considering the Brownian motion and thermophoresis were employed and different nanoparticles volume fractions were analyzed. It was found that the heat transfer trend reached a peak value at maximum particle loading. Using these results, new correlations were developed for different parameters such as the optimal particle loading and a maximum value of the heat transfer augmentation.

Goodarzi et al. [51] investigated both laminar and turbulent mixed convection of Cu-water nanofluid inside a rectangular shallow cavity. The upper movable lid of the cavity was considered at a lower temperature, compared to the bottom wall. FLUENT commercial code was utilized to solve the problem, along with some modifications in governing equations by developing user-defined function (UDF) codes. The results showed that the impact of the volume fraction on turbulent kinetic energy, turbulence intensity, skin friction, and wall shear stress is insignificant. However, under similar conditions, lower Richardson number leads to higher wall shear stress and turbulence kinetics energy.

The single- and two-phase models were employed by Naphon and Nakharintr [55] to investigate the 3-D laminar convection heat transfer of nanofluids inside a mini-channel heat sink. Some experiments were also carried out for validation purpose. The research outcomes demonstrated that two-phase mixture model is in better agreement with experimental results, compared to single-phase model.

Recently, Siavashi et al. [56] investigated the application of nanofluids and porous media to enhance the heat transfer inside an annular pipe. The simulation was conducted to investigate the effects of different parameters such as the Darcy and Reynolds numbers as well as porous medium radius and its position on heat transfer enhancement, heat loss, and entropy generation. Two-phase mixture model along with Darcy-Brinkman-Forchheimer equation was employed for nanofluid flow simulation in porous media. A FVM code was developed to solve the governing equations.

The results showed that the geometry, nanoparticle concentration, and magnitude of the Reynolds number have considerable effects on both the performance and entropy generation numbers.

By reviewing the literature, it can be seen that among different multiphase approaches, the mixture model is the most popular model for nanofluids modeling. This popularity can be due to some facts such as accuracy, simplicity in both theory and implementation, and low computational cost. However, for using this model there are some limitations and requirements, which were addressed in detail by Moraveji and Ardehali [49], Bahiraei [50], and Goodarzi et al. [51].

2.2.2.3. EULERIAN MODEL

In this model, pressure is assumed to be equal for all the phases, while other governing equations are solved separately for primary and secondary phases. The volume of the two phases is estimated by integrating the volume fraction on solution domain, where the aggregate of volume fractions totality becomes one [50]. The Eulerian model corresponding equations can be expressed as follows [42]:

Continuity:

$$\nabla.(\varphi_z \rho_z \vec{u}_z) = 0 \tag{33}$$

Where $\vec{u}_z = \int_V \varphi_z \, dv \;, \quad \sum_{z=1}^{n} \varphi_z = 1$.

Conservation of momentum (z^{th} phase):

$$\nabla.(\varphi_z \rho_z \vec{u}_z \vec{u}_z) = \sum_{p=1}^{n} \vec{R}_{pz} + F_{lift,z} + \varphi_z \nabla.(\mu_z \nabla \vec{u}) + \varphi_z \rho_z \vec{g} - \varphi_z \nabla P \tag{34}$$

Where $\sum_{p=1}^{n} \vec{R}_{pz} = \sum_{p=1}^{n} S_{pz}(\vec{u}_p - \vec{u}_z)$ denotes the interaction force between phases, $S_{pz} = (\varphi_z \varphi_z \rho_z f) / \tau_p$, $\tau_p = (\rho_p d_p^2)/(18\mu_z)$, and f indicates the drag friction, which is computed by Schiller and Naumann [57] recommendation:

$$f = \frac{C_D Re_p}{24} \tag{35}$$

where

$$C_D = \begin{cases} \dfrac{24(1+0.15Re_p^{0.687})}{Re_p} & Re \leq 1000 \\ 0.44 & Re > 1000 \end{cases} \tag{36}$$

The nanoparticle Reynolds number (Re_p) in Eq. (36) and lift force in Eq. (34) [58] are, however, based on particle-fluid relative velocity, which is extremely small for nanoparticles.

$$Re_p = \frac{\rho_q \, |\vec{u}_p - \vec{u}_z| \, d_p}{\mu_z} \cong 0 \tag{37}$$

$$F_{lift,z} = -\frac{1}{2}\rho_z \varphi_Z (\vec{u}_p - \vec{u}_z) \times (\nabla \times \vec{u}_z) \cong 0 \tag{38}$$

Considering Eq. (37), the first two terms on the right side of Eq. (34) should be ignored.

Conservation of energy:

$$\nabla.(\varphi_Z \rho_Z \vec{u}_Z h_z) = -\nabla.(k_Z \nabla T_Z) - \tau_q : \nabla \vec{u}_Z + \sum_{p=1}^{n} \vec{Q}_{PZ} \tag{39}$$

Where $\vec{Q}_{pz} = h\,(\vec{u}_p - \vec{u}_Z)$ is the heat exchange coefficient and $h = \dfrac{6k_q \varphi_q \varphi_p N u_p}{d_p^2}$. Also, Nu_p is calculated from Eq. (21).

Kalteh et al. [59] investigated the laminar-forced convection heat transfer of Cu-water nanofluid inside a microchannel. The Eulerian model utilized for flow simulation and governing equations was solved by FVM. The results demonstrated that the nanoparticles are distributed uniformly inside the solution domain. The two-phase model also presented a higher heat transfer augmentation compared to the single-phase model.

Laminar- and turbulent-forced convection of nanofluids inside small tubes were investigated by Chen et al. [60]. The multiphase flow was simulated using both mixture and Eulerian models and the results were compared with experimental data as well as the correlations from the literature. The obtained results for two models were quite similar, although mixture approach results showed better agreement with experimental results.

Thermal behavior and nanofluid flow at the entrance region of a pipe under constant heat flux were modeled by Göktepe et al. [61]. The results demonstrated that two-phase models predict heat transfer coefficient and friction factor with a higher accuracy at the entrance region, compared to the single-phase model. The authors also suggested that more suitable relations for nanoparticles are required to enhance the forecast accuracy of the Eulerian model.

Recently, Sadeghi et al. [62] studied nanoparticle aggregation effect on laminar convection heat transfer of alumina-water nanofluid in a circular tube. The Eulerian model was implemented according to nanoclusters Brownian motion and their fractal structure. The governing equations were solved using ANSYS CFX commercial software. The results revealed that nanoparticles size and concentration as well as fractional structure have undeniable effects on heat transfer phenomenon of nanofluid. Also, it was noted that Brownian motion can affect the convective heat transfer of nanofluids significantly.

All in all, it can be concluded that the main advantage of the Eulerian model in comparison to single-phase model is that there is no need to apply effective property models for the nanofluids [59]. However, it may not be as precise as the mixture model [17, 60].

2.3. Other Approaches

2.3.1. Lattice-Boltzmann Method

Lately, Lattice-Boltzmann method or Thermal Lattice-Boltzmann method has become an attractive alternative to simulate the nanofluids flow. The gap between microscopic and macroscopic phenomena is removed by employing Lattice-Boltzmann method since it considers molecular dynamics [50]. In Lattice-Boltzmann method, the conservation equations are resolved by the assumption that the nanoparticles are microscopically located in a chain of lattices where their distributions are determined based on Boltzmann method. In the paper of Succi [63], microscopic interaction between the nanoparticles was numerically modeled utilizing a collision model and microscopic and macroscopic quantities of components were joined together. Also in [64, 65], two more different methods were employed, namely D2Q9 (two-dimensional and 9-velocity) square and D3Q19 (three-dimensional and 19-velocity) cube lattice structures. In Lattice-Boltzmann method, it is easy to deal with the complex boundaries; also, the other advantages of this method include physical representation of microscopic interactions and the existence of uniform algorithms to solve the multiphase flows [65].

For the first time, Xuan and Yao [66] proposed LBM for simulating flow and energy transport of the nanofluids. After this study, the use of this method was rapidly increased. Considering interaction forces such as Brownian, gravity-buoyancy, drag, and interaction potential forces between two phases, Qi et al. [67] studied the free convection of nanofluid using a two-phase Lattice-Boltzmann model. It was found that while Brownian, gravity-buoyancy, and interaction potential forces have positive impacts on the augmentation of free convection, drag force has a negative impact.

Karimipour et al. [68] studied laminar-forced convective heat transfer of copper-water nanofluid inside a microchannel using double-population LBM-BGK method. The obtained results of this study were in a fair agreement with previous studies, which shows that LBM could be utilized to simulate forced convection for the nanofluids flow inside microsized configurations.

Recently, by employing a 2-D double multiple-relaxation-time (MRT) thermal Lattice-Boltzmann model, Zhang and Che [69] simulated the magneto-hydrodynamic (MHD) flow and heat transfer of copper water in an inclined cavity with four heat sources. The governing equations were solved using D2Q9- and D2Q5-MRT models, which was validated by previous investigations. The results showed that the inclination angle has a considerable effect on flow fields, the temperature patterns, and the local Nusselt number distributions. Moreover, it was concluded that MRT Lattice-Boltzmann method is competent for solving heat transfer of nanofluids in enclosures affected by a magnetic field.

In the end, LBM has been widely used for natural, forced, and mixed convection of nanofluids, which can be found in details [70, 71]. The results of this model have higher accuracy than the results of conventional CFD approaches. However, it seems that more research may be needed in order to find out to what extent LBM is applicable in the simulation of nanofluids flow and characteristics.

2.3.2. Non-Homogeneous Two-Component Model (Buongiorno Model)

Buongiorno [72] investigated the effects of seven different slip mechanisms between the base fluid and nanoparticles: gravity, thermophoresis, Brownian diffusion, inertia, Magnus effect, fluid drainage and diffusiophoresis, in the absence of turbulent effects. It was demonstrated that thermophoresis and Brownian diffusion are the most influential mechanisms on nanofluids flow and heat transfer, which can affect nanoparticle concentration variations. Under such conditions, the four coupled governing equations were proposed as follows [73, 74]:

Conservation of mass:

$$\nabla.(\rho_{nf}\vec{u}) = 0 \tag{40}$$

Conservation of momentum:

$$\nabla.(\rho_{nf}\vec{u}\vec{u}) = -\nabla P + \nabla.(\mu_{nf}\nabla\vec{u}) \tag{41}$$

Conservation of energy:

$$\nabla.((\rho c_p)_{nf}\vec{u}T) = \nabla.(k_{nf}\nabla T) - c_{p_{nf}}\vec{J}_{np}.\nabla T \tag{42}$$

Conservation of nanoparticles:

$$\vec{u}.\nabla\varphi = -\frac{1}{\rho_{np}}\nabla.\vec{J}_{np} \tag{43}$$

Where $\vec{J}_{np}$ is nanoparticles flux and is defined as

$$\vec{J}_{np} = \vec{J}_{np_{Br}} + \vec{J}_{np_T} \tag{44}$$

The aforementioned terms can be calculated as follows [75]:

$$\vec{J}_{np_{Br}} = \frac{k_{Br}T}{3\pi\mu_f d_{np}} \tag{45}$$

$$\vec{J}_{np_T} = 0.26(\frac{k_f}{2k_f + k_{np}}\frac{\mu_f}{\rho_f}\varphi) \tag{46}$$

where *D* represents the diffusion coefficient.

TABLE 1. Some recent studies on modified Buongiorno model.

Authors	Geometry of study	Type of nanofluid	Properties	Remarks
Malvandi [79]	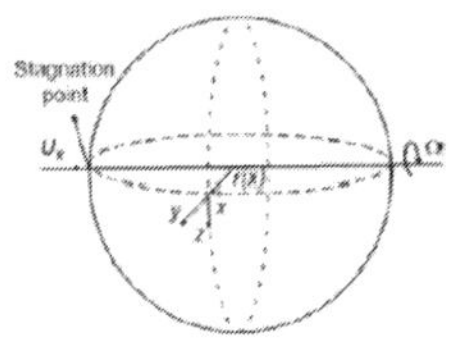	Non-homo geneous mixtures	$1\,nm \le d_p \le 100\,nm$ $0 \le \phi \le 0.2$	Increasing the thermophoresis is found to decrease heat transfer and concentration rates. This decrease suppresses for higher thermophoresis number. In addition, it was observed that unlike the heat transfer rate, a rise in Brownian motion leads to an increase in concentration rate
Malvandi et al. [80]	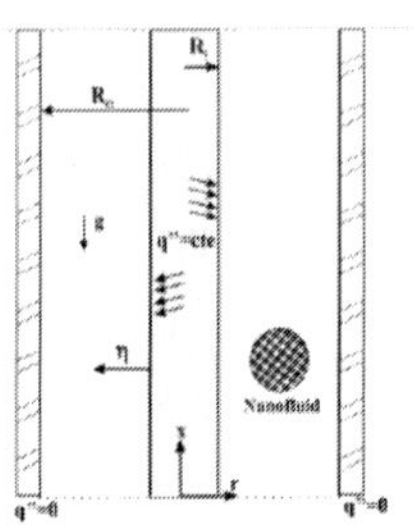	Al_2O_3 and TiO_2/water-based nanofluids	$1\,nm \le d_p \le 100\,nm$ $0 \le \phi \le 0.1$ $0.1 \le N_{BT} \le 10$	The buoyancy has negative effect on the efficiency of nanoparticle inclusion; however, slip velocity at the surface enhances both the heat transfer rate and the efficiency
Authors	**Geometry of study**	**Type of nanofluid**	**Properties**	**Remarks**
Malvandi, Ganji [81]	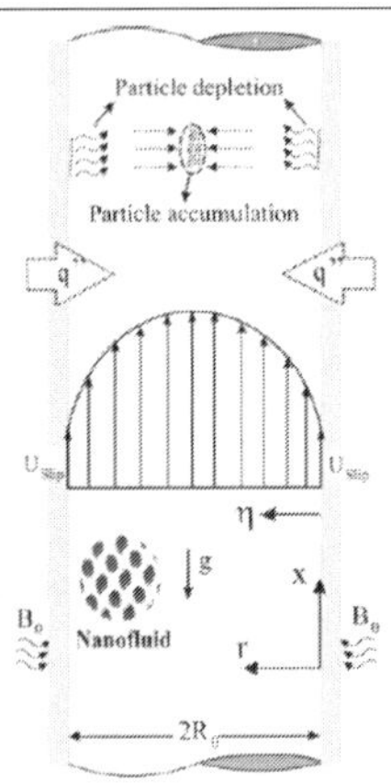	Al_2O_3 /water nanofluid	$1\,nm \le d_p \le 100\,nm$ $0 \le \phi \le 0.1$ $0.1 \le N_{BT} \le 10$ $0 \le Ha \le 10$	The heat transfer rate is enhanced by the presence of the magnetic field especially for the smaller nanoparticles. Moreover, as the magnetic field strength (Ha) intensifies, the peak of the velocity profile near the walls is increased; however, the peak of the velocity profile at the core region is decreased.

Authors	Geometry of study	Type of nanofluid	Properties	Remarks
Malvandi, Ganji [82]	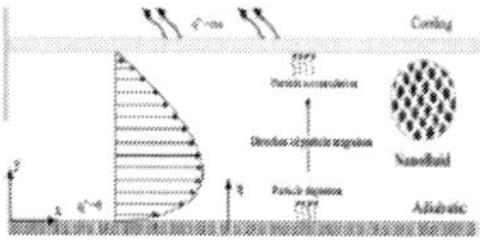	Al_2O_3 /water nanofluid	$1\,nm \le d_p \le 100\,nm$ $0 \le \phi \le 0.1$ $4 \le N_{BT} \le 10$	The concentration of nanoparticles is higher near the cold wall (nanoparticles accumulation), while it is lower near the adiabatic wall (nanoparticles depletion). Also, there is an optimum value for the bulk mean of nanoparticle volume fraction in which the heat transfer rate is maximum. This optimum value decreases for smaller nanoparticles.
Malvandi et al. [83]	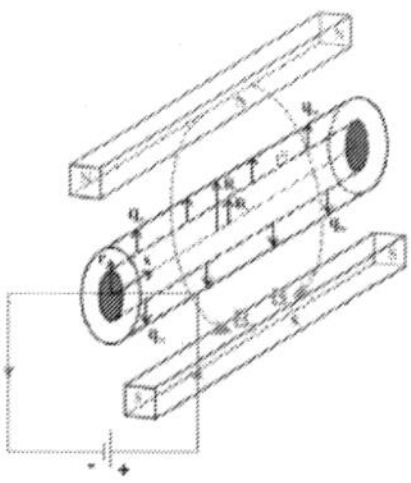	Al_2O_3 /water nanofluid	$1\,nm \le d_p \le 100\,nm$ $0 \le \phi \le 0.1$ $0.1 \le N_{BT} \le 10$ $0 \le Ha \le 15$	In the presence of the magnetic field, the velocity gradients near the wall grow, which increases the slip velocity at boundaries and thus the heat transfer rate intensifies. What is more? Increasing Ha (intensifying the magnetic field) leads to an increase in the Lorentz force (a retarding force to the transport phenomena), which tends to resist the fluid flow and thus reduces the flow's velocity.
Malvandi, Ganji [84]	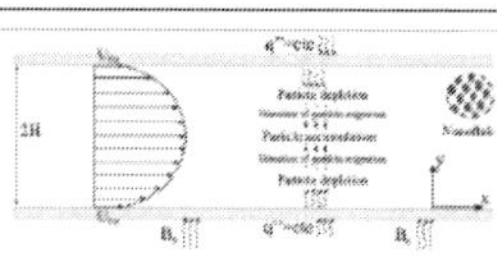	Al_2O_3 /water nanofluid	$1\,nm \le d_p \le 100\,nm$ $0 \le \phi \le 0.1$ $0.2 \le N_{BT} \le 10$ $0 \le Ha \le 10$	Obtained results indicated that nanoparticles move from the heated walls (nanoparticles depletion) toward the core region of the channel (nanoparticles accumulation) and construct a non-uniform nanoparticle distribution. Moreover, in the presence of the magnetic field, the near-wall velocity gradients increase, enhancing the slip velocity and thus the heat transfer rate and pressure drop increase.
Malvandi, Ganji [85]	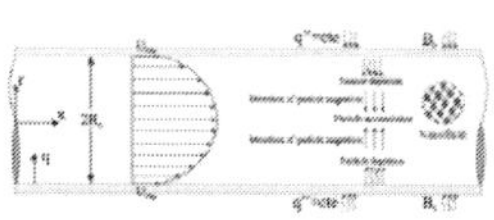	Al_2O_3 /water nanofluid	$1\,nm \le d_p \le 100\,nm$ $0 \le \phi \le 0.1$ $0.2 \le N_{BT} \le 10$ $0 \le Ha \le 10$	Nanoparticles concentration is higher in the core region of the microchannel (nanoparticles accumulation) while taking its minimum values closer to the

				heated wall (nanoparticles depletion). That is to say, nanoparticles move from the heated wall toward the core region and construct a non-uniform nanoparticle distribution.
Moshizi et al. [86]	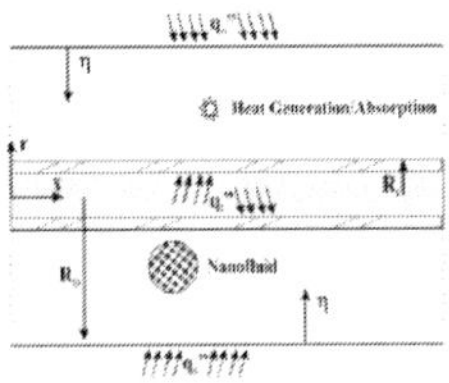	Al_2O_3 /water nanofluid	$1\,nm \le d_p \le 100\,nm$ $0 \le \phi \le 0.1$ $0.2 \le N_{BT} \le 10$	In the case of heat absorption, by imposing heatflux at both walls, the dimensional temperature profile becomes to be more uniform. The variations on the heat transfer coefficient enhancement in the case of heat absorption are smaller than in the case of heat generation, for a moderate range of NBT. Furthermore, the heat absorption boosts the pressure drops of nanofluid.

Authors	Geometry of study	Type of nanofluid	Properties	Remarks
Malvandi, Ganji [87]	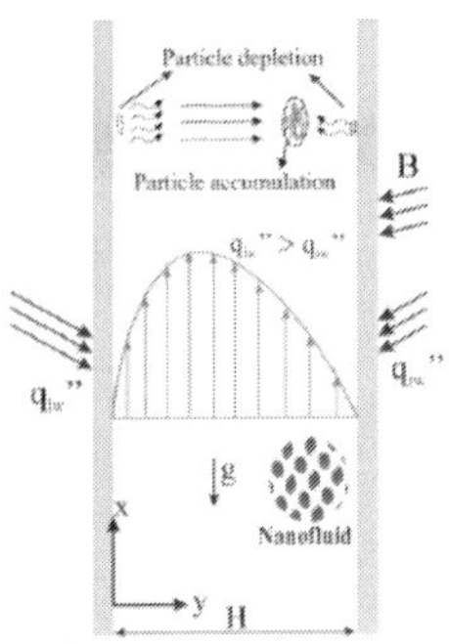	Al_2O_3 /water nanofluid	$1\,nm \le d_p \le 100\,nm$ $0 \le \phi \le 0.1$ $0.2 \le N_{BT} \le 10$ $0 \le Ha \le 10$	It is shown that nanoparticles eject themselves from the heated walls, construct a depleted region, and accumulate in the core region, but they are more likely to accumulate toward the wall with the lower heat flux.
Malvandi et al. [88]	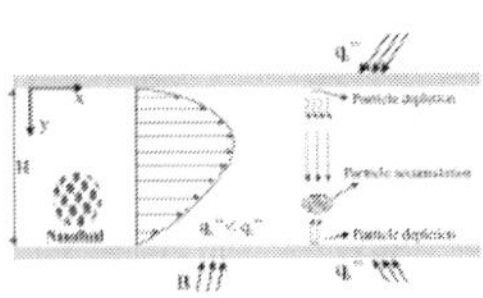	Al_2O_3 /water nanofluid	$1\,nm \le d_p \le 100\,nm$ $0 \le \phi \le 0.1$ $0.2 \le N_{BT} \le 10$ $0 \le Ha \le 10$	The non-uniform nanoparticle distribution makes the velocities move toward the wall with the higher heat flux and enhances the heat transfer rate there. In addition, it is shown that the advantage of nanoparticle inclusion is increased in the presence of a magnetic field, though heat transfer enhancement is decreased.

Authors	Geometry of study	Type of nanofluid	Properties	Remarks
Malvandi et al. [89]	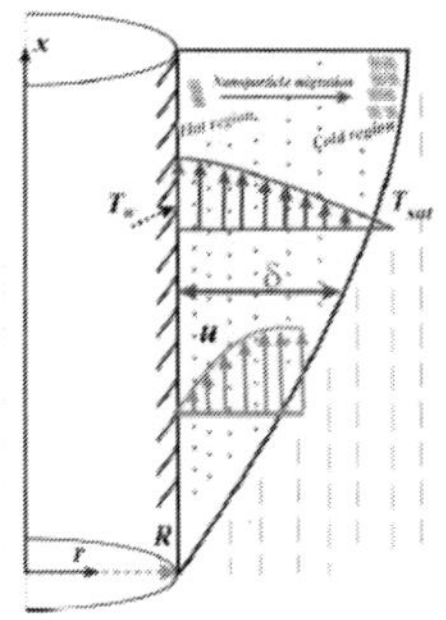	Al_2O_3 and TiO_2/water-based nanofluids	$1\,nm \le d_p \le 100\,nm$ $0 \le \phi \le 0.06$ $0.2 \le N_{BT} \le 10$	The heat transfer enhancement of titania-water nanofluids is completely insignificant relative to such enhancement for alumina-water nanofluid. Therefore, alumina-water nanofluid exhibits a better performance compared to titania-water nanofluids.
Malvandi [90]	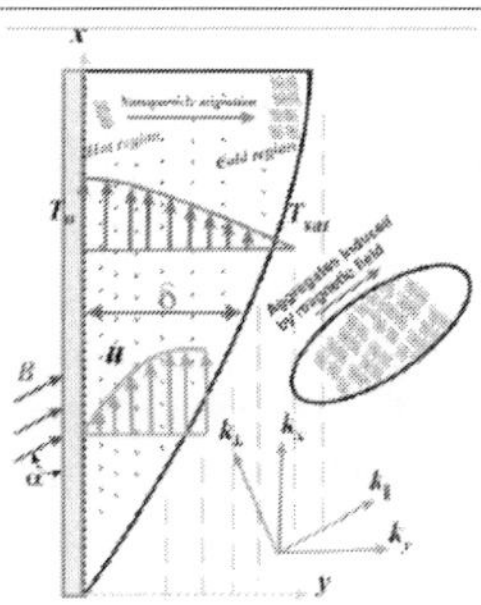	Fe_3O_4 -water nanofluid	$1\,nm \le d_p \le 100\,nm$ $0 \le \phi \le 0.06$ $0.2 \le N_{BT} \le 10$ $0 \le Ha \le 5$ $45° < \alpha < 90°$	A closed-form expression for the distribution of nanoparticle volume fraction has been obtained. It has been shown that the heat transfer rate is improved further when an external magnetic field exerts in the direction of the temperature gradient.

Sheikhzadeh et al. [76] studied the effects of Brownian motion, thermophoresis, and Dufour (transport model) on laminar-free convection heat transfer of alumina-water nanofluid flow in a square enclosure. Variable thermophysical properties utilized for fluid characterization and the governing equations were discretized using FVM. The results illustrated that the Dufour effect on heat transfer is not significant. In addition, a comparison between experimental data and numerical results revealed that the transport model is in better agreement with experimental results, compared to single-phase model.

Using the same method, Bahiraei et al. [77] studied the laminar convection heat transfer of alumina-water nanofluid inside a circular tube, considering particle migration effects. The results showed that with the Reynolds number or volume fraction augmentation, the average heat transfer coefficient enhances. In addition, it was reported that by considering the particle migration effect, higher heat transfer coefficient would be achieved.

Using modified Buongiorno model, Malvandi et al. [78] investigated MHD mixed convection heat transfer for Al_2O_3-water nanofluid inside a vertical annular pipe. The governing equations reduced to two-point O.D.E.s, which were solved by means of the Runge-Kutta-Fehlberg scheme. The obtained results indicated that the excellence of using nanofluids for heat transfer enhancement purpose is diminished by the presence of a magnetic field. Moreover, it was noted that the imposed thermal asymmetry may change the direction of nanoparticle migration, and, hence, alters the velocity, temperature, and nanoparticle concentration profiles. Table 1 shows some new works on modified Buongiorno model.

2.3.3. Other Approaches

In some other studies, novel numerical approaches have been employed to solve the governing equations of nanofluids. SPH method has been used by Mansour and Bakier [91] to study free convection within an enclosed cavity filled with Al_2O_3 nanoparticles. The left and right walls of the cavity had a complex-wavy geometry while upper and lower walls were both flat and insulated. Complex-wavy walls were modeled as the superposition of two sinusoidal functions. The results revealed that heat transfer performance may be optimized by tuning the wavy-surface geometry parameter in accordance with the Rayleigh number. Using optimal homotopy analysis method (OHAM), Nadeem et al. [92] examined 2-D stagnation point flow of a non-Newtonian Casson nanofluid over a convective-stretching surface. The governing non-linear partial differential equations were converted into non-linear ordinary differential equations and solved analytically using OHAM. The results showed that heat transfer rate is an increasing function of the stretching parameter, Prandtl and Biot numbers and it decreases with an increase in non-Newtonian parameter, Brownian motion, and thermophoresis.

The laminar axisymmetric flow of a nanofluid over a non-linearly stretching sheet was studied by Mustafa et al. [93], both numerically and analytically. The simultaneous effects of Brownian motion and thermophoretic diffusion of nanoparticles were taken into account. The numerical solution was computed by employing implicit finite difference scheme known as Keller-Box method. The results obtained from both solutions were in excellent agreement with each other. The results demonstrated that the effect of Brownian motion on fluid temperature and wall heat transfer rate is insignificant. Moreover, it was reported that increases in Schmidt number lead to a thinner nanoparticle volume fraction boundary layer.

3. CONCLUSION

A comprehensive review of popular methods in the simulation of the nanofluids was carried out. Different CFD approaches including single-phase, multiphase, and other methods were reviewed. For each model, the governing equations and recent literature were studied.

Conventional single-phase model was the most common method to study the convective heat transfer of nanofluids. This can be due to the fact that this model simplifies the simulation and in comparison to other models has the lowest computational cost. However, the results obtained from this model may have some deviation from the experimental data. For instance, it was reported in many studies that homogeneous model underestimates the heat transfer coefficient and Nusselt number, when compared to the dispersion and two-phase models. However, it was also revealed that using the temperature-dependent thermophysical properties in homogeneous model can lead to more realistic results. On the other hand, dispersion model for both constant and temperature-dependent properties showed promising results, compared with experimental data. This model requires less computational time compared to two-phase model. In addition, the model takes into account thermal dispersion effect, which leads to more reliable results in comparison with the homogeneous model.

Nanofluids are inherently multiphase fluids; therefore, employing two-phase model taking into account the slip velocity, Brownian motion, thermophoresis, and so forth, can lead to more appropriate results. Most of the publications confirmed that different two-phase models predict more accurate results than the homogeneous model. Also, higher values of the heat transfer coefficient were reported for two-phase models, compared to conventional single-phase model. A vast number of studies utilized the mixture and Eulerian models, and to smaller extent VOF and Lagrangian-Eulerian models. Some publications noted that among all two-phase models, mixture model predicts more precise results compared with experimental data. However, this model has some limitation and cannot be applied in some cases. On the other hand, since VOF and Lagrangian-Eulerian models are employed less than other two-phase models, it seems that further research might be needed to assess their precision in nanofluids simulation.

In the end, LBM and non-homogeneous two-component models are rather novel approaches, used in several cases. The results predicted by these approaches showed a promising accordance with the results obtained from previous studies. Moreover, according to literature, these methods may present some well-known advantages in the modeling of nanofluids. Obviously, more attempts should be made to find the flow characteristics of nanofluids in various systems and in the presence of different modes of heat transfer in order to examine the aforementioned approaches.

4. ACKNOWLEDGEMENTS

The authors gratefuly acknowledge High Impact Research Grant UM.C/HIR/MOHE/ENG/23 and the Faculty of Engineering, University of Malaya, Malaysia, for support in conducting this research work.

REFERENCES

1. Yarmand H, Gharehkhani S, Kazi SN, Sadeghinezhad E, Safaei MR. Numerical investigation of heat transfer enhancement in a rectangular heated pipe for turbulent nanofluid. The Scientific World Journal. 2014;2014(Article ID 369593):1–9. doi:10.1155/2014/369593.
2. Mojumder S, Sourav S, Sumon S, Mamun M. Combined effect of Reynolds and Grashof numbers on mixed convection in a lid-driven T-shaped cavity filled with water-Al2O3 nanofluid. Journal of Hydrodynamics, Ser B. 2015;27(5):782–94.
3. Abu-Nada E, Chamkha AJ. Mixed convection flow in a lid-driven inclined square enclosure filled with a nanofluid. European Journal of Mechanics-B/Fluids. 2010;29(6):472–82.
4. Rostamani M, Hosseinizadeh S, Gorji M, Khodadadi J. Numerical study of turbulent forced convection flow of nanofluids in a long horizontal duct considering variable properties. International Communications in Heat and Mass Transfer. 2010;37(10):1426–31.
5. Pak BC, Cho YI. Hydrodynamic and heat transfer study of dispersed fluids with submicron metallic oxide particles. Experimental Heat Transfer an International Journal. 1998;11(2):151–70.
6. Maiga SEB, Palm SJ, Nguyen CT, Roy G, Galanis N. Heat transfer enhancement by using nanofluids in forced convection flows. International Journal of Heat and Fluid Flow. 2005;26(4):530–46.
7. Abu-Nada E. Rayleigh-Bénard convection in nanofluids: effect of temperature dependent properties. International Journal of Thermal Sciences. 2011;50(9):1720–30.
8. Bouhalleb M, Abbassi H. Numerical Investigation Of Heat Transfer By CuO–Water Nanofluid In Rectangular Enclosures. Heat Transfer Engineering. 2016;37(1):13–23.
9. Togun H, Safaei MR, Sadri R, Kazi SN, Badarudin A, Hooman K et al. Numerical simulation of laminar to turbulent nanofluid flow and heat transfer over a backward-facing step. Applied Mathematics and Computation. 2014;239:153–70.
10. Yarmand H, Ahmadi G, Gharehkhani S, Kazi SN, Safaei MR, Alehashem MS et al. Entropy generation during turbulent flow of zirconia-water and other nanofluids in a square cross section tube with a constant heat flux. Entropy. 2014;16(11):6116–32.
11. Togun H, Ahmadi G, Abdulrazzaq T, Shkarah AJ, Kazi S, Badarudin A et al. Thermal performance of nanofluid in ducts with double forward-facing steps. Journal of the Taiwan Institute of Chemical Engineers. 2015;47:28–42.
12. Selimefendigil F, Öztop HF, Chamkha AJ. Fluid-structure-magnetic field interaction in a nanofluid filled lid-driven cavity with flexible side wall. European Journal of Mechanics – B/Fluids. doi:http://dx.doi.org/10.1016/j.euromechflu.2016.03.009.
13. Selimefendigil F, Öztop HF. Conjugate natural convection in a cavity with a conductive partition and filled with different nanofluids on different sides of the partition. Journal of Molecular Liquids. 2016;216:67–77. doi:http://dx.doi.org/10.1016/j.molliq.2015.12.102.

14. Selimefendigil F, Öztop HF, Chamkha AJ. Mhd mixed convection and entropy generation of nanofluid filled lid driven cavity under the influence of inclined magnetic fields imposed to its upper and lower diagonal triangular domains. Journal of Magnetism and Magnetic Materials. 2016;406:266–81. doi:http:// dx.doi.org/10.1016/j.jmmm.2016.01.039.
15. Buddakkagari V, Kumar M. Transient boundary layer laminar free convective flow of a nanofluid over a vertical cone/plate. International Journal of Applied and Computational Mathematics. 2015;1(3):427–48.
16. Jahanbin AH, Javaherdeh K. Numerical investigation of CuO nanoparticles effect on forced convective heat transfer inside a mini-channel: comparison of different approaches. Life Science Journal. 2013;10(8s):183–189.
17. Lotfi R, Saboohi Y, Rashidi A. Numerical study of forced convective heat transfer of nanofluids: comparison of different approaches. International Communications in Heat and Mass Transfer. 2010;37(1):74–8.
18. Ding Y, Wen D. Particle migration in a flow of nanoparticle suspensions. Powder Technology. 2005;149(2):84–92.
19. Nikkhah Z, Karimipour A, Safaei MR, Forghani-Tehrani P, Goodarzi M, Dahari M et al. Forced convective heat transfer of water/functionalized multi-walled carbon nanotube nanofluids in a microchannel with oscillating heat flux and slip boundary condition. International Communications in Heat and Mass Transfer. 2015;68:69–77.
20. Safaei M, Togun H, Vafai K, Kazi S, Badarudin A. Investigation of heat transfer enhancement in a forward-facing contracting channel using FMWCNT nanofluids. Numerical Heat Transfer, Part A: Applications. 2014;66(12):1321–40.
21. Xuan Y, Roetzel W. Conceptions for heat transfer correlation of nanofluids. International Journal of heat and Mass transfer. 2000;43(19):3701–7.
22. Plumb O, editor. The effect of thermal dispersion on heat transfer in packed bed boundary layers. ASME/JSME Thermal Engineering Joint Conference; 1983.
23. Hunt M, Tien C. Effects of thermal dispersion on forced convection in fibrous media. International Journal of Heat and Mass Transfer. 1988;31(2):301–9.
24. Khaled A-R, Vafai K. Heat transfer enhancement through control of thermal dispersion effects. International Journal of Heat and Mass Transfer. 2005;48(11):2172–85.
25. Mojarrad MS, Keshavarz A, Shokouhi A. Nanofluids thermal behavior analysis using a new dispersion model along with single-phase. Heat and Mass Transfer. 2013;49(9):1333–43.
26. Zeinali Heris S, Noie SH, Talaii E, Sargolzaei J. Numerical investigation of Al2O3/water nanofluid laminar convective heat transfer through triangular ducts. Nanoscale Research Letters. 2011;6(1):1.
27. Zeinali Heris S, Kazemi-Beydokhti A, Noie S, Rezvan S. Numerical study on convective heat transfer of Al2O3/water, CuO/water and Cu/water nanofluids through square cross-section duct in laminar flow. Engineering Applications of Computational Fluid Mechanics. 2012;6(1):1–14.
28. Heris SZ, Etemad SG, Esfahany MN. Experimental investigation of oxide nanofluids laminar flow convective heat transfer. International Communications in Heat and Mass Transfer. 2006;33(4):529–35.
29. Zeinali Heris S, Esfahany MN, Etemad SG. Experimental investigation of convective heat transfer of Al2O3/water nanofluid in circular tube. International Journal of Heat and Fluid Flow. 2007;28(2):203–10.
30. Kumar S, Prasad SK, Banerjee J. Analysis of flow and thermal field in nanofluid using a single phase thermal dispersion model. Applied Mathematical Modelling. 2010;34(3):573–92.
31. Akbaridoust F, Rakhsha M, Abbassi A, Saffar-Avval M. Experimental and numerical investigation of nanofluid heat transfer in helically coiled tubes at constant wall temperature using dispersion model. International Journal of Heat and Mass Transfer. 2013;58(1):480–91.

32. Safaei M, Mahian O, Garoosi F, Hooman K, Karimipour A, Kazi S et al. Investigation of micro-and nanosized particle erosion in a 90° pipe bend using a two-phase discrete phase model. The Scientific World Journal. 2014;2014(Article ID 740578):1–12. doi:10.1155/2014/740578.
33. Minkowycz W, Sparrow EM, Murthy J. Handbook of numerical heat transfer. Wiley Online Library; New York 1988.
34. Mirzaei M, Saffar Avval M, Naderan H. Heat transfer investigation of laminar developing flow of nanofluids in a microchannel based on Eulerian–Lagrangian approach. The Canadian Journal of Chemical Engineering. 2014;92(6):1139–49.
35. Oosthuizen PH, Carscallen WE. Introduction to compressible fluid flow. Boca Raton, FL: CRC Press; 2013.
36. Bahremand H, Abbassi A, Saffar-Avval M. Experimental and numerical investigation of turbulent nanofluid flow in helically coiled tubes under constant wall heat flux using Eulerian–Lagrangian approach. Powder Technology. 2015;269:93–100.
37. Moraveji MK, Esmaeili E. Comparison between single-phase and two-phases CFD modeling of laminar forced convection flow of nanofluids in a circular tube under constant heat flux. International Communications in Heat and Mass Transfer. 2012;39(8):1297–302.
38. Tahir S, Mital M. Numerical investigation of laminar nanofluid developing flow and heat transfer in a circular channel. Applied Thermal Engineering. 2012;39:8–14.
39. Behroyan I, Ganesan P, He S, Sivasankaran S. Turbulent forced convection of Cu-water nanofluid: CFD model comparison. International Communications in Heat and Mass Transfer. 2015;67:163–72.
40. Vanaki SM, Ganesan P, Mohammed H. Numerical study of convective heat transfer of nanofluids: a review. Renewable and Sustainable Energy Reviews. 2016;54:1212–39.
41. Xu P, Wu Z, Mujumdar A, Yu B. Innovative hydrocyclone inlet designs to reduce erosion-induced wear in mineral dewatering processes. Drying Technology. 2009;27(2):201–11.
42. Akbari M, Galanis N, Behzadmehr A. Comparative analysis of single and two-phase models for Cfd studies of nanofluid heat transfer. International Journal of Thermal Sciences. 2011;50(8):1343–54.
43. Akbari M, Galanis N, Behzadmehr A. Comparative assessment of single and two-phase models for numerical studies of nanofluid turbulent forced convection. International Journal of Heat and Fluid Flow. 2012;37:136–46.
44. Behzadmehr A, Saffar-Avval M, Galanis N. Prediction of turbulent forced convection of a nanofluid in a tube with uniform heat flux using a two phase approach. International Journal of Heat and Fluid Flow. 2007;28(2):211–9.
45. Hejazian M, Moraveji MK, Beheshti A. Comparative numerical investigation on TiO2/water nanofluid turbulent flow by implementation of single phase and two phase approaches. Numerical Heat Transfer, Part A: Applications. 2014;66(3):330–48.
46. Naphon P, Nakharintr L. Turbulent two phase approach model for the nanofluids heat transfer analysis flowing through the minichannel heat sinks. International Journal of Heat and Mass Transfer. 2015;82:388–95.
47. Hanafizadeh P, Ashjaee M, Goharkhah M, Montazeri K, Akram M. The comparative study of single and two-phase models for magnetite nanofluid forced convection in a tube. International Communications in Heat and Mass Transfer. 2015;65:58–70.
48. El-Batsh H, Doheim M, Hassan A. On the application of mixture model for two-phase flow induced corrosion in a complex pipeline configuration. Applied Mathematical Modelling. 2012;36(11):5686–99.
49. Moraveji MK, Ardehali RM. Cfd modeling (comparing single and two-phase approaches) on thermal performance of Al2O3/water nanofluid in mini-channel heat sink. International Communications in Heat and Mass Transfer. 2013;44:157–64.

50. Bahiraei M. A comprehensive review on different numerical approaches for simulation in nanofluids: traditional and novel techniques. Journal of Dispersion Science and Technology. 2014;35(7):984–96.
51. Goodarzi M, Safaei M, Vafai K, Ahmadi G, Dahari M, Kazi S et al. Investigation of nanofluid mixed convection in a shallow cavity using a two-phase mixture model. International Journal of Thermal Sciences. 2014;75:204–20.
52. Bianco V, Manca O, Nardini S. Numerical investigation on nanofluids turbulent convection heat transfer inside a circular tube. International Journal of Thermal Sciences. 2011;50(3):341–9.
53. Shariat M, Moghari RM, Akbarinia A, Rafee R, Sajjadi S. Impact of nanoparticle mean diameter and the buoyancy force on laminar mixed convection nanofluid flow in an elliptic duct employing two phase mixture model. International Communications in Heat and Mass Transfer. 2014;50:15–24.
54. Corcione M, Cianfrini M, Quintino A. Two-phase mixture modeling of natural convection of nanofluids with temperature-dependent properties. International Journal of Thermal Sciences. 2013;71:182–95.
55. Naphon P, Nakharintr L. Numerical investigation of laminar heat transfer of nanofluid-cooled mini-rectangular fin heat sinks. Journal of Engineering Physics and Thermophysics. 2015;88(3):666–75.
56. Siavashi M, Bahrami HRT, Saffari H. Numerical investigation of flow characteristics, heat transfer and entropy generation of nanofluid flow inside an annular pipe partially or completely filled with porous media using two-phase mixture model. Energy. 2015;93:2451–66.
57. Schiller L, Naumann Z. A drag coefficient correlation. Vdi Zeitung. 1935;77(318):51.
58. Drew D, Lahey R. Analytical modeling of multiphase flow. In: Particulate two-phase flow. Butterworth–Heinemann Boston 1993:509–66.
59. Kalteh M, Abbassi A, Saffar-Avval M, Harting J. Eulerian–Eulerian two-phase numerical simulation of nanofluid laminar forced convection in a microchannel. International journal of heat and fluid flow. 2011;32(1):107–16.
60. Chen Y-j, Li Y-y, Liu Z-h. Numerical simulations of forced convection heat transfer and flow characteristics of nanofluids in small tubes using two-phase models. International Journal of Heat and Mass Transfer. 2014;78:993–1003.
61. Göktepe S, Atalık K, Ertürk H. Comparison of single and two-phase models for nanofluid convection at the entrance of a uniformly heated tube. International Journal of Thermal Sciences. 2014;80:83–92.
62. Sadeghi R, Haghshenasfard M, Etemad SG, Keshavarzi E. Theoretical investigation of nanoparticles aggregation effect on water-alumina laminar convective heat transfer. International Communications in Heat and Mass Transfer. 2016;72:57–63.
63. Succi S. Lattice Boltzmann equation: failure or success? Physica A: Statistical Mechanics and its Applications. 1997;240(1):221–8.
64. Karimipour A, Esfe MH, Safaei MR, Semiromi DT, Jafari S, Kazi S. Mixed convection of copper–water nanofluid in a shallow inclined lid driven cavity using the Lattice Boltzmann method. Physica A: Statistical Mechanics and its Applications. 2014;402:150–68.
65. Zhang J. Lattice Boltzmann method for microfluidics: models and applications. Microfluidics and Nanofluidics. 2011;10(1):1–28.
66. Xuan Y, Yao Z. Lattice Boltzmann model for nanofluids. Heat and mass transfer. 2005;41(3):199–205.
67. Qi C, He Y, Yan S, Tian F, Hu Y. Numerical simulation of natural convection in a square enclosure filled with nanofluid using the two-phase Lattice Boltzmann method. Nanoscale Research Letters. 2013;8(1):1–16.
68. Karimipour A, Nezhad AH, D'Orazio A, Esfe MH, Safaei MR, Shirani E. Simulation of copper–water nanofluid in a microchannel in slip flow regime using the Lattice Boltzmann method. European Journal of Mechanics-B/Fluids. 2015;49:89–99.

69. Zhang T, Che D. Double Mrt thermal Lattice Boltzmann simulation for MHD natural convection of nanofluids in an inclined cavity with four square heat sources. International Journal of Heat and Mass Transfer. 2016;94:87–100.
70. Kamyar A, Saidur R, Hasanuzzaman M. Application of computational fluid dynamics (CFD) for nanofluids. International Journal of Heat and Mass Transfer. 2012;55(15–16):4104–15.
71. Sidik NAC, Razali SA. Lattice Boltzmann method for convective heat transfer of nanofluids–a review. Renewable and Sustainable Energy Reviews. 2014;38:864–75.
72. Buongiorno J. Convective transport in nanofluids. Journal of Heat Transfer. 2006;128(3):240–50.
73. Malvandi A, Ganji D. Effects of nanoparticle migration on water/alumina nanofluid flow inside a horizontal annulus with a moving core. Journal of Mechanics. 2015;31(03):291–305.
74. Hedayati F, Malvandi A, Kaffash M, Ganji D. Fully developed forced convection of alumina/water nanofluid inside microchannels with asymmetric heating. Powder Technology. 2015;269:520–31.
75. Malvandi A, Moshizi SA, Ganji DD. Two-component heterogeneous mixed convection of alumina/water nanofluid in microchannels with heat source/sink. Advanced Powder Technology. 2016;27(1):245–54. doi:http://dx.doi.org/ 10.1016/ j.apt. 2015.12.009.
76. Sheikhzadeh GA, Dastmalchi M, Khorasanizadeh H. Effects of nanoparticles transport mechanisms on Al2O3–water nanofluid natural convection in a square enclosure. International Journal of Thermal Sciences. 2013;66:51–62.
77. Bahiraei M, Mostafa Hosseinalipour S, Hangi M. Prediction of convective heat transfer of Al2O3-water nanofluid considering particle migration using neural network. Engineering Computations. 2014;31(5):843–63.
78. Malvandi A, Safaei M, Kaffash M, Ganji D. Mhd mixed convection in a vertical annulus filled with Al2O3–water nanofluid considering nanoparticle migration. Journal of Magnetism and Magnetic Materials. 2015;382:296–306.
79. Malvandi A. The unsteady flow of a nanofluid in the stagnation point region of a time-dependent rotating sphere. Thermal Science. 2015;19(5):1603–12.
80. Malvandi A, Moshizi S, Soltani EG, Ganji D. Modified Buongiorno's model for fully developed mixed convection flow of nanofluids in a vertical annular pipe. Computers & Fluids. 2014;89:124–32.
81. Malvandi A, Ganji DD. Magnetohydrodynamic mixed convective flow of Al2O3–water nanofluid inside a vertical microtube. Journal of Magnetism and Magnetic Materials. 2014;369(0):132–41. doi:http://dx.doi.org/10.1016/j.jmmm.2014.06.037.
82. Malvandi A, Ganji D. Effects of nanoparticle migration on force convection of alumina/water nanofluid in a cooled parallel-plate channel. Advanced Powder Technology. 2014;25(4):1369–75.
83. Malvandi A, Moshizi SA, Ganji DD. Effect of magnetic fields on heat convection inside a concentric annulus filled with Al2O3–water nanofluid. Advanced Powder Technology. doi:http://dx.doi.org/10.1016/j.apt.2014.07.013 2016; *25*(6):1817-1824..
84. Malvandi A, Ganji DD. Magnetic field effect on nanoparticles migration and heat transfer of water/alumina nanofluid in a channel. Journal of Magnetism and Magnetic Materials. 2014;362:172–9. doi:http://dx.doi.org/ 10.1016/ j.jmmm. 2014.03.014.
85. Malvandi A, Ganji DD. Brownian motion and thermophoresis effects on slip flow of alumina/water nanofluid inside a circular microchannel in the presence of a magnetic field. International Journal of Thermal Sciences. 2014;84:196–206. doi:http://dx.doi.org/10.1016/j.ijthermalsci.2014.05.013.
86. Moshizi SA, Malvandi A, Ganji DD, Pop I. A two-phase theoretical study of Al2O3–water nanofluid flow inside a concentric pipe with heat generation/absorption. International Journal of Thermal Sciences. 2014;84:347–57. doi:http://dx.doi.org/10.1016/j.ijthermalsci.2014.06.012.

87. Malvandi A, Ganji D. Effects of nanoparticle migration on hydromagnetic mixed convection of alumina/water nanofluid in vertical channels with asymmetric heating. Physica E: Low-dimensional Systems and Nanostructures. 2015;66:181–96.
88. Malvandi A, Ganji D, Kaffash M. Magnetic field effects on nanoparticle migration and heat transfer of alumina/water nanofluid in a parallel-plate channel with asymmetric heating. The European Physical Journal Plus. 2015;130(4):1–21.
89. Malvandi A, Heysiattalab S, Ganji D. Thermophoresis and Brownian motion effects on heat transfer enhancement at film boiling of nanofluids over a vertical cylinder. Journal of Molecular Liquids. 2016;216:503–9.
90. Malvandi A. Film boiling of magnetic nanofluids (MNFs) over a vertical plate in presence of a uniform variable-directional magnetic field. Journal of Magnetism and Magnetic Materials. 2016;406:95–102.
91. Mansour M, Bakier M. Free convection heat transfer in complex-wavy-wall enclosed cavity filled with nanofluid. International Communications in Heat and Mass Transfer. 2013;44:108–15.
92. Nadeem S, Mehmood R, Akbar NS. Optimized analytical solution for oblique flow of a casson-nano fluid with convective boundary conditions. International Journal of Thermal Sciences. 2014;78:90–100.
93. Mustafa M, Khan JA, Hayat T, Alsaedi A. Analytical and numerical solutions for axisymmetric flow of nanofluid due to non-linearly stretching sheet. International Journal of Non-Linear Mechanics. 2015;71:22–9.

CHAPTER 14

Open-Source Tools for Dynamical Analysis of Liley's Mean-Field Cortex Model

Kevin R. Green*, Lennaert van Veen

Faculty of Science, University of Ontario Institute of Technology, 2000 Simcoe Street North, Oshawa, L1H 7K4 Ontario, Canada

ABSTRACT

Mean-field models of the mammalian cortex treat this part of the brain as a two-dimensional excitable medium. The electrical potentials, generated by the excitatory and inhibitory neuron populations, are described by nonlinear, coupled, partial differential equations that are known to generate complicated spatio-temporal behaviour. We focus on the model by Liley et al. (Network: Computation in Neural Systems 13 (2002) 67–113). Several reductions of this model have been studied in detail, but a direct analysis of its spatio-temporal dynamics has, to the best of our knowledge, never been attempted before. Here, we describe the implementation of implicit time-stepping of the model and the tangent linear model, and solving for equilibria and time-periodic solutions, using the open-source library PETSc. By using domain decomposition for parallelization, and iterative solving of linear problems, the code is capable of parsing some dynamics of a macroscopic slice of cortical tissue with a sub-millimetre resolution.

Keywords: Mean-field modeling Hyperbolic partial differential equations Numerical partial differential equations 35Q92 65Y05

1. INTRODUCTION

Models of cortical dynamics come in two main families: neuronal network models and mean-field models. The former describe many interacting neurons, each with their own dynamical rules, while the latter describe electrical potentials, generated collectively by many neurons, as continuous in space and time. These potentials can be thought of as averages over a number of macrocolumns, groups of hundreds of thousands of neurons in columnar structures at the surface of the cortex. Both of these modelling approaches can be classified as *forward*: they attempt to predict the future state of the cortex, given the current state and a set of physiological parameter values. A complementary approach, that can be called *backward*, is to divide the cortex into interacting components that can be regarded as functional units, and compute the strength of coupling between these units. The backward approach is often based on functional Magnetic Resonance Imaging (fMRI) experiments. A promising modelling strategy is to describe each functional component with a neuronal network or mean-field model, and then have them interact according to empirically determined coupling, thus combining the forward and backward approaches [1].

When considering the forward modelling of a macroscopic piece of cortical tissue, a fundamental difference between the neuronal network and mean-field families is apparent. A model of the former kind should describe billions of neurons, and many times more connections between them. As demonstrated by recent publications, such as by Izhikevich and Edelman [2] or by the Blue Brain team [3], progress in super computing allows for the simulation of ever larger neuronal networks, that reflect actual brain dynamics. However, it is hard to see how the output of such models can be analyzed, other than by purely statistical techniques. Models of the latter kind, in contrast, can be analyzed as smooth, infinite-dimensional dynamical systems.

An added advantage of the mean-field approach is that the electrical potentials, which appear as dependent variables, are observable, macroscopic quantities. An indirect measurement of these fields is provided by the electroencephalograph (EEG) [4]. The EEG is usually measured with electrodes on the scalp or, in exceptional circumstances, directly on the surface of the brain. In either case, the measured signal is not that of individual neurons, but that of many neurons, spread out over a few square centimetres or millimetres. Thus, the way the signals of individual neurons are smeared out by the spatial averaging of mean-field modelling is similar to the way they are mixed up in EEG measurements. Because of the link between the local mean potential and the EEG, mean-field models are sometimes called EEG models (e.g. [5,6]). The geometry of the cortical surface, however, is not taken into consideration in doing so. This surface is folded, and electrocortical activity will result in different EEG signals depending on the location and orientation of the generating tissue. A more direct link between the model variables and measurements may be given by the Local Field Potential (LFP), which is typically measured *in vitro* or under anaesthesia.

The origin of mean-field modelling lies in the 1970s, when pioneers like Freeman [7], Wilson and Cowan [8] and Lopes da Silva et al. [9] started to model components of the human cortex with continuous fields. Over the past four decades, mean-field models have been used to study a range of open questions in neuroscience, such as the generation of the alpha rhythm, 8–13 Hz oscillations in the EEG (see, e.g., [9,5]), epilepsy (see, e.g., [10–12]) and anaesthesia [6]. In a different context, they are used in models for sensorimotor control, pattern discrimination and target tracking [13]. As discussed above, mean-field models also appear as components of combined forward-backward models that aim to capture the functioning of the cortex as a whole, such as in Honey et al. [14].

Although mean-field models have been used in all these settings, little analysis has been done on their behaviour as spatially extended dynamical systems. In part, this is due to their staggering complexity. The Liley model [15] considered here, for instance, consists of fourteen coupled Partial Differential Equations (PDEs) with strong nonlinearities, imposed by coupling between the mean membrane potentials and the mean synaptic inputs. The model can be reduced to a system of Ordinary Differential Equations (ODEs) by considering only spatially homogeneous solutions, and the resulting system has been examined in detail using numerical bifurcation analysis (see [16] and references therein). In order to compute equilibria, periodic orbits and such objects for the PDE model, we need a flexible, stable simulation code for the model and its linearization that can run in parallel to scale up to a domain size of about 2500 cm^2, the size of a full-grown human cortex. We also need efficient, iterative solvers for linear problems with large, sparse matrices. In this paper, we will show that all this can be accomplished in the open-source software package PETSc [17]. Our implementation consists of a number of functions in C that are available publicly [18].

The goal of this computational work is to parse the spatio-temporal dynamics of a full-fledged mean-field model. We will present the numerical implementation of algorithms for the computation of equilibria and time-periodic solutions and study their stability and parameter dependence. Thus, our goal is similar to that of Coombes et al., who analyzed “spots”: rotationally symmetric, localized solutions in a model of a single neuron population in two dimensions [19]. The challenge lies in giving up the restriction to a single population, a single space dimension, or solutions with a fixed spatial symmetry.

1.1. Liley's model

The model we use was first proposed by Liley et al. [15]. The dependent variables are the mean inhibitory and excitatory membrane potential, h_i and h_e, the four mean synaptic inputs, originating from either population and connecting to either, I_{ee}, I_{ei}, I_{ie} and I_{ii}, and the excitatory axonal activity in long-range fibres, connecting to either population, ϕ_{ee} and ϕ_{ei}. The model equations are:

$$\tau_k \frac{\partial h_k(\vec{x}, t)}{\partial t} = h_k^r - h_k(\vec{x}, t) + \frac{h_{ek}^{eq} - h_k(\vec{x}, t)}{\left|h_{ek}^{eq} - h_k^r\right|} I_{ek}(\vec{x}, t) + \frac{h_{ik}^{eq} - h_k(\vec{x}, t)}{\left|h_{ik}^{eq} - h_k^r\right|} I_{ik}(\vec{x}, t) \quad (1)$$

$$\left(\frac{\partial}{\partial t} + \gamma_{ek}\right)^2 I_{ek}(\vec{x}, t) = e\Gamma_{ek}\gamma_{ek}\{N_{ek}^{\beta} S_e[h_e(\vec{x}, t)] + p_{ek} + \phi_{ek}(\vec{x}, t)\} \quad (2)$$

$$\left(\frac{\partial}{\partial t} + \gamma_{ik}\right)^2 I_{ik}(\vec{x}, t) = e\Gamma_{ik}\gamma_{ik}\{N_{ik}^{\beta} S_i[h_i(\vec{x}, t)] + p_{ik}\} \quad (3)$$

$$\left[\left(\frac{\partial}{\partial t} + v\Lambda\right)^2 - \frac{3}{2}v^2\nabla^2\right] \phi_{ek}(\vec{x}, t) = N_{ek}^{\alpha} v^2 \Lambda^2 S_e[h_e(\vec{x}, t)] \quad (4)$$

$$S_k[h_k] = S_k^{max}\left(1 + \exp\left[-\sqrt{2}\frac{h_k - \mu_k}{\sigma_k}\right]\right)^{-1} \quad (5)$$

where index $k = \{e, i\}$ denotes excitatory or inhibitory. The meaning of the parameters, along with some physiological bounds and the values used in our tests, are given in Table 1. A detailed description of these equations can be found in Refs. [15,16]. Here, we will focus on the aspects of the model most relevant for the numerical implementation.

Table 1. Meaning, ranges and values for the model parameters. The values used for the tests presented in Section 7 are taken from Ref. [20]. We make note that p_{ie} and p_{ii} are not listed here, as they are set to 0.

Parameter	Definition	Minimum	Maximum	Value	Units
her	Resting excitatory membrane potential	−80	−60	−72.293	mV
Hir	Resting inhibitory membrane potential	−80	−60	−67.261	mV
τ_e	Passive excitatory membrane decay time	5	150	32.209	ms
τ_i	Passive inhibitory membrane decay time	5	150	92.260	ms
heeeq	Excitatory reversal potential	−20	10	7.2583	mV
heieq	Excitatory reversal potential	−20	10	9.8357	mV
hieeq	Inhibitory reversal potential	−90	hkr−5	−80.697	mV
hiieq	Inhibitory reversal potential	−90	hkr−5	−76.674	mV
Γ_{ee}	EPSP peak amplitude	0.1	2.0	0.29835	mV
Γ_{ei}	EPSP peak amplitude	0.1	2.0	1.1465	mV
Γ_{ie}	IPSP peak	0.1	2.0	1.2615	mV

	amplitude				
Γ_{ii}	IPSP peak amplitude	0.1	2.0	0.20143	mV
γ_{ee}	EPSP characteristic rate constant	100	1000	122.68	s^{-1}
γ_{ei}	EPSP characteristic rate constant	100	1000	982.51	s^{-1}
γ_{ie}	IPSP characteristic rate constant	10	500	293.10	s^{-1}
γ_{ii}	IPSP characteristic rate constant	10	500	111.40	s^{-1}
Neeα	No. of cortico-cortical synapses, target excitatory	2000	5000	3228.0	–
Neiα	No. of cortico-cortical synapses, target inhibitory	1000	3000	2956.9	–
Neeβ	No. of excitatory intracortical synapses	2000	5000	4202.4	–
Neiβ	No. of excitatory intracortical synapses	2000	5000	3602.9	–
Nieβ	No. of inhibitory intracortical synapses	100	1000	443.71	–
Niiβ	No. of	100	1000	386.43	–

	inhibitory intracortical synapses				
v	Axonal conduction velocity	100	1000	116.12	cm s^{-1}
1/Λ	Decay scale of cortico-cortical connectivity	1	10	1.6423	cm
Semax	Maximum excitatory firing rate	50	500	66.433	s^{-1}
Simax	Maximum inhibitory firing rate	50	500	393.29	s^{-1}
μ_e	Excitatory firing threshold	−55	−40	−44.522	mV
μ_i	Inhibitory firing threshold	−55	−40	−43.086	mV
σ_e	Standard deviation of excitatory firing threshold	2	7	4.7068	mV
σ_i	Standard deviation of inhibitory firing threshold	2	7	2.9644	mV
p_{ee}	Extracortical synaptic input rate	0	10,000	2250.6	s^{-1}
p_{ei}	Extracortical synaptic input rate	0	10,000	4363.4	s^{-1}

There are two sources of nonlinearity, related to the coupling of the synaptic inputs to the membrane potential and vice versa. The former connection is quadratically nonlinear, while the latter is given by the sigmoidal function S_k, which describes the onset of firing as the potential exceeds the threshold value μ_k. These nonlinearities tend to form sharp transitions of the potentials across the domain. That is one reason why we opted for a finite-difference discretization over a pseudo-spectral approach. Spectral accuracy would be of limited value in the presence of steep gradients and the finite-difference scheme can be parallelized much more efficiently. The second reason is that we would like to be able to change the geometry of the domain and the boundary conditions in future work. The finite-difference scheme is more flexible in that respect.

The only spatial derivatives in the model are those in the equations for the long-range connections. These are damped wave equations. We will discretize the Laplacian using a five-point stencil on a rectangular grid. In previous work, Bojak and Liley chose a second-order centred difference scheme for the time derivatives [6]. A disadvantage of this approach is that the stability condition of this scheme dictates that we set the time step inversely proportional to the grid spacing. In practice, they used a time step of 0.05 ms. To avoid this obstacle, we want to use implicit timestepping, and have currently implemented the unconditionally stable implicit Euler method, as described in Section 3.

Following earlier work on this model (e.g. [6,20]), we adopt periodic boundary conditions in both dimensions. This is a common choice in the study of mean-field models, and is partially justified by the observation, that each part of the cortex is connected to each other part. A discussion of this argument can be found in chapter 11 of Nunez and Srinivasan [4]. A result of this choice is that the model PDEs will be equivariant under translations and reflections. This equivariance has consequences for the behaviour of the model. An in-depth discussion of these consequences is beyond the scope of the current paper, but in Section 6 we will decribe how to compute periodic solutions for the equivariant system.

Other authors have used this model with an additional diffusive term in the equations for the membrane potentials to model gap junctions [21]. Inclusion of these terms can drastically change the bifurcation behaviour, as they can cause Turing transitions to space-dependent equilibria. Without the additional terms, a Hopf bifurcation from a spatially homogeneous equilibrium to a space dependent periodic orbit or a saddle-node bifurcation of the equilibrium often appear to be the primary instability. The gap junction terms can readily be included in our implementation, and in Section 5 we will describe how to solve for equilibrium states that may depend on space.

We will test our implementation by comparing to, and extending, the computations of oscillations with a 40 Hz component by Bojak and Liley [20]. Oscillations with this frequency are called gamma oscillations, and have been hypothesized to aid in the communication between groups of neurons [22]. Both simulations and experiments indicate that gamma oscillations occur in subjects performing cognitive tasks (see, e.g. [23] and references therein). Gamma band activity was found in the Liley model despite the fact that it was in no way tuned or formulated to produce this behaviour.

The parameter values for this experiment are listed in Table 1. The 40 Hz oscillations arise spontaneously if the number of local inhibitory-to-inhibitory connections is changed slightly. We introduce a scaling parameter *r* by replacing Niiβ→rNiiβ. This is the only parameter that will be varied in our tests.

1.2. PETSc overview

Rather than creating our code from scratch, we opted to work with the Portable, Extensible Toolkit for Scientific Computation (PETSc): an open-source, object oriented library that is designed for the scalable solution and analysis of PDEs [24,17]. PETSc is written in the C language, and is usable from C/C++ as well as Fortran and Python. We use PETSc in conjunction with the Scalable Library for Eigenvalue Problem Computations (SLEPc) [25], for the computation of eigenspectra of equilibrium and periodic solutions. Since our implementation uses some features of PETSc that are recent additions and are still being modified, we use the development version of both projects.

PETSc is split up into multiple components to address the various problems associated with solving PDEs numerically. For our purposes, we treat the DMcomponent, which handles the topology of the discretization, as the most fundamental, from which we can easily derive memory allocation and communication for distributed vectors (Vec) and matrices (Mat). With vectors and matrices, we can now solve linear systems, such as those that arise in Newton iteration for implicit time-stepping and the computation of equilibria and periodic orbits. PETSc's component for this is called KSP, and it has numerous iterative solvers implemented, as well as preconditioners, (PC), to increase convergence rates. For implicit time-stepping, for example, we use GMRES, preconditioned with incomplete LU (ILU) factorization, combined with the block Jacobi method [26,27]. On top of the linear solvers come the nonlinear solvers, PETSc's SNEScomponent, which implements a few different methods, such as globally convergent Newton iteration with line search [28]. Finally, PETSc provides a timestepping component, TS, to obtain time dependent solutions. Implemented here are numerous explicit and implicit schemes such as adaptive stepsize Runge–Kutta and implicit Euler. The implicit schemes make use of the SNEScomponent. A schematic of the hierarchy discussed here can be found in Fig. 1.

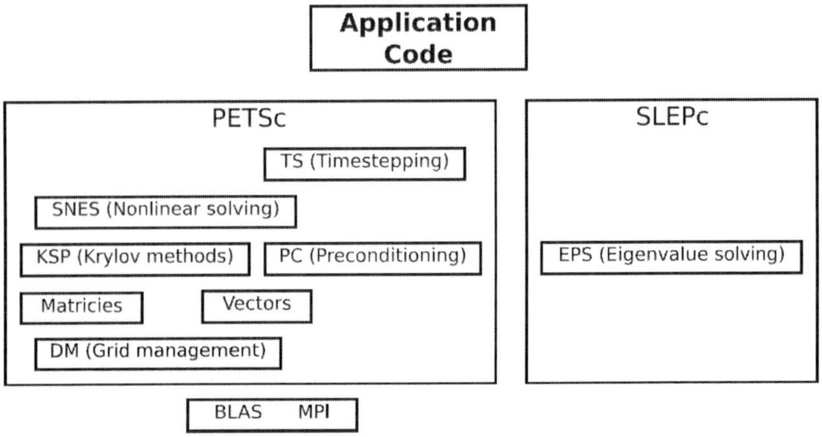

Figure. 1. Schematic representation of the components of PETSc and SLEPc used in our code, and their relative hierarchy.

For our dynamical systems calculations we will frequently need to compute specific eigenvalues and eigenvectors for system-sized matrices. For this end, we use SLEPc, which implements iterative eigenvalue solvers using PETSc Vecand Mat distributed data structures. The component of SLEPc that we use is EPS, which has a few algorithms for iteratively solving eigenproblems. Its default algorithm is Krylov–Schur iteration.

2. MODEL IMPLEMENTATION

2.1. Geometry

Following earlier work by Bojak and Liley (e.g. [6,20]) we consider the PDEs on a rectangular domain with periodic boundary conditions. On this domain, we use a rectangular grid of N_x by N_y points. In the tests presented in Section 7, the domain and the grid are square. PETSc allows for more complicated domain shapes and grids, that can be encoded in the DM component, independent of the higher-level components. This choice of periodic boundary conditions is one of computational convenience. As we have no formulation of boundary conditions that come from the physiology, we adopt periodic boundaries and look at phenomena that are on length scales below that of the system size. A brief demonstration of this can be seen in Fig. 2 and its caption.

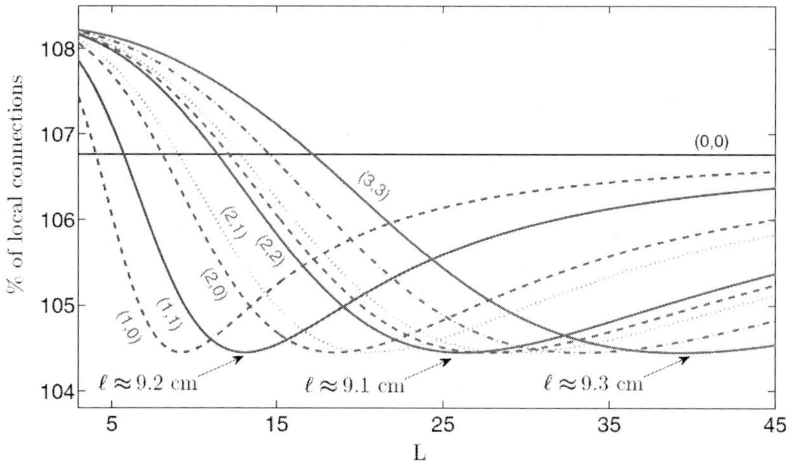

Figure. 2. Neutral stability curve for the spatially homogeneous equilibrium of the Liley model with parameters set according to Table 1. Shown is the scaling parameter, r, versus the linear domain size, L, and wave numbers $\boldsymbol{k} = (k_x, k_y)$ are shown in parenthesis. When varying r, only for very small domains the primary instability is spatially homogeneous. For domain sizes over 12.5 cm × 12.5 cm^2 the location of the primary instability approaches $r = 1.045$ and the length scale of the leading instability approaches $L/\|\boldsymbol{k}\| = 9.3$ cm.

Within DM, PETSc provides a simpler subcomponent, DMDA, for working with finite differences on structured grids such as our rectangle. If we specify a stencil to use for the spatial derivatives in the DMDA, PETSc will automatically handle numerous things for parallel execution, such as memory allocation and the communication setup for distributed vectors and for the distributed Jacobian matrix.

2.2. Fields

To make use of PETSc's solvers, the model must be written as a system of equations that is first order in time. This we achieve by introducing new states J_{jk} and ψ_{ek} according to

$$\frac{\partial I_{jk}}{\partial t} = J_{jk} - \gamma_{jk} I_{jk} \tag{6}$$

$$\frac{\partial J_{jk}}{\partial t} = e\Gamma_{jk}\gamma_{jk}\{N_{jk}^{\beta} S_j[h_j] + \phi_{jk} + p_{jk}\} - \gamma_{jk} J_{jk} \tag{7}$$

$$\frac{\partial \phi_{ek}}{\partial t} = \psi_{ek} - v^2 \Lambda^2 \phi_{ek} \tag{8}$$

$$\frac{\partial \psi_{ek}}{\partial t} = v^2 \Lambda^2 N_{ek}^{\alpha} S_e[h_e] + \frac{3}{2} v^2 \nabla^2 \phi_{ek} - v^2 \Lambda^2 \psi_{ek} \tag{9}$$

with indices $j, k = \{e, i\}$.

We opted to use a struct, seen in Code 2.1, to store the fields, rather than a triply indexed array.

Code 2.1

Struct for the fields.

```
typedef struct Field{
PetscReal h e, h i,
        I ee, J ee,
        I ie, J ie,
                I ei, J ei,
        I ii, J ii,
        phi ee, psi ee,
        phi ei, psi ei;
} Field;
```

This allows the code to be more readable in the function and Jacobian evaluation routines. For example, one accesses the ϕ_{ee} component at grid point (x_i, y_j) simply as u[j][i].phi_ee, provided that the elements of the array (Field **u;) are stored on the processor in which the call is made.

2.3. Parameters

All of the model parameters are stored in a struct designated as the application context. The application context is how PETSc gets problem related parameters into the user-defined functions needed by its solvers.

Code 2.2

Application context struct with the model parameters.

```
typedef struct AppCtx{
PassiveReal hr e, hr i,
        tau e, tau i,
        heq ee, heq ie,
        heq ei, heq ii,
        Gamma ee, Gamma ie,
        Gamma ei, Gamma ii,
        gamma ee, gamma ie,
        gamma ei, gamma ii,
        Nalpha ee, Nalpha ei,
        Nbeta ee, Nbeta ie,
        Nbeta ei, Nbeta ii,
        v, Lambda,
        Smax e, Smax i,
        mu e, mu i,
        sigma e, sigma i,
        p ee, p ei,
        p ie, p ii;
        . . .
} AppCtx;
```

Similar to the fields, this allows readable code for the parameters. For example, one accesses the Γ_{ie} parameter as user->Gamma_ie, if user is defined as the pointer AppCtx *user;. How the parameters show up in our struct for the application context is shown in Code 2.2.

2.4. User supplied functions

In addition to the structs listed above, we need to provide PETSc with (at least) a C function that computes the vector field for a given state. We call this function FormFunction, and from this PETSc is capable of approximating the Jacobian with various finite difference methods. However, we also supply a C function to explicitly compute the Jacobian, named FormJacobian, because this allows for more efficient calculations, especially when looking at stepping the variational equations in Section 4.

3. TIMESTEPPING

We currently use the implicit Euler method to time-step the discretized equations. As mentioned in Section 1.1, this allows us to take larger time steps than feasible with explicit methods. Since we are aiming to compute periodic orbits, rather than to generate long time series, the first order accuracy of the method is not an issue. Once a periodic orbit is computed, the time-step size can be reduced to increase accuracy. Another option is to use Richardson extrapolation to increase the order of accuracy, using the same nonlinear solving as described below.

3.1. Mathematical basis

We symbolically write the dynamical system as

$$\dot{u} = f(u), \mathrm{f} : \mathbb{R}^N \to \mathbb{R}^N \tag{10}$$

where N is the total number of unknowns after discretization, in our case $14 \times N_x \times N_y$. The implicit Euler scheme for time integration is given by

$$u_{n+1} = u_n + d_t f\,(u_{n+1}) \tag{11}$$

where the subscript represents the step number, dt the step size, and u_0 the initial conditions. This nonlinear equation is solved by Newton iteration:

$$u_{n+1}^{k+1} = u_{n+1}^{k} + \mathrm{d}u^k \,, \tag{12}$$

where the superscript denotes the Newton iterate, and du^k is the solution to the linear system

$$\left(\mathbb{I} - \mathrm{d}t \left.\frac{\partial f}{\partial u}\right|_{u_{n+1}^k}\right) \mathrm{d}u^k = \mathrm{d}t f(u_{n+1}^k) - u_{n+1}^k + u_n^k, \tag{13}$$

where $\partial f/\partial u$ denotes the $N \times N$ Jacobian matrix. Provided that the initial approximation, u_{n+1}^0, is close enough to the actual solution of Eq. (11), this iteration should converge quadratically. This is achieved by making the initial approximation the result of an explicit Euler step

$$u_{n+1}^0 = u_n + \mathrm{d}t f(u_n) \tag{14}$$

As we scale up the size of our problems, it becomes the linear solve in Eq. (13)that takes most of the time. This problem is handled by using GMRES to solve the linear system. For large time steps, the spectrum of the matrix in Eq. (13) is spread out, and we need to precondition it for iterative solving. We make use ILU, which has shown to be reliable for this type of problem [29,30]. If we use more than one processor, PETSc uses distributed storage for the matrix, and combines ILU with block Jacobi preconditioning.

3.2. Implementation

PETSc provides a simple interface for timestepping in its TS component. The basic code required to set up a TS is given in Code 3.1. With a TS set up like this, the timestepping parameters are set from command line arguments at run time. For example, to do implicit Euler timestepping for 40.67 ms with a time step of 0.1 ms, one needs to provide the arguments

```
-ts_type beuler -ts_dt 0.1 -ts_final_time 40.67.
```

In this specific case, since the final time is not an integer number of timesteps, PETSc will step past it, and interpolate at the desired time.

Code 3.1 PETSc code for setting up and running the timestepping. FormFunctionTS and FormJacobianTS are user provided functions that compute the rhs of Eq. (10), and its Jacobian respectively. J is an appropriately allocated matrix to hold the Jacobian, and u a vector to hold the solutions.

```
TS ts;

TSCreate(PETSC COMM WORLD,&ts);

TSSetProblemType(ts,TS NONLINEAR);

TSSetExactFinalTime(ts);
```

```
TSSetRHSFunction(ts,PETSC NULL,FormFunctionTS,&user);
TSSetRHSJacobian(ts,J,J,FormJacobianTS,&user);
TSSetFromOptions(ts);
TSSolve(ts,u);
```

4. STEPPING OF THE VARIATIONAL EQUATIONS

4.1. Mathematical basis

The variational equations for the dynamical system are written as

$$\dot{v} = \left.\frac{\partial f}{\partial u}\right|_{u} v, \quad v \in \mathbb{R}^N \tag{15}$$

and must be integrated simultaneously with the dynamical system (10). Solving the variational equations allow us to compute the stability of solutions, and is also an essential ingredient for the treatment of boundary value problems such as those that arise in the computation of periodic orbits.

Performing implicit Euler timestepping on the variational Eq. (15) requires solutions of the linear problems

$$\left(\mathbb{I} - \mathrm{dt} \left.\frac{\partial f}{\partial u}\right|_{u_{n+1}} \right) v_{n+1} = v_n \tag{16}$$

Since we already have the Jacobian of the dynamical system at timestep $n+1$, stepping the variational equations requires only one additional $N \times N$ linear solve per time step.

4.2. Implementation

In PETSc, we implement the timestepping of the variational equations as a MATSHELL, effectively viewing it as a matrix-free multiplication. Within a MATSHELL, one needs to provide a context for storing the relevant data and write functions for the desired matrix operation(s). For example, we point the operation MATOP_MULT to a function that takes the initial state of the variational system $v(0)$as input, and outputs the result $v(T)$ at the end of the timestepping. The context we use for the time stepping of the variational equations is shown in Code 4.1. The function we provide for MATOP_MULT works by first taking a step of the TS, then loading the Jacobian computed from that step and solving Eq. (16). This is repeated until the TS reaches its end.

Code 4.1 The MATSHELL context for timestepping of the variational equations. The TS holds the relevant info for stepping the dynamical system.

```
typedef struct PeriodIntegrationCtx{
// timestepping of the original eqn
TS ts;
Mat tsJac;
Vec initState,endState,fullSol;
// additional requirements for variational eqn
Mat J,eye;
KSP ksp;
} PeriodIntegrationCtx;
```

The MATSHELL thus defined can be used by SLEPc for the iterative computation of eigenpairs. In particular, we will use this approach to compute the Floquet multipliers of periodic orbits.

5. EQUILIBRIA

Having set up the function FormFunction for the right hand side of the dynamical system, and its Jacobian computation FormJacobian, also used for time integration, we can set up equilibrium calculations using PETSc's SNEScomponent with very little effort.

5.1. Mathematical basis

Equilibrium solutions to the dynamical system (10) are solutions that satisfy

$$f(u) = 0 \tag{17}$$

To solve this, we can set up a Newton iteration scheme

$$u^{k+1} = u^k + du^k \tag{18}$$

with du coming from the solution of the linear system

$$\left.\frac{\partial f}{\partial u}\right|_{u^k} \mathrm{d}u^k = -f(u^k) \tag{19}$$

As with the timestepping, if the initial guess is good enough this will converge quadratically provided that $(\partial f/\partial u)|_u k$ is nonsingular. Unlike the case of time stepping, though, we do not always have a way to produce an initial approximation that is good enough. For stable equilibrium solutions, we can use timestepping to get close to an equilibrium, but this will not work for unstable equilibria. One possible solution is using globally convergent Newton methods. Using such methods we can find equilibria from very coarse initial data, at the cost of computing many iterations. The line search algorithm and the trust region approach (see, e.g. [28]) are implemented in the SNES component.

Stability of equilibrium solutions follows from the spectrum of the Jacobian. Due to discrete symmetries of a domain, these can appear in groups. On a square domain, for instance, a single eigenvalue will be associated with up to eight eigenvectors, with wavenumbers $(\pm k_x, \pm k_y)$ and $(\pm k_y, \pm k_x)$.

As discussed in Section 1.1, the model is also equivariant under translatations in both dimensions. In the presence of this symmetry, it is more natural to search for *relative* equilibria, also called travelling waves. This leads to the introduction of two extra unknowns, that can be thought of as the wave velocities, into system (17), and an extension by two equations of the associated linear system (19). However, since we have so far only observed spatially homogeneous equilibrium states, we will discuss this adjustment in Section 6 on periodic solutions.

5.2. Implementation

Setting up and using a nonlinear solver within PETSc is straightforward, as shown in Code 5.1. The default algorithm used by SNES is Newton's method with line search.

Code 5.1 Code snippet for solving for equilibria. Vectors r and u are preallocated, with ubeing the initial approximation, and J a preallocated matrix for the Jacobian.

```
SNES snes;

SNESCreate(PETSC COMM WORLD,&snes);

SNESSetFunction(snes,r,FormFunctionSNES,&user);
```

```
SNESSetJacobian(snes,J,J,FormJacobianSNES,&user);
SNESSetFromOptions(snes);
SNESSolve(snes,PETSC NULL,u);
```

6. PERIODIC SOLUTIONS

The primary instability in the Liley model is often a Hopf bifurcation, and periodic orbits have been shown to play an important role in the dynamics of ODE reductions of the model (e.g. [16,31]). However, space dependent periodic orbits have not previously been computed and studied. Using PETSc data structures for bordered matrices, in conjunction with a MATSHELL, we can solve for periodic orbits based on the time stepping described in Sections 3 and 4.

6.1. Mathematical basis

Relative periodic orbits solve the boundary value problem

$$F(u,t) = \phi(t,u) - T_{ab}u = 0, \tag{20}$$

where ϕ is the flow of the dynamical system (10), t is the period, and $T_{ab}u(x,y) = u(x-a, y-b)$ the translation operator. Our strategy for solving this equation is essentially that of Sanchez et al. [32], namely Newton iterations combined with unconditioned GMRES iteration. Linearizing Eq. (20) gives

$$(D_u\phi(u,t) - \mathbb{I})\mathrm{d}u + f(\phi(u,t))\mathrm{d}t + T_{ab}\frac{\partial u}{\partial x}\mathrm{d}a + T_{ab}\frac{\partial u}{\partial y}\mathrm{d}b$$
$$= -F(u,t), \tag{21}$$

where $D_u\phi$ is a matrix of derivatives of the flow with respect to its initial condition. Upon convergence, this is the monodromy matrix of the periodic orbit. The result is N equations in $N + 3$ unknowns, which must be closed by phase conditions. For the temporal phase, we opted to handle this with a one dimensional Poincaré section, which gives a constraint on the Newton update step:

$$[D_u\phi(u,t)]_{k,.}\mathrm{d}u + f_k(\phi(u,t))\mathrm{d}t = C - \phi_k(u,t), \tag{22}$$

where $[D_u\phi(u,t)]_{k,.}$ denotes the kth row of the matrix $D_u\phi$, and C the desired value of the flow on the Poincaré section. For the spatial phase, we restrict the update step to be orthogonal to the generators of spatial translations:

$$\begin{aligned}\frac{\partial u}{\partial x}\mathrm{d}u &= 0\\ \frac{\partial u}{\partial y}\mathrm{d}u &= 0\end{aligned} \tag{23}$$

These choices give the bordered system

$$\begin{bmatrix} (D_u\phi(u,t)-T_{ab}) & f(\phi(u,t)) & T_{ab}\frac{\partial u}{\partial x} & T_{ab}\frac{\partial u}{\partial y}\\ [D_u\phi(u,t)]_{k,.} & f_k(\phi(u,t)) & 0 & 0\\ \frac{\partial u}{\partial x} & 0 & 0 & 0\\ \frac{\partial u}{\partial y} & 0 & 0 & 0 \end{bmatrix} \begin{bmatrix} \mathrm{d}u\\ \mathrm{d}t\\ \mathrm{d}a\\ \mathrm{d}b \end{bmatrix} = \begin{bmatrix} -F(u,t)\\ C-\phi_k(u,t)\\ 0\\ 0 \end{bmatrix} \tag{24}$$

the solution to which can be used to update the approximate solution

$$\begin{bmatrix} u^{n+1}\\ t^{n+1}\\ a^{n+1}\\ b^{n+1} \end{bmatrix} = \begin{bmatrix} u^{n}\\ t^{n}\\ a^{n}\\ b^{n} \end{bmatrix} + \begin{bmatrix} \mathrm{d}u\\ \mathrm{d}t\\ \mathrm{d}a\\ \mathrm{d}b \end{bmatrix} \tag{25}$$

The matrix $D_u\phi$ is dense, so we should avoid calculating and storing it explicitly. Iterative solving of the linear problem, (24), requires the computation of matrix-vector products, which are constructed from the integration of the variational Eq. (15) with $v(0) = d_u$ and the vector field $f(\phi(u,t))$ at the end point of the approximately periodic orbit. Since the governing PDE is dissipative, most of the eigenvalues of the monodromy matrix are clustered around zero. This aids the convergence of GMRES, without any preconditioning. Sanchez et al. [32]provide bounds for the number of GMRES iterations for the Navier–Stokes equation, and the convergence we observe for the Liley model is qualitatively similar.

6.2. Implementation

The problem of creating a bordered matrix system in a distributed environment is not a trivial one. The specific case that we have is one vector, *u*, that is sparsely connected and distributed among processors, and three parameters *t*, *a*, and *b* that must exist and be synchronized across all processors.

PETSc's DM module has some recently introduced functionality that allows us to handle this in a straightforward way, letting us make use of the DMDA already used in the other types of calculations.

DMRedundant can be used for the *a*, *b*, *t* components of our extended system, as it has the precise behaviour that we require. Next, we use a DMComposite to join together the DMDA of the grid with the DMRedundant of the period and translations. We can then derive vectors from this DMComposite, and use these vectors for PETSc's iterative linear solvers. PETSc code that illustrates this idea is shown in Code 6.1.

Code 6.1 Additional DM pieces for extended vectors as in Eq. (25), assuming that da is the DM associated with the grid structure. The numerical arguments in DMRedundantCreate represent the processor where the redundant entries live (in global vectors), and the number of redundant entries respectively. Note that the spatial and the time symmetries are taken to be in separate DMRedundant entries, this is just a matter of preference.

```
DM packer, redT, redC;

DMCompositeCreate(PETSC COMM WORLD,&packer);

DMRedundantCreate(PETSC COMM WORLD,0,1,&redT);

DMRedundantCreate(PETSC COMM WORLD,0,2,&redC);

DMCompositeAddDM(packer,da);

DMCompositeAddDM(packer,redT);

DMCompositeAddDM(packer,redC);
```

The matrix multiplication is done through a MATSHELL, and the struct that holds the relevant data is found in Code 6.2.

Code 6.2 For finding periodic solution, we need a method for integrating the variational equations (the MATSHELL discussed in Section 4), additional DMs, and space for holding *f* evaluated at the state at the end of the integration.

```
typedef struct PeriodFindCtx{
```

```
Mat *linTimeIntegration;
DM packer,redT,redC;
Vec endState,f at endState;
} PeriodFindCtx;
```

7. EXAMPLE CALCULATIONS

In this section, we present some computations that serve to validate our implementation and to investigate its efficiency. All tests are based on the parameter set in Table 1, and the scaling of the number of local inhibitory-to-inhibitory connections, r, is varied around the first bifurcation from an equilibrium to more complicated, spatio-temporal behaviour.

Fig. 2 shows the neutral stability curve for the spatially homogeneous equilibrium, which is the unique attractor of the model at small values of r. The primary transition is a Hopf bifurcation with spatial wave numbers that depend on the system size. For systems smaller than $2 \times 2\ \text{cm}^2$, the emerging periodic orbit is spatially homogeneous. For larger systems, space dependent orbits emerge, and their typical length scale converges to about 9.3 cm for large system sizes. These stability curves were computed by solving small eigenvalue problems for each combination of wavenumbers, independent from the PETSc implementation. The eigenvalues computed by Krylov–Schur iteration in SLEPc, presented in Section 7.2, are in good agreement.

A partial bifurcation diagram, for spatially homogeneous solutions only, is shown in Fig. 3. In this diagram, the Hopf bifurcation is subcritical, and time series analysis indicates that the Hopf bifurcations associated with nonzero wave numbers are, too. The time series presented in Section 7.1 was generated by starting from the equilibrium at $r = 1$ and adding a finite-size perturbation in the least stable direction, with wave number $|k_x| = |k_y| = 1$.

7.1. Timestepping

For the timestepping demonstration, we used a system size of $12.8 \times 12.8\ \text{cm}^2$ with 0.5 mm resolution, resulting in a 256×256 grid, and $N = 917{,}504$ unknowns in total. Setting the parameter $r = 1.0$, we initialize with the stable equilibrium solution perturbed by its least stable eigenmode, shown in Fig. 7. Since the equilibrium solution is stable, small perturbations just decay, but sufficiently large perturbations grow. The snapshots of Fig. 4 were taken after a transient time of 600 ms. The membrane potentials show behaviour that is nearly periodic, with a dominant period of 40 Hz, as demonstrated by the power spectrum shown in the last panel. The power spectrum is computed from a spatial average over the black box in the other panels, which was done to mimic the smearing of the signal observed from the scalp. Averaging over other regions of the same size produces qualitatively similar power spectra.

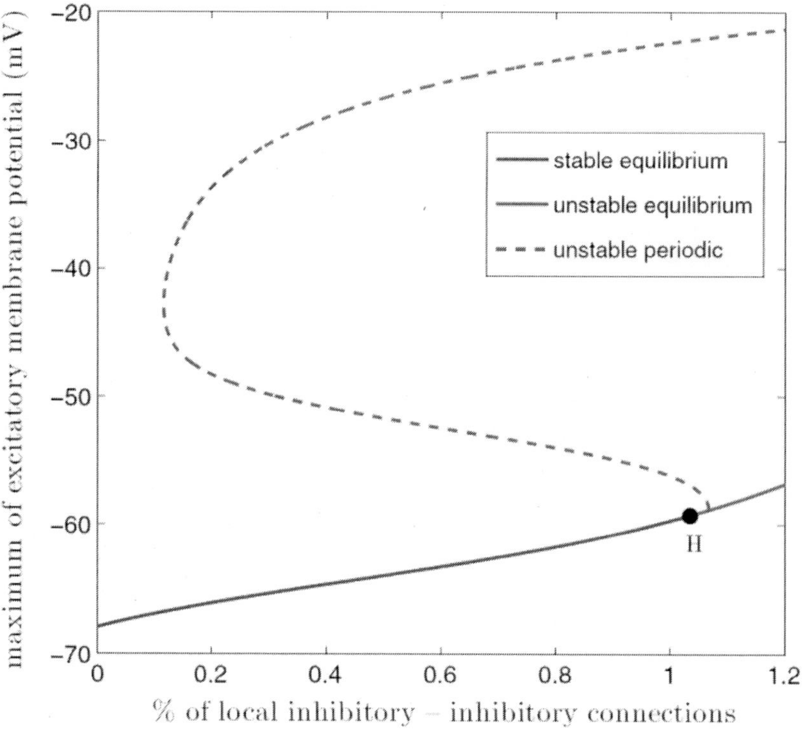

Figure. 3. Partial bifurcation diagram showing the primary transition from a spatially homogeneous equilibrium to a space and time dependent periodic orbit. On the vertical axis the scaling parameter *r* is plotted, and on the vertical axis the (maximum of) the excitatory membrane potential. The branch of periodic solutions shown with a dashed line is a spatially homogeneous branch that is unstable to space-dependent perturbations.

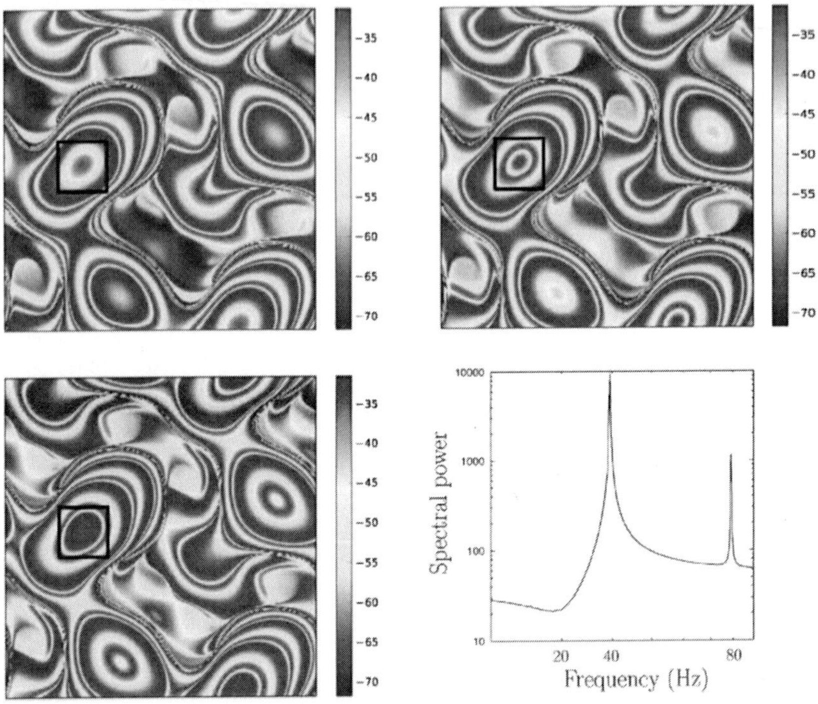

Figure. 4. Three snapshots of the excitatory membrane potential (colour bars in mV), 6 ms apart, of a solution at $r = 1$, near the primary Hopf bifurcation. The domain size is $12.8 \times 12.8\ cm^2$, the resolution is 0.5 mm and the time-step size 1 ms. The fourth panel shows the power spectrum of h_e, averaged over the region inside the black square. These images of the membrane potential are reminiscent of observations made of spontaneous cortical activity as in Kenet et al. [33].

Since the time-stepping code lies at the core of the periodic orbit solver, we also investigated its scaling with an increasing number of processors. Doubling the domain size, while keeping the grid spacing fixed, gives a dynamical system with $N = 3{,}670{,}016$ degrees of freedom. We time-stepped this system on varying numbers of Intel Xeon processors with Infiniband interconnects (saw.sharcnet.ca) and the results are displayed in Fig. 5. Scaling for few processors is slightly hindered by the change in preconditioner that is used, namely ILU on a single processor versus block Jacobi plus ILU in parallel. Except for this irregularity, the scaling is shown to be linear up to 256 processors.

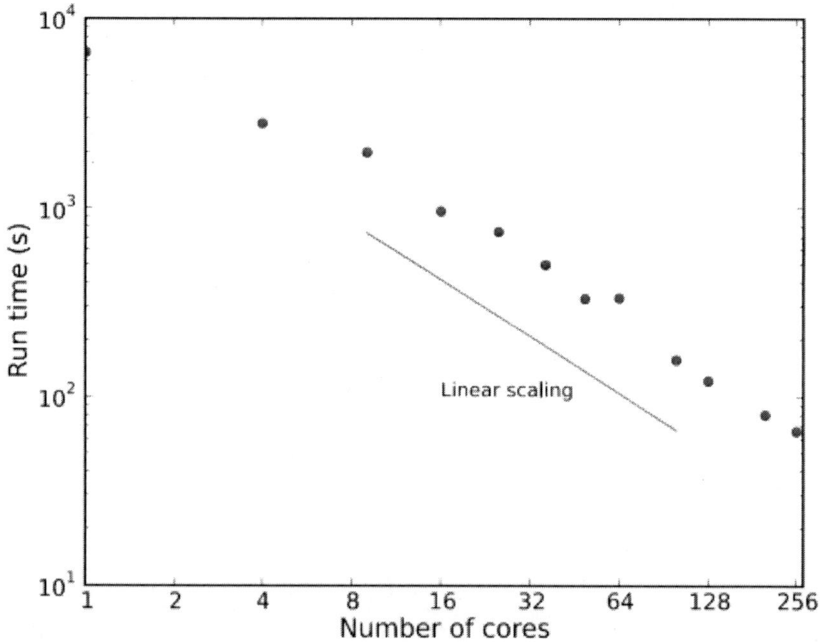

Figure. 5. Wall time for the computation of 100 time steps of 0.1 ms each on a 25.6 × 25.6 cm^2 domain with 0.5 mm resolution. The fully implicit Euler steps are computed with Newton iterations, each of which is solved for by GMRES, preconditioned with a combination of block Jacobi and ILU. The initial guess is given by an explicit Euler step. Two or three Newton iterations are sufficient to reduce the residual by a factor of 10^8. About 100 Krylov vectors are computed by GMRES to bring the relative residual down to 10^{-5}. The number of unknowns is N = 3,670,016.

At 256 processors, we are at 14,336 unknowns per process. As per PETSc's recommendations of not going below 10,000 unknowns per process [24], we do not expect efficient scaling to many more processors.

7.2. Equilibrium

We computed the whole equilibrium curve of Fig. 3 through parameter continuation, which is a trivial extension of the algorithm for computing equilibria, presented in Section 5. For each computed equilibrium solution, we took the Jacobian and used SLEPc to compute the eigenvalues with the largest real parts. The result is shown in Fig. 6. As predicted by the neutral stability curve computation, the (1, 1) mode turns unstable first, immediately followed by the (1, 0) mode. Around $r = 1.08$, the (0, 2) mode crosses the (0, 1) mode and proceeds to become the most unstable mode for larger values of r. The least stable eigenmode for $r = 1.046$, just after its eigenvalue has crossed zero, is shown in Fig. 7.

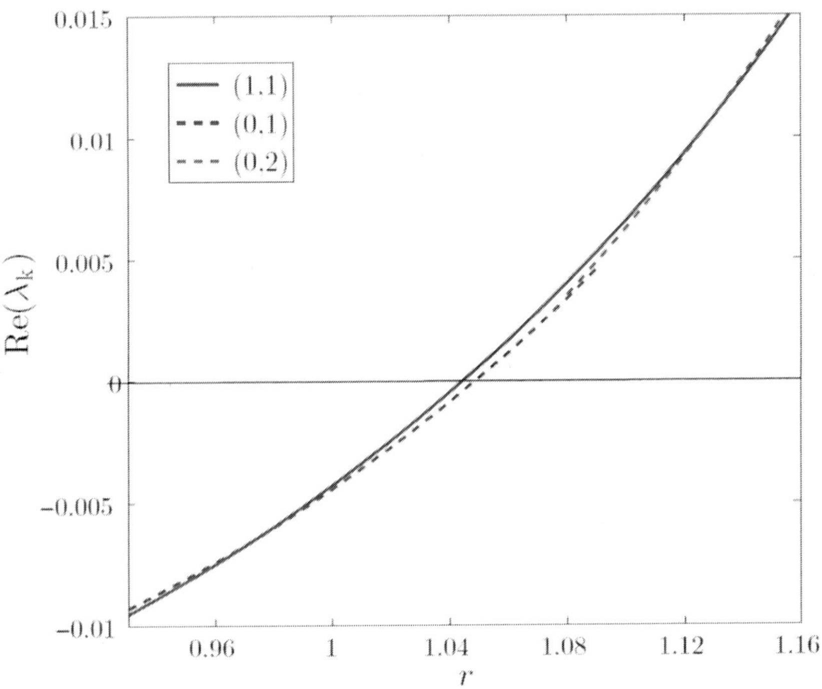

Figure. 6. The real parts of the leading two eigenvalue pairs of the spatially homogeneous equilibrium tracked in the scaling parameter r, for system size $L = 12.8$ cm. The primary transition is tied to wave numbers $|k_x| = |k_y| = 1$. The other curves shown are for wave numbers $k_x = 0, k_y = \pm 1$ and $k_x = \pm 1, k_y = 0$ and for $k_x = 0, k_y = \pm 2$ and $k_x = \pm 2, k_y = 0$.

7.3. Periodic solutions

We tested the computation of periodic orbits on a smaller grid, namely 16×16 points, still with 0.5 mm resolution, and with $r = 1.2$. The primary Hopf bifurcation is sub critical, so there is no easy way to compute the branch of space-dependent periodic solutions. Instead, we computed one of the spatially homogeneous orbits, for which an approximate solution can readily be obtained from analysis of the ODE reduction of the model. In fact, the upper part of the branch of periodic orbits shown in Fig. 3 is stable to all spatially homogeneous perturbations.

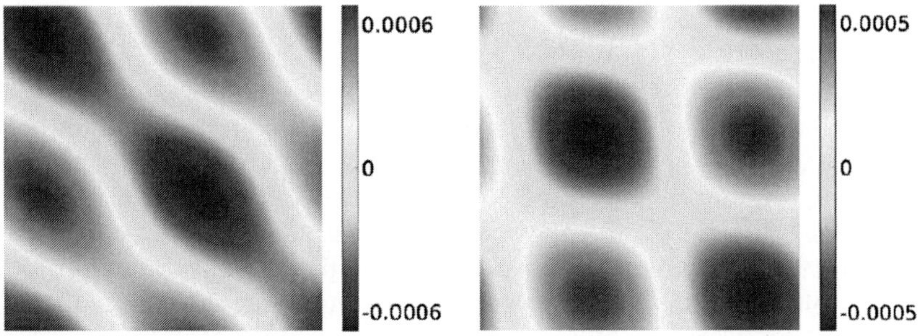

Figure. 7. The real part of the least stable eigenmode of the stable equilibrium located at $r = 1.046$. Displayed are the excitatory (left) and inhibitory (right) membrane potentials. The eigenvector, with wave numbers (1, 1), was computed by Arnoldi iteration and is scaled to have unit L_2 norm.

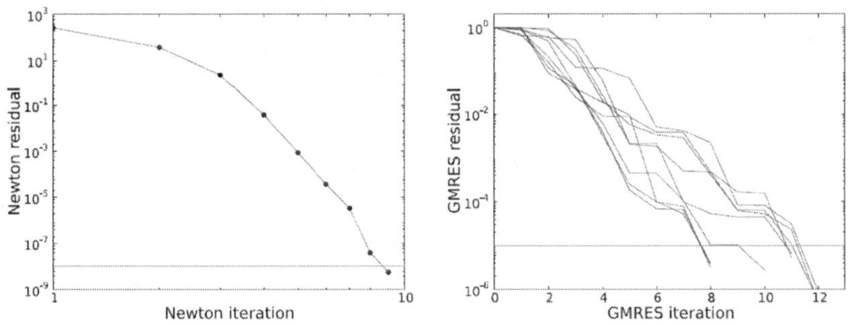

Figure. 8. Residuals of the Newton iteration (left) and the corresponding GMRES iterations (right). The latter is normalized by the norm of the right hand side of Eq. (24), i.e. the Newton residual. The tolerance was set at 10^{-8} for the Newton iteration and to 10^{-5} for the GMRES iteration. Note the super linear convergence of the former.

Starting from a coarse initial approximation, the Newton iterations converged faster than linear, and each Newton step took between 8 and 12 GMRES iterations, out of a maximum of $N + 3 = 3587$. The nonlinear and linear residuals are shown in Fig. 8. Subsequently, we computed the most unstable multipliers, using SLEPC with the MATSHELL for stepping the variational system as described in Section 4.2. The most unstable multiplier is $\mu_1 = 1.111$ and corresponds to a wave number (1, 1) perturbation.

8. CONCLUSION AND FUTURE IMPROVEMENTS

In the current paper, we have presented the basic implementation of the model and example computations to validate it and test its performance. The code will be available publicly [18]. As it is built on top of PETSc, the user has access to a range of nonlinear and linear solvers and preconditioners, which can be used to solve the boundary value problems that typically arise in dynamical systems analysis. The periodic orbit computation, presented in Section 7.3, is a simple example of such a boundary value problem, that has all the ingredients: a module for time-stepping the system and perturbations and a representation of user-specified, bordered matrices.

The next step in the development of the code is the implementation of pseudo-arclength continuation of periodic orbits. This will enable us, for instance, to complement the bifurcation diagram of the current test case, Fig. 3, with the branches of space-dependent periodic solutions that actually regulate the observed dynamics, in contrast to the highly unstable spatially homogeneous periodic orbits computed from the ODE reduction of the model.

We finish with emphasizing that there are some propagating modes within the cortex that can be described without considering its coupling to, say, the thalamus, as seen in Muller and Destexhe [34]. Although we did not attempt to tune the model parameters or initial conditions for the purpose, a comparison of the model output in Fig. 4 to the voltage sensitive dye experiments presented in [34] (see their Fig. 3) shows qualitative agreement with respect to the spatial extent of the patterns.

This implementation will be useful to studying such dynamics within the Liley model, and is easily modified to similar mean-field models of the cortex; examples being those that include the effects of gap junctions [21], and models that incorporate different properties for the long range connections.

ACKNOWLEDGEMENTS

L.v.V. was supported by NSERC Grant no. 355849-2008. Some of the computations were made possible by the facilities of the Shared Hierarchical Academic Research Computing Network (SHARCNET: www.sharcnet.ca) and Compute/Calcul Canada.

REFERENCES

1. G. Deco, V. Jirsa, A. McIntosh **Emerging concepts for the dynamical organization of resting-state activity in the brain** Nature Reviews Neuroscience, 12 (2011), pp. 43-56
2. E. Izhikevich, G. Edelman **Large-scale model of mammalian thalamocortical systems** Proceedings of the National Academy of Sciences, 105 (2008), pp. 3593-3598
3. H. Markram **The blue brain project** Nature Reviews Neuroscience, 7 (2006), pp. 153-160
4. P. Nunez, R. Srinivasan **Electric Fields of the Brain. The Neurophysics of EEG** (2nd ed.), Oxford University Press, Oxford (2006)
5. P. Nunez **The brain wave equation: a model for the EEG** Mathematical Biosciences, 21 (1974), pp. 279-297
6. Bojak, D. Liley **Modelling the effects of anaesthesia on the electroence phal ogram** Physical Review E, 71 (2005), pp. 1-22
7. W. Freeman**Mass Action in the Nervous System** Academic Press, New York (1975)
8. H.R. Wilson, J.D. Cowan **Excitatory and inhibitory interactions in localized populations of model neurons** Biophysical Journal, 12 (1972), pp. 1-24
9. F. Lopes da Silva, A. Hoeks, H. Smits, L. Zetterberg **Model of brain rhythmic activity** Kybernetik, 15 (1974), pp. 27-37
10. M. Breakspear, J. Roberts, J. Terry, S. Rodrigues, M.N.P. Robinson **A unifying explanation of primary generalized seizures through nonlinear brain modelling and bifurcation analysis** Cerebral Cortex, 16 (2006), pp. 1296-1313
11. M. Kramer, A. Szeri, J. Sleigh, H. Kirsch **Mechanisms of seizure propagation in a cortical model** Journal of Computational Neuroscience, 22 (2007), pp. 63-80
12. Blenkinsop, A. Valentin, M. Richardson, J. Terry **The dynamic evolution of focal-onset epilepsies - combining theoretical and clinical observations** European Journal of Neuroscience, 36 (2012), pp. 2188-2200
13. J. Quinton, B. Girau **Spatiotemporal pattern discrimination using predictive dynamic neural fields** BMC Neuroscience, 13 (Suppl. 1) (2012), p. O16
14. C. Honey, R. Kötter, M. Breakspear, O. Sporns **Network structure of cerebral cortex shapes functional connectivity on multiple time scales** Proceedings of the National Academy of Sciences of the United States of America, 104(2007), pp. 10240-10245
15. D.T.J. Liley, P.J. Cadusch, M.P. Dafilis **A spatially continuous mean field theory of electrocortical activity** Network-Computation in Neural Systems, 13 (2002), pp. 67-113
16. F. Frascoli, L. van Veen, I. Bojak, D. Liley**Metabifurcation analysis of a mean field model of the cortex** Physica D, 240 (2011), pp. 949-962
17. S. Balay, J. Brown, K. Buschelman, V. Eijkhout, W.D. Gropp, D. Kaushik, M.G. Knepley, L.M. Curfman, B.F. Smith, H. Zhang **PETSc Users Manual. Tech. rep.** Argonne National Laboratory (2012)
18. Source code available from. http://bitbucket.org/kegr/mfm.
19. S. Coombes, P. Graben, R. Potthast, J.J. Wright (Eds.), Neural Field Theory. Spots: Breathing, Drifting and Scattering in a Neural Field Model, Springer Verlag (2012) (Chapter)
20. Bojak, D. Liley **Self-organized 40 Hz synchronization in a physiological theory of EEG** Neurocomputing, 70 (2007), pp. 2085-2090
21. M.L. Steyn-Ross, D.A. Steyn-Ross, J.W. Sleigh, M.T. Wilson **A mechanism for ultra-slow oscillations in the cortical default network** Bulletin of Mathematical Biology, 73 (2011), pp. 398-416

22. F. Varela, J.-P. Lachaux, E. Rodriguez, J. Martineri **The brainweb: phase synchronization and large-scale integration** Nature Reviews Neuroscience, 2 (2001), pp. 229-239
23. E. Rolls, T. Webb, G. Deco **Communication before coherence** Cognitive Neuroscience, 36 (2012), pp. 2689-2709
24. S. Balay, J. Brown, K. Buschelman, W.D. Gropp, D. Kaushik, M.G. Knepley, L.M. Curfman, B.F. Smith, H. Zhang, PETSc Web page (2012). http://www. mcs.anl. gov/petsc.
25. V. Hernandez, J.E. Roman, V. Vidal **SLEPc: a scalable and flexible toolkit for the solution of eigenvalue problems** ACM Transactions on Mathematical Software, 31 (3) (2005), pp. 351-362
26. Y. Saad, M.H. Schultz **GMRES: a generalized minimal residual algorithm for solving nonsymmetric linear systems** SIAM Journal on Scientific and Statistical Computing, 7 (1986), pp. 856-869
27. Y. Saad **Iterative Methods for Sparse Linear Systems** (2nd ed.), SIAM, Philadelphia, PA (2003)
28. J.E. Dennis, R.B. Schnabel **Numerical Methods for Unconstrained Optimization and Nonlinear Equations** SIAM, Philadelphia, PA (1996)
29. J. Sanchez, F. Marques, J.M. Lopez**A continuation and bifurcation technique for Navier–Stokes flows** Journal of Computational Physics, 180 (2002), pp. 78-98
30. Y. Saad **Preconditioned Krylov Subspace Methods for CFD Applications. Tech. rep.** Minnesota Supercomputer Institute, Minneapolis (1994)
31. L. van Veen, D. Liley **Chaos via Shilnikov's saddle-node bifurcation in a theory of the electroencephalogram** Physical Review Letters, 97 (2006), p. 208101
32. J. Sanchez, M. Net, B. Garcia-Archilla, C. Simó **Newton–Krylov continuation of periodic orbits for Navier–Stokes flows** Journal of Computational Physics, 201 (2004), pp. 13-33
33. T. Kenet, D. Bibitchkov, M. Tsodyks, A. Grinvald, A. Arieli**Spontaneously emerging cortical representations of visual attributes** Nature, 425 (2003), pp. 954-956
34. L. Muller, A. Destexhe **Propagating waves in thalamus, cortex and the thalamocortical system: experiments and models** Journal of Physiology, Paris, 106 (2012), pp. 222-238

CHAPTER 15

Esoteric Twist: An Efficient in-Place Streaming Algorithmus for the Lattice Boltzmann Method on Massively Parallel Hardware

Martin Geier* and Martin Schönherr*

Institute for Computational Modeling in Civil Engineering, TU Braunschweig, 38106 Braunschweig, Germany

ABSTRACT

We present and analyze the Esoteric Twist algorithm for the Lattice Boltzmann Method. Esoteric Twist is a thread safe in-place streaming method that combines streaming and collision and requires only a single data set. Compared to other in-place streaming techniques, Esoteric Twist minimizes the memory footprint and the memory traffic when indirect addressing is used. Esoteric Twist is particularly suitable for the implementation of the Lattice Boltzmann Method on Graphic Processing Units.

Keywords: lattice Boltzmann; EsoTwist; in place streaming; indirect addressing; GPGPUs

1. INTRODUCTION

The lattice Boltzmann method (LBM) is a simple algorithm used in computational fluid dynamics for solving the Navier–Stokes equations. The efficient implementation of the lattice Boltzmann method has received considerable attention from the computer scientist community [1,2,3]. The LBM has several properties that make it interesting from a performance oriented algorithmic point of view. The ratio of floating point operations to memory access is relatively low so that the performance of the method is usually bandwidth limited. The LBM uses more variables than mathematically necessary for solving the Navier–Stokes equation, which adds to the relatively large data traffic. Still, the LBM is often considered to be efficient if implemented correctly. The algorithm is perfectly suitable for massively parallel implementation [4]. The time integration is second order accurate even though it depends only on the previous time step [5]. A particular feature of the LBM in distinction to finite difference schemes is that the number of input variables of the local (i.e., node-wise) time integration scheme equals the number of output variables. This is important as it implies that each input datum is required only once. In a finite difference scheme, functional values usually have to be gathered in a finite neighborhood of a grid node and the same datum is required to update several grid nodes. Most algorithmic and hardware optimizations of recent years focused on the reuse of data. Spatial blocking and scan-line algorithms are prominent examples developed for accelerating finite differences methods [6]. Caching is a generic hardware optimization meant to improve performance for repeatedly used data. It is interesting to note that the LBM does not benefit from such optimizations because it does not reuse data in a single time step. The LBM benefits from hardware cache only indirectly or when the cache is so large that it can hold the state variables until they are reused in the next time step. This zero-data-redundancy is not necessarily a disadvantage of the LBM. For example, it has been argued by Asinari et al. [7] that the LBM was inefficient due to the utilization of the distribution functions which increase the memory requirements and should intuitively increase traffic between the CPU and the main memory. Asinari et al. [8] proposed replacing the distribution functions with macroscopic variables, which reduced the memory footprint but counter-intuitively increased the memory traffic since data had to be read repeatedly for the computation of finite differences [7]. In addition, the savings in memory turned out to be smaller than naively assumed: the LBM can overwrite each input datum with the output datum while the finite difference method requires a source and a destination array to avoid the overwriting of data that is still required for neighboring grid nodes. In addition to these purely algorithmic differences, there are differences in accuracy between finite difference and LBM implementations. Depending on the details of the model and the validation example, either the finite difference method [9] or the LBM [10] excels over the other. In the current paper, we will not discuss the numerical modeling underlying the LBM. Instead, we focus entirely on its efficient implementation. By doing so, we exploit the features of the method and arrive at a result that is very different from what one should do to implement a finite difference method efficiently.

2. LATTICE BOLTZMANN METHOD

In this section, we describe the LBM from a purely algorithmic point of view without considering the details of the modeling. We direct the reader to various text books for an introduction to the method [11,12,13,14].

The LBM is a computational method to solve the lattice Boltzmann equation which can be written in three dimensions as:

$$f_{ijk(x+i)(y+j)(z+k)(t+1)} = f_{ijkxyzt} + \Omega_{ijkxyzt}. \tag{1}$$

Here, the indexes i, j and k are integer values indicating the direction in which the distribution fijkxyzt moves on a Cartesian lattice with the nodes being located at positions x, y and z. The time step is indicated by t. The so-called collision operator Ωijkxyzt is a function of all local distributions fijkxyzt but neither depends on the state of any other lattice node nor on the state of the same node at any time other than t. It is hence seen that communication between lattice nodes happen only on the left-hand side of (1), the so-called streaming. The right-hand side, the so-called collision, is entirely local. It is also seen that the lattice Boltzmann equation is a local mapping from Q→Q variables. Each datum is used only once.

The LBM can be implemented for different velocity sets where the indexes i, j and k take different values. Without loss of generality, we restrict us here to the technically most relevant case with 27 velocities and i, j, k∈{−1,0,1}.

3. IMPLEMENTATION OF THE LATTICE BOLTZMANN METHOD

The LBM has a superficial algorithmic similarity to a finite difference time domain technique and it is admissible to implement it in a likewise fashion. A naive approach would utilize two data arrays $f_{ijkxyzt}$ and $f^*_{ijkxyzt}$ and rewrite (1) as:

$$f^*_{ijk(x+i)(y+j)(z+k)t} = f_{ijkxyzt} + \Omega_{ijkxyzt}. \tag{2}$$

After completing one time step, the post-collision distributions might be redefined as the new pre-collision distributions:

$$f_{ijkxyz(t+1)} = f^*_{ijkxyzt}. \tag{3}$$

This updated algorithm is denoted as AB-pattern [15,16]. In a finite difference time domain method, such an update was necessary in order to avoid the overwriting of data at the neighboring node. In the LBM, this is not necessary because the collision maps a set of inputs to the same number of outputs. It is hence admissible to overwrite the input directly by the output and a secondary array is never required. A naive implementation of this concept would require moving the distributions in a separate step:

$$f^*_{ijkxyzt} = f_{ijkxyzt} + \Omega_{ijkxyzt}, \tag{4}$$

$$f_{ijk(x+i)(y+j)(z+k)(t+1)} = f^*_{ijkxyzt}. \tag{5}$$

It is desired to implement the LBM in such a way that only one array is required while both streaming and collision are combined in a single step, i.e., we are looking for an algorithm that requires only a single array to hold the distributions and reads and writes each datum only once in each time step. Such a technique is called in-place streaming and several methods have been proposed in literature. An early example of an in-place streaming algorithm is the swap algorithm of Mattila et al. [17] and Latt [18]. The swap algorithm combines the streaming and the collision in a single step, but it does not overwrite the input of the local collision by the output. As a result, the swap algorithm requires a certain order in which the lattice nodes have to be processed. This is a severe disadvantage, as it limits the application of the algorithm to serial computations with a clear order of the instructions. The method is not applicable to massively parallel computations.

An alternative approach to in-place streaming is the compressed grid that approximately saves half the memory [19]. The idea behind the compressed grid is to extend a rectangular simulation domain by one line of nodes in all directions and keep a certain order in the processing of the grid nodes, which, after the update of the first line of nodes, are written to spare line of nodes. Once the nodes in the first line are processed, they are no longer required and the second line of nodes can be written to the first line of nodes and so on until all lines have been processed. The compressed grid method works only for a rectangular domain and requires a fixed order of updating the nodes.

Bailey [16] studied the implementation of the LBM on Graphics Cards where lattice nodes are updated in random order by thousands of threads in parallel. He proposed an algorithm called AA-pattern that overwrites the input with the output by distinguishing between odd and even time steps (see Figure 1). In the odd time step, the distributions are not streamed and the local distribution are directly overwritten. In the even time step, the distributions are fetched from the neighboring nodes and written back to the neighboring nodes in an opposite direction after collision. In such a way, the distributions move two lattice spacings every second time step. The AA-pattern requires the code to be implemented twice, once for the even and once for the odd time steps.

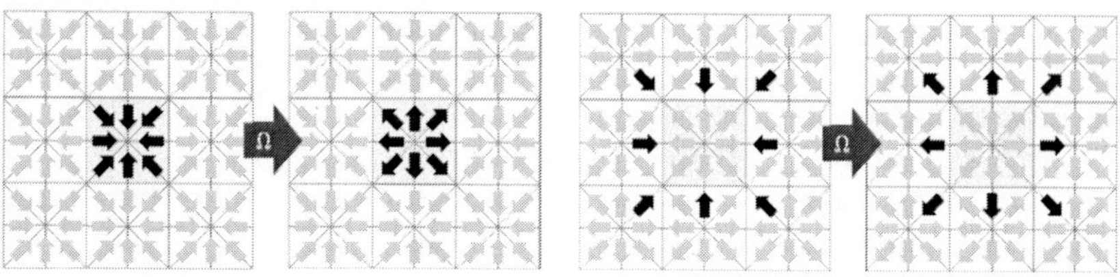

Figure 1. AA-pattern in 2D. (**left**) Odd time step; (**right**) Even time step.

The AA-pattern solves the concurrency problem of the swap algorithm, but it was designed to run on a full matrix implementation. Real world applications of computational fluid dynamics usually require a higher geometrical flexibility and local grid-refinement. For this purpose, it is usually desired to implement the LBM as a sparse matrix method. The AA-pattern permits an implementation on sparse matrices, but it requires pointers to all neighboring nodes, which implies a relatively large memory footprint. In the reminder of this paper, we present and discuss the properties of the Esoteric Twist data structure that solves the concurrency problem similar to the AA-pattern but leads to substantial savings in memory when indirect addressing is used.

4. ESOTERIC TWIST

The Esoteric Twist, or short EsoTwist, algorithm combines the following desired features:

- Streaming and collision are combined in one step.
- Each datum is read once and written once in each time step.
- The method is thread safe; all nodes can be processed in arbitrary order or entirely in parallel.
- The method is well suitable for indirect addressing, i.e., has a small memory footprint.

The swap algorithm fulfills all but the last two features, and the AA-pattern fulfills all but the last feature. The AA-pattern has to access all neighboring nodes in every second time step, which implies that it first has to read the pointer to all neighbors from an index array if indirect addressing is used. We show below that the EsoTwist method is more economical in terms of memory footprint and memory reads for indirect addressing. In this section, we first introduce the EsoTwist method for the general case.

The name Esoteric Twist originates from the unintuitive (esoteric) observation that the streaming step can be eliminated if the distributions are written back in opposite (twisted) order compared to the reading before collision. As in the case of the AA-pattern, the key to a thread safe in-place streaming is that the collision operator writes only to the memory from which it draws its inputs. We assume that modern computers have access to a slow but vast memory that holds all the distributions of all nodes and a fast but small memory in which all operations associated with a collision at a single node can be executed. We will call the former the main memory and the latter the registers. We assume, as in the AA-pattern, that the distributions belonging to one node are transferred from the main memory to the registers before collision and written back to the main memory after collision. We assume that the transfer between main memory and registers is expensive and should be minimized. Since the distributions ought to move, they either need to be drawn from the neighboring nodes (pull-scheme) before collision or they have to be sent to the neighboring nodes after collision (push-scheme). In the EsoTwist approach, all distributions that move in negative direction with $i, j, k<0$ are pulled while all distributions moving in positive directions with $i, j, k>0$ are pushed. The procedure is depicted in Figure 2 for the two-dimensional case. The yellow node in Figure 2 is the one for which we consider one update step. The collision on the node depends on all Q incoming distributions and it returns Q outgoing distributions. Only the incoming distributions moving in positive x- and y-directions are read from the yellow node itself. The distributions moving in negative directions are read from the respective neighbors in the positive directions as depicted in the figure. After collision, the distributions are written back to the places where the distributions moving in the opposite direction were read from. As is the case in the AA-pattern, each distribution swaps places with the distribution moving in the opposite direction after collision. For the movement of the distribution to happen, the method has to distinguish between odd and even time steps, as depicted in Figure 2. This can be realized in different ways. Unlike in the case of the AA-pattern, EsoTwist does not necessarily require two implementations of the method for odd and even time steps. A much simpler realization of the method is obtained by storing the distributions in a structure of arrays, i.e., by grouping the distributions in arrays according to the direction in which they are moving. After one time step is completed, all pointers to the arrays are swapped with the pointers to the arrays moving in the opposite direction. Then, the next time step is executed as before. However, a structure of arrays is not necessarily desired. The EsoTwist method can also be implemented like the AA-pattern by explicitly distinguishing between odd and even time steps. Algorithm 1 shows the concept of the EsoTwist algorithm in one dimension. The complete algorithm in three dimensions is shown in the Appendix A.

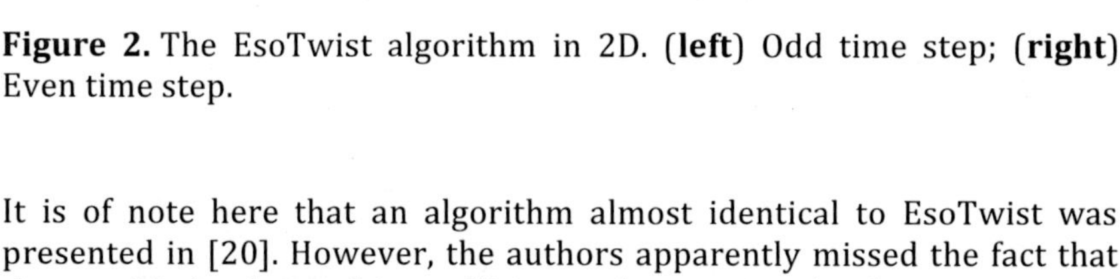

Figure 2. The EsoTwist algorithm in 2D. (**left**) Odd time step; (**right**) Even time step.

It is of note here that an algorithm almost identical to EsoTwist was presented in [20]. However, the authors apparently missed the fact that they could simply combine collision and swapping the directions into a single step. Instead, they swapped the distributions in a separate step, which unnecessarily doubles the data traffic.

Algorithm 1: The basic EsoTwist Algorithm for a one-dimensional LBM with three speeds in a structure of arrays format. The asterisk marks the post-collision distribution. The capital F denotes a register to which the distribution is read from the distributions in main memory f for the collision. The distribution f1 moving in a positive direction and the stationary distribution f0 are read at the index n while the distribution f−1 moving in a negative direction is read from the neighbor in a positive direction. After collision, the distribution moving in positive direction F∗1 is written to the neighbor in a positive direction at the slot where the distribution going to the opposite direction was originally read. The distribution going in negative direction F∗−1 is now written to location n. In that way, both the distribution going in a negative direction and the distribution going in a positive direction moved a distance of one node during the time step. After all nodes have been collided, it is sufficient to swap the pointers to the arrays f−1 and f1 to execute the streaming of all nodes.

for *all time steps* **do**
    for *all nodes n* **do**
        $n_x \leftarrow neighborX(n);$
        $F_{-1} \leftarrow f_{-1}(n_x);$
        $F_0 \leftarrow f_0(n);$
        $F_1 \leftarrow f_1(n);$
        $F^* \leftarrow collision(F);$
        $f_{-1}(n_x) \leftarrow F_1^*;$
        $f_0(n) \leftarrow F_0^*;$
        $f_1(n) \leftarrow F_{-1}^*;$
    end
    $swapPointer(f_{-1}, f_1);$
end

4.1. Indirect Addressing

In almost all real-world applications of computational fluid dynamics, it would be unpractical to discretize the computational domain with a rectangular Cartesian grid. To discretize domains of arbitrary shape, sparse matrices are the most flexible choice. In a sparse matrix implementation, each node has to access its neighbors via a pointer, or equivalently via an index which has to be stored in an additional array.

For example, if a three-dimensional LBM with 27 velocities is implemented on a sparse matrix using the AA-pattern, each node has to find the 26 neighbors of the node in the even time step. The AA-pattern is efficient in the sense that it has to find the neighbors only every other time step. Still, it has to store the pointer to these neighbors. A pointer usually occupies the same amount of memory as a scalar datum such that 26 pointers basically occupy almost as much memory as the distributions themselves. This memory can be substantially reduced by applying pointer chasing. For example, using the AA-pattern, it would be sufficient to store pointers in the six axial directions and access the diagonal neighbors by querying, for example, the neighbor in the positive x-direction for its neighbor in the negative y-direction. Applying this technique reduces the number of pointers per node from 26 to 6. This number can be reduced further to four if the nodes are aligned in one of the cardinal directions along the array. However, each of these optimizations reduce the flexibility of the method. The alignment of the grid along one of the cardinal axes in connection with indirect addressing usually means that the grid cannot be modified after its generation, which is an obstacle for adaptive simulations. That is to say, the feature that nodes on a sparse matrix gird can be inserted and deleted ad libitum is lost once the nodes are aligned along one dimension in order to eliminate the pointers in this direction. We will hence not consider the alignment of the data along one of the cardinal directions in our analysis for sparse matrix grids.

Using pointer chasing, the AA-pattern has to store links to six neighbors at each node ($6\times N$ data values for N nodes). The total memory requirement without further meta data for an LBM with 27 discrete speeds is $33\times N$ data values. Since the AA-pattern streams data only every other time step, it requires $(27+26/2)\times N=40\times N$ reads and $27\times N$ writes from and to the main memory in every time step with $27\times N$ reads in the odd and $53\times N$ reads in the even time steps. It is interesting to note that, while pointer chasing reduces the memory footprint, it does not reduce the number of reads. All 26 pointers to all neighbors have to be read every other time step.

In comparison to the AA-pattern, EsoTwist reduces both the memory footprint and the number of reads. This is due to EsoTwist being asymmetric in space. Only neighbors in positive directions in either dimension have to be accessed. In three dimensions, using 27 speeds, all distributions accessed by a node are stored at the indexes of the node and the seven neighbors in the upper octant of the surrounding cube. Thus, by applying pointer chasing, $3\times N$ links have to be stored and each time step requires $(27+7)\times N=34\times N$ reads and $27\times N$ writes. The total memory requirement is hence $30\times N$ (9% less then for the AA-pattern).

The data reads are even 25% less than for the AA-pattern and the data writes are the same. In their analysis, Wittmann et al. [3] conclude that the AA-pattern and EsoTwist have the lowest data traffic for the known streaming algorithm, but their analysis ignores that EsoTwist requires fewer pointers and fewer reads than the AA-pattern.

In addition to those direct savings in memory, there is also an indirect savings compared to the AA-pattern due to the fact that the fewer ghost nodes are required when indirect addressing is used (see Figure 3). Both EsoTwist and the AA-pattern store part of the distributions associated with each node at some neighboring nodes. The memory for these nodes has to be allocated even if the node itself is not part of the simulation domain. In the AA-pattern, all 26 neighbors of an existing node have to be allocated while, in EsoTwist, only the seven neighbors in the upper octant of the surrounding cube have to be allocated. Depending on the complexity of the geometry, this can result in considerable savings through a smaller number of ghost nodes.

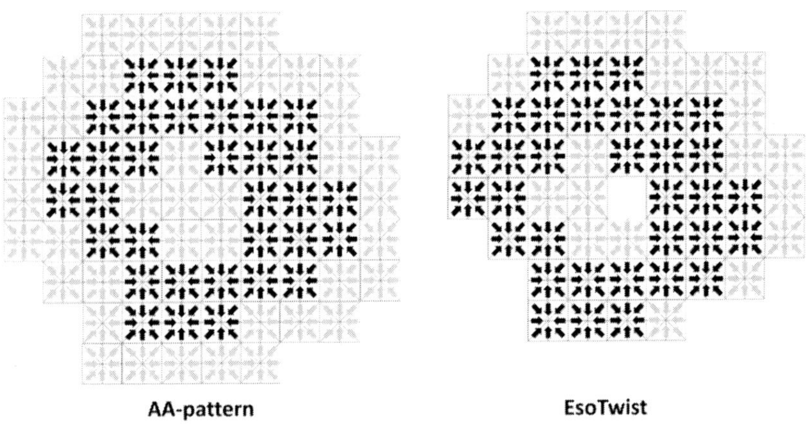

Figure 3. Ghost nodes in the AA-pattern and EsoTwist when indirect addressing is used. Since both methods store part of the distributions at neighboring nodes the memory for these ghost nodes (in gray) have to be allocated even though they are not part of the simulation domain. EsoTwist has an advantage over the AA-pattern in that it requires these neighbors only in positive directions. Thus, fewer ghost nodes are required for EsoTwist than for the AA-pattern.

It should be noted that the above analysis ignores the effect of caching for the number of reads and writes such that no general conclusion on performance can be drawn at this stage. Modern CPUs usually read complete cache lines at once and this enhances performance when data is aligned. However, since, in the current subsection, we consider sparse matrix grids, we have no general guarantee that our nodes are arranged in any favorable ordering and that we would benefit from caching. The above analysis is valid for the worst case scenario of completely unaligned data.

4.2. Variants

Even though pointers are an indispensable element of all software on the machine level, several popular programming languages (e.g., Fortran) do not support pointers explicitly. This is very unfortunate since a simple operation that is efficiently executed by the hardware in nano-seconds has to be mimicked by more complicated and certainly less efficient operations. Still, it is possible to implement EsoTwist without pointers. One possibility is to combine the arrays for distributions moving in opposite directions into one array of twice the length. Staying with our example of the D3Q27 velocity set, we would have one array of length N for the distributions f000 and thirteen arrays of length 2N for the other distributions. For example, let g1−10[n] be the array for the distributions f1−10 and f−110. Now, let odd=1 and even=0for odd time steps and odd=0 and even=1 for even time steps. It is possible to access the distributions without pointer exchange trough:

$$f_1{-}10[n] = g1{-}10[n{+}odd{*}N], \tag{6}$$

$$f{-}110[n]{=}g1{-}10[n{+}even{*}N]. \tag{7}$$

Thus, it is also possible to implement EsoTwist in a programming language that does not support pointers.

Another consideration concerns the structure of arrays data type. On massively parallel hardware like GPGPUs that apply single instruction multiple data operations, it is beneficial to store the data in different arrays according to the directions. However, this is not the optimal data layout for serial computing since all 27 distributions are stored in different locations of the main memory and none of them would share a single cache line. On those machines, storing all data belonging to a node together would be optimal. Geller [21] proposed a compromise that enhanced the performance of EsoTwist on serial computers. He suggested to use two containers, one for the local distributions and one for the distributions at neighboring nodes. In this way, half the data required for the collision could always be accessed together. Only the static distribution was stored separately.

EsoTwist has also been implemented successfully for block structured grids [22,23]. Block structuring can be used to increase locality of the data and is useful for load balancing parallel codes. Inside a block, no pointers to neighboring nodes are required. In connection with EsoTwist, each block needs only to access its seven neighboring blocks in the upper octant of the surrounding cube for data exchange. Hence, each block would require only three links to neighboring blocks if pointer chasing is used. Block structured grids might increase performance through high data locality, and they largely simplify the parallelization of the method, but they also have some disadvantages compared to the application of EsoTwist on sparse matrix grids. The data transfer at the interfaces between the blocks does not happen automatically and requires an explicit step of data exchange. Usually, also at least one additional layer of nodes is required to facilitate the exchange of data. In terms of memory occupation for a cubic block in three dimensions with N×N×N nodes, (N+1)3−N3ghost nodes are usually required. Using pointer chasing, only three links to neighboring blocks are required per block. In contrast, sparse matrices require three links per node and ghost cells only on the domain boundaries. Assuming that each datum (distributions and pointers) occupies the same size in memory, we can determine the breakeven point in memory consumption when block structured grids occupy less memory than a sparse matrix representation. For a method with 27 speeds, we have to equate the three pointers with the average amount of additional data required for the block structured grids:

$$3 = \frac{27((N+1)^3 - N^3) + 3}{N^3}. \quad (8)$$

This solves to $N \approx 27.98$. We hence see that a block size of at least 28×28×28 nodes is required to occupy less memory on a block structured grid compared to a sparse matrix. This analysis does not take into account that block structured grids usually contain nodes which do not belong to the domain. In principle, a block has to be allocated completely whenever at least one node in the block is located in the fluid domain. It is hence seen that memory economy is usually a poor motivation for the use of block structured grids, at least when combined with EsoTwist.

The EsoTwist data structure has also been implemented in a variant that secures memory alignment of all distributions [24]. This was historically important for the implementation of the lattice Boltzmann method on Nvidia GPUs of older generations [25] where data was always read in form of sixteen adjacent four byte values. The trick in this EsoStripe called variant is that the distributions of adjacent points are stored in a distance of sixteen, such that whenever sixteen nodes are processed simultaneously by the vector processor, the data access is completely aligned. While this leads to more than a 25% performance increase on

GeForce GTX 465, no performance gain was observed on newer hardware when compared to non-aligned reads and writes [24]. In that case, data alignment was found to not be essential for newer hardware.

4.3. Implicit Bounce Back

EsoTwist has an interesting feature concerning lattice nodes which are excluded from the computation. If a lattice node is not touched, i.e., if no distributions are read and no distributions are written, the direction in which the populations on this node move is reversed through the pointer exchange. In the LBM, a reversal of the directions of the distributions is used to model solid walls in the so-called simple bounce back scheme. Using EsoTwist bounce back is implicitly applied to all nodes skipped during the update, and these nodes will thus behave like solid walls. Applying simple bounce back in EsoTwist is entirely for free since the respective nodes do not even need to be touched. This way of imposing bounce back by omission of the node is called implicit bounce back. This property is shared by other swap methods like the one proposed by Mattila et al. [17] and Latt [18] and by the AA-pattern [16].

5. RESULTS

The EsoTwist data structure has been successfully implemented both on parallel CPU systems [22,23] and on GPUs [26,27,28]. Here, we focus on the implementation for GPUs in connection with indirect addressing.

5.1. Performance Model

Our performance model is based on the observation that the LBM is typically a memory bandwidth limited algorithm [3,29,30,31,32]. The EsoTwist method for a 27 speed lattice requires 27 distributions that have to be read and written in every time step. Even though only three pointers to neighbors are stored with every node, we still need to read seven such pointers per node and time step through pointer chasing. One additional number is used to indicate the type of the node. Since this indicator is also used as a pointer to an array of boundary conditions (each node can have its individual boundary condition), this number must also have at least four bytes. In total, our method requires 27 + 7 + 1 = 35 reads and 27 writes per node and time step which sums to #Bytes = 62 × 4 Bytes = 248 Bytes. The performance of our code is measured in Million Node Updates Per Second (MNUPS). The bandwidth occupancy P is calculated by:

$$P = \frac{\#MNUPS \times \#Bytes}{bandwidth} \times 100\%. \tag{9}$$

When comparing results of EsoTwist to results from a full matrix code, we have to take into account that the full matrix code does not require any pointers such that it has only 28 reads and 27 writes in every time step per node.

5.2. Comparison to Full Matrix Implementation

Here, we compare the performance of EsoTwist on sparse matrices with a full matrix implementation. In order to make the setup comparable, we chose a cubic domain with 128×128×128 nodes. In the first example, we simulate duct flow with implicit bounce back boundary condition and use the Bhatnagar Gross Krook collision model [33]. This example is run on an old Nvidia Tesla C1060 GPU with a bandwidth of 102 GBytes/s. The full matrix implementation obtains 205 MNUPS (41.2% bandwidth) and the sparse matrix implementation obtains 191 MNUPS (43.2% bandwidth).

In our second example, we compare the performance of a simulation with the same grid size but with a cascaded collision kernel [34] on a GeForce GTX Titan X (Maxwell GM200), which is a much more recent GPU than the Tesla C1060. The theoretical peak bandwidth of the GeForce GTX Titan X is 336.6 GBytes/s. The full matrix version of the code uses the AB-pattern [15,16]. Despite the fact that the cascaded kernel is computationally more intensive, we obtain much better performance figures on this newer hardware. The full matrix version runs with 999.425 MNUPS (60.8% bandwidth), and the sparse matrix version runs with 993.514 MNUPS (68.2% bandwidth).

It is observed in both cases that, while the performance on the full matrix is higher in terms of lattice updates, the sparse matrix version makes better use of the bandwidth. The differences in execution speed are marginal. The sparse matrix version obtains 93.2% of the execution speed of the full matrix version on the outdated Tesla C1060 and 99.4% on the more recent GeForce GTX Titan X. This result is a further confirmation of the observation by Linxweiler [24] that progress in hardware reduces the advantage of highly optimized full matrix codes over more flexible codes.

5.3. Isotropy

We probe the isotropy of the EsoTwist method by comparing the execution speed of a simulation of the same geometry under three different orientations. This is done because the nodes on this grid are filled along one of the cardinal directions first, depending on whether the orientation of the geometry neighboring nodes are either close together or far apart. The problem hence probes how the method reacts to different levels of data locality. The geometry is shown in Figure 4. It is composed of three rectangular pipes with different cross sections. Each pipe is 128 grid nodes long. The cross-sections of the pipes are 5×5, 10×10 and 20×20 grid nodes, respectively. Simulations are executed on a Nvidia GeForce GTX Titan GPU running with device driver version 331.82 and the CUDA (Compute Unified Device Architecture) toolkit version 6.0. The theoretical bandwidth of this GPU is 288.4 GBytes/s [35]. The computational model used was the BGK collision kernel in single precision. We observe similar performance regardless of the orientation of the domain in space. For the larger pipe oriented along the x-axis, we obtain 738.5 MNUPS (59.1% peak bandwidth); for the larger pipe oriented along the y-axis, we obtain 729.8 MNUPS (58.4% peak bandwidth) and, for the z direction, we obtain 727.3 MNUPS (58.2% peak bandwidth). Our performance figures compare to between 407 and 684 MNUPS reported for the GeForce GTX Titan by Tomczak and Szafran [36] using a lattice Boltzmann method with nineteen speeds (D3Q19 lattice) in double precision in a tiling layout (similar to block structuring) and two data sets (AB-pattern), which corresponds to between 48% and 72.6% bandwidth (the numbers are taken directly from [36] as our performance model does not apply directly to their code). The relatively wide range in the performance of [36] is due to differences in the data locality in the different test cases.

5.4. Porous Material

Our implementation of the LBM using EsoTwist with indirect addressing on GPUs participated in a comparative study between different Computational Fluid Dynamics (CFD) methods. Simulation results were presented in [27]. Our method was found to be the most efficient when compared to the competing methods to solve the Navier–Stokes equations in porous media. Here, we present only performance values.

We discretize a random packing of 6864 mono-disperse spheres (beads) with radius 0.5 mm as depicted in Figure 5. The bead pack is discretized with Cartesian grids using two different resolutions: 40 μm (Figure 6)

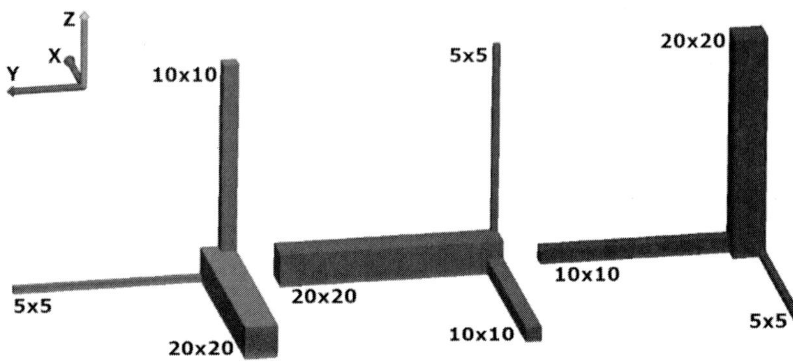

Figure 4. Different orientations of the same test geometry to probe the isotropy of EsoTwist with indirect addressing. The performance in terms of Million Node Updates Per Second is seen to depend only weakly on the orientation. Picture reproduced from [37].

and 20 μm (Figure 7). The full matrix representation of the grids would contain 1.74×107 and 1.39×108 grid nodes for the 40 μm and the 20 μm case, respectively. By the use of indirect addressing, the number of nodes is reduced to 9.05×106 (52%) and 6.465×107(46.5%), respectively. We hence see that indirect addressing eliminates about half of the grid nodes in this case. It is interesting to note that, in the case of the AA-pattern without pointer chasing, there would hardly be any advantage of the sparse data structure over the full matrix in this case since the number of pointers required for indirect addressing is almost as large as the memory required to store the distributions. In the case of EsoTwist combined with pointer chasing, only three pointers have to be stored with each grid node. The 20 μm case was simulated for 72 s in real time, which translates to 1,600,000 time steps. Other than in the previous simulation, the current test case applied second order accurate boundary conditions to capture the curvature of the beads and the simulation domain was decomposed and distributed over several GPUs. On two Nvidia K40c GPUs, we obtained a performance of 525.2 MNUPS and on six Tesla C1060 we obtained 448.6 MNUPS. The 40 μm test case was simulated on a single K40c GPU and run for 400.000 time steps. It obtained a performance of 233.3 MNUPS. However, these performance figures do not include the fact that the simulation of a porous medium is boundary dominated and that the application of the second order accurate interpolation boundary condition [26] is done in a separate step. In addition, it must be taken into account that the complex cumulant collision kernel [26] was used. The application of the second order boundary conditions produce about the same data traffic as one collision. In the case of the 40 µm simulation, we had about 4.26×106 boundary nodes such that the bandwidth occupancy including the application of the boundary conditions is calculated as 27.5%. These performance figures hence show that, even though the optimal performance is never obtained in real-world applications with complex boundary conditions, the performance remains comparable to the optimal case.

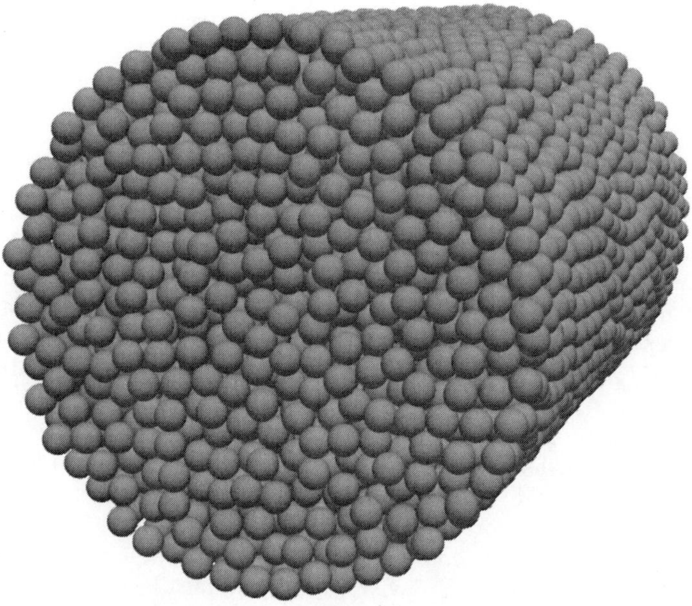

Figure 5. The random packing of 6864 spheres is discretized with a sparse matrix. Picture reproduced from [37].

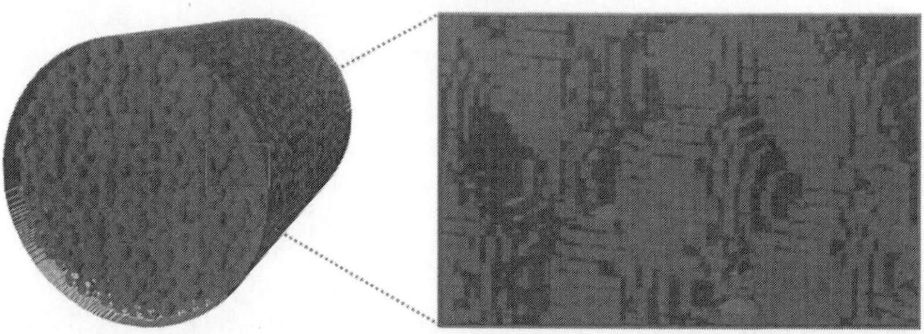

Figure 6. The discretization of the random packing of spheres with a lattice spacing of 40 µm. Picture reproduced from [37].

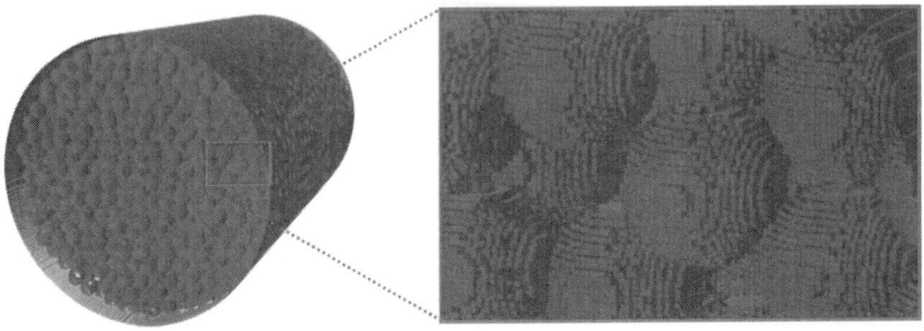

Figure 7. The discretization of the random packing of spheres with a lattice spacing of 20 μm. Picture reproduced from [37].

6. CONCLUSIONS

We presented the Esoteric Twist data structure for efficient thread safe execution of the LBM on massively parallel hardware. EsoTwist requires only a single read and write operation for each datum in each time step and only a single place in main memory. When combined with indirect addressing, EsoTwist requires the smallest number of pointers to neighboring nodes of any LBM algorithm using sparse matrices known to us. To our knowledge, EsoTwist has the smallest memory footprint of any sparse matrix LBM implementations, and it requires the smallest number of reads of any known LBM algorithm using sparse matrices. These claims are also confirmed by comparing to a recent analysis of Wittmann et al. [3]. Ignoring that EsoTwist requires, in fact, less pointers and reads than the AA-pattern, Wittmann et al. concluded that the AA-pattern and EsoTwist have the lowest data traffic of the known streaming algorithms. While being efficient, EsoTwist is also very flexible and is easily applied to real-world applications with complex boundaries. It also combines very favorably with local grid refinement techniques as shown in several publications [22,23,26,28].

Our results show comparable performance in terms of memory bandwidth (up to ~68.2% peak bandwidth) to a block structured code with two distributions on recent Nvidia GPUs [36] and to full matrix codes. In terms of execution time, the sparse matrix EsoTwist method was found to obtain 99.4% of the performance of the AB-pattern on a recent GPU. It is of note that the ratio on older hardware was found to be inferior (93.2% for a Tesla C1060). It is hence seen that progress in hardware allows for more flexibility of the code without severe performance penalty.

ACKNOWLEDGMENTS

This work was funded by the TU-Braunschweig.

APPENDIX A. ESOTWIST IN 3D

Here, we present some pseudocode of the EsoTwist algorithm. We present separately the routine for read (Algorithm A1) and write (Algorithm A2) that are used in the EsoTwist algorithm in three dimensions (Algorithm A3).

Algorithm A1: Reading the data from the main memory to the registers. An efficient implementation of the method writes out all for-loops explicitly.

Function $readNode(n, n_x, n_y, n_z, n_{xy}, n_{xz}, n_{yz}, n_{xyz})$
  for $\{i,j,k\} \in \{\{0,0,0\},\{0,0,1\},\{0,1,0\},\{0,1,1\},\{1,0,0\},\{1,0,1\},\{1,1,0\},\{1,1,1\}\}$ **do**
    $F_{ijk} \leftarrow f_{ijk}[n];$
  end
  for $\{i,j,k\} \in \{\{-1,0,0\},\{-1,0,1\},\{-1,1,0\},\{-1,1,1\}\}$ **do**
    $F_{ijk} \leftarrow f_{ijk}[n_x];$
  end
  for $\{i,j,k\} \in \{\{0\bar{1},0\},\{0,\bar{1},1\},\{1,\bar{1},0\},\{1,\bar{1},1\}\}$ **do**
    $F_{ijk} \leftarrow f_{ijk}[n_y];$
  end
  for $\{i,j,k\} \in \{\{0,0,-1\},\{0,1,-1\},\{1,0,-1\},\{1,1,-1\}\}$ **do**
    $F_{ijk} \leftarrow f_{ijk}[n_z];$
  end
  for $\{i,j,k\} \in \{\{-1,-1,0\},\{-1,-1,1\}\}$ **do**
    $F_{ijk} \leftarrow f_{ijk}[n_{xy}];$
  end
  for $\{i,j,k\} \in \{\{-1,0,-1\},\{-1,1,-1\}\}$ **do**
    $F_{ijk} \leftarrow f_{ijk}[n_{xz}];$
  end
  for $\{i,j,k\} \in \{\{0,-1,-1\},\{1,-1,-1\}\}$ **do**
    $F_{ijk} \leftarrow f_{ijk}[n_{yz}];$
  end
  $F_{-1-1-1} \leftarrow f_{-1-1-1}[n_{xyz}];$

Algorithm A2: Writing data back to main memory. An efficient implementation of the method writes out all for-loops explicitly. Note that the direction of the distributions has been swapped compared to the Algorithm A1.

Function *writeNode*($n, n_x, n_y, n_z, n_{xy}, n_{xz}, n_{yz}, n_{xyz}$)

for $\{i,j,k\} \in \{\{0,0,0\},\{0,0,1\},\{0,1,0\},\{0,1,1\},\{1,0,0\},\{1,0,1\},\{1,1,0\},\{1,1,1\}\}$ **do**
  $f_{ijk}[n] \leftarrow F^*_{-i-j-k}$;
end
for $\{i,j,k\} \in \{\{-1,0,0\},\{-1,0,1\},\{-1,1,0\},\{-1,1,1\}\}$ **do**
  $f_{ijk}[n_x] \leftarrow F^*_{-i-j-k}$;
end
for $\{i,j,k\} \in \{\{0,-1,0\},\{0,-1,1\},\{1,-1,0\},\{1,-1,1\}\}$ **do**
  $f_{ijk}[n_y] \leftarrow F^*_{-i-j-k}$;
end
for $\{i,j,k\} \in \{\{0,0,-1\},\{0,1,-1\},\{1,0,-1\},\{1,1,-1\}\}$ **do**
  $f_{ijk}[n_z] \leftarrow F^*_{-i-j-k}$;
end
for $\{i,j,k\} \in \{\{-1,-1,0\},\{-1,-1,1\}\}$ **do**
  $f_{ijk}[n_{xy}] \leftarrow F^*_{-i-j-k}$;
end
for $\{i,j,k\} \in \{\{-1,0,-1\},\{-1,1,-1\}\}$ **do**
  $f_{ijk}[n_{xz}] \leftarrow F^*_{-i-j-k}$;
end
for $\{i,j,k\} \in \{\{0,-1,-1\},\{1,-1,-1\}\}$ **do**
  $f_{ijk}[n_{yz}] \leftarrow F^*_{-i-j-k}$;
end
$f_{-1-1-1}[n_{xyz}] \leftarrow F^*_{111}$;

Algorithm A3: EsoTwist in 3D: the algorithm requires three arrays neighborX, neighborY and neighborZ with links to the neighboring node of *n*. The functions readNode() (Algorithm A1) and write Node() (Algorithm A2) transfer the distributions fijk to the registers Fijk and back.

```
for all time steps do
    for all nodes n do
        n_x ← neighborX[n];
        n_y ← neighborY[n];
        n_z ← neighborZ[n];
        n_xy ← neighborY[n_x];
        n_xz ← neighborZ[n_x];
        n_yz ← neighborZ[n_y];
        n_xyz ← neighborZ[n_xy];
        readNode(n, n_x, n_y, n_z, n_xy, n_xz, n_yz, n_xyz);
        F* ← collideNode(F);
        writeNode(n, n_x, n_y, n_z, n_xy, n_xz, n_yz, n_xyz);
    end
    for i ∈ −1...1 ; j,k ∈ 0...1 ; j + k ≠ 0 do
        swapPointer(f_ijk, f_−i−j−k);
    end
    swapPointer(f_100, f_−100);
    swapPointer(f_10−1, f_−101);
    swapPointer(f_1−10, f_−110);
    swapPointer(f_0−11, f_01−1);
end
```

REFERENCES

Axner, L.; Bernsdorf, J.; Zeiser, T.; Lammers, P.; Linxweiler, J.; Hoekstra, A. Performance evaluation of a parallel sparse lattice Boltzmann solver. *J. Comput. Phys.* **2008**, *227*, 4895–4911.

Wittmann, M.; Zeiser, T.; Hager, G.; Wellein, G. Comparison of different propagation steps for lattice Boltzmann methods. *Comput. Math. Appl.* **2013**, *65*, 924–935.

Wittmann, M.; Zeiser, T.; Hager, G.; Wellein, G. Modeling and analyzing performance for highly optimized propagation steps of the lattice Boltzmann method on sparse lattices. *arXiv*, 2014; arXiv:1410.0412.

Schönherr, M.; Kucher, K.; Geier, M.; Stiebler, M.; Freudiger, S.; Krafczyk, M. Multi-thread implementations of the lattice Boltzmann method on non-uniform grids for CPUs and GPUs. *Comput. Math. Appl.* **2011**, *61*, 3730–3743.

Dellar, P.J. An interpretation and derivation of the lattice Boltzmann method using Strang splitting. *Comput. Math. Appl.* **2013**, *65*, 129–141.

Hager, G.; Wellein, G.; Wittmann, M.; Zeiser, T.; Fehske, H. Efficient Temporal Blocking for Stencil Computations by Multicore-Aware Wavefront Parallelization. In Proceedings of the 2009 33rd Annual IEEE International Computer Software and Applications Conference (COMPSAC 2009), Seattle, WA, USA, 20–24 July 2009; Volume 1, pp. 579–586.

Obrecht, C.; Asinari, P.; Kuznik, F.; Roux, J.J. Thermal link-wise artificial compressibility method: GPU implementation and validation of a double-population model. *Comput. Math. Appl.* **2016**, *72*, 375–385.

Asinari, P.; Ohwada, T.; Chiavazzo, E.; Rienzo, A.F.D. Link-wise artificial compressibility method. *J. Comput. Phys.* **2012**, *231*, 5109–5143.

Ohwada, T.; Asinari, P. Artificial compressibility method revisited: Asymptotic numerical method for incompressible Navier-Stokes equations. *J. Comput. Phys.* **2010**, *229*, 1698–1723.

Dubois, F.; Lallemand, P.; Obrecht, C.; Tekitek, M.M. Lattice Boltzmann model approximated with finite difference expressions. *Comput. Fluids* **2016**.

Thorne, D.T.; Michael, C. *Lattice Boltzmann Modeling: An Introduction for Geoscientists and Engineers, 2nd. ed.*; Springer: Berlin/Heidelberg, Germany, 2006.

Mohamad, A.A. *Lattice Boltzmann Method: Fundamentals and Engineering Applications with Computer Codes*; Springer Science & Business Media: Berlin, Germany, 2011.

Guo, Z.; Shu, C. *Lattice Boltzmann Method and Its Applications in Engineering*; World Scientific: Singapore, 2013; Volume 3.

Krüger, T.; Kusumaatmaja, H.; Kuzmin, A.; Shardt, O.; Silva, G.; Viggen, E.M. *The Lattice Boltzmann Method: Principles and Practice*; Springer: Cham, Switzerland, 2016.

Wellein, G.; Zeiser, T.; Hager, G.; Donath, S. On the single processor performance of simple lattice Boltzmann kernels. *Comput. Fluids* **2006**, *35*, 910–919.

Bailey, P.; Myre, J.; Walsh, S.; Lija, D.J.; Saar, M.O. Accelerating lattice Boltzmann fluid flow simulations using graphics processors. In Proceedings of the 2009 International Conference on Parallel Processing, Vienna, Austria, 22–25 September 2009; pp. 550–557.

Mattila, K.; Hyväluoma, J.; Rossi, T.; Aspnäs, M.; Westerholm, J. An efficient swap algorithm for the lattice Boltzmann method. *Comput. Phys. Commun.* **2007**, *176*, 200–210.

Latt, J. *How to Implement Your DdQq Dynamics with Only q Variables Per Node (Instead of 2q)*; Technical Report; Tufts University: Medford, MA, USA, 2007.

Pohl, T.; Kowarschik, M.; Wilke, J.; Iglberger, K.; Rüde, U. Optimization and profiling of the cache performance of parallel lattice Boltzmann codes. *Parallel Process. Lett.* **2003**, *13*, 549–560.

Neumann, P.; Bungartz, H.J.; Mehl, M.; Neckel, T.; Weinzierl, T. A Coupled Approach for Fluid Dynamic Problems Using the PDE Framework Peano. *Commun. Comput. Phys.* **2012**, *12*, 65.

Geller, S.; ICON Technology & Process Consulting Ltd., Braunschweig, Germany. Personal communication, 2016.

Far, E.K.; Geier, M.; Kutscher, K.; Krafczyk, M. Simulation of micro aggregate breakage in turbulent flows by the cumulant lattice Boltzmann method. *Comput. Fluids* **2016**, *140*, 222–231.

Far, E.K.; Geier, M.; Kutscher, K.; Krafczyk, M. Distributed cumulant lattice Boltzmann simulation of the dispersion process of ceramic agglomerates. *J. Comput. Methods Sci. Eng.* **2016**, *16*, 231–252.

Linxweiler, J. Ein Integrierter Softwareansatz zur Interaktiven Exploration und Steuerung von Strömungssimulationen auf Many-Core-Architekturen. Ph.D. Thesis, TU Braunschweig, Braunschweig, Germany, June 2011. (In German)

Tölke, J.; Krafczyk, M. TeraFLOP computing on a desktop PC with GPUs for 3D CFD. *Int. J. Comput. Fluid Dyn.* **2008**, *22*, 443–456.

Geier, M.; Schönherr, M.; Pasquali, A.; Krafczyk, M. The cumulant lattice Boltzmann equation in three dimensions: Theory and validation. *Comput. Math. Appl.* **2015**, *70*, 507–547.

Yang, X.; Mehmani, Y.; Perkins, W.A.; Pasquali, A.; Schönherr, M.; Kim, K.; Perego, M.; Parks, M.L.; Trask, N.; Balhoff, M.T.; et al. Intercomparison of 3D pore-scale flow and solute transport simulation methods. *Adv. Water Resour.* **2016**, *95*, 176–189.

Pasquali, A.; Schönherr, M.; Geier, M.; Krafczyk, M. Simulation of external aerodynamics of the DrivAer model with the LBM on GPGPUs. In Proceedings of the ParCo2015, Edinburgh, UK, 1–4 September 2015.

Zeiser, T.; Wellein, G.; Hager, G.; Donath, S.; Deserno, F.; Lammers, P.; Wierse, M. *Optimized Lattice Boltzmann Kernels as Testbeds for Processor Performance*; Regional Computing Center of Erlangen (RRZE): Erlangen, Germany, 2004; Volume 1.
Welleina, G.; Lammersb, P.; Hagera, G.; Donatha, S.; Zeisera, T. Towards optimal performance for lattice Boltzmann applications on terascale computers. In Proceedings of the Parallel CFD Conference, Busan, Korea, 15–18 May 2006; Elsevier: Amsterdam, The Netherlands, 2006; pp. 31–40.
Williams, S.; Oliker, L.; Carter, J.; Shalf, J. Extracting ultra-scale lattice Boltzmann performance via hierarchical and distributed auto-tuning. In Proceedings of the 2011 International Conference for High Performance Computing, Networking, Storage and Analysis (SC), Seattle, WA, USA, 12–18 November 2011; pp. 1–12.
Feichtinger, C.; Habich, J.; Köstler, H.; Rüde, U.; Aoki, T. Performance modeling and analysis of heterogeneous lattice boltzmann simulations on cpu–gpu clusters. *Parallel Comput.* **2015**, *46*, 1–13.
Qian, Y.H.; d'Humières, D.; Lallemand, P. Lattice BGK Models for Navier-Stokes Equation. *EPL (Europhys. Lett.)* **1992**, *17*, 479.
Geier, M.; Greiner, A.; Korvink, J.G. A factorized central moment lattice Boltzmann method. *Eur. Phys. J. Spec. Top.***2009**, *171*, 55–61.
Jeong, H.; Lee, W.; Pak, J.; Choi, K.J.; Park, S.H.; Yoo, J.S.; Kim, J.H.; Lee, J.; Lee, Y.W. Performance of Kepler GTX Titan GPUs and Xeon Phi System. *arXiv*, 2013; arXiv:1311.0590.
Tomczak, T.; Szafran, R.G. Memory layout in GPU implementation of lattice Boltzmann method for sparse 3D geometries. *arXiv*, 2016; arXiv:1611.02445.
Schönherr, M. Towards Reliable LES-CFD Computations Based on Advanced LBM Models Utilizing (Multi-) GPGPU Hardware. Ph.D. Thesis, TU Braunschweig, Braunschweig, Germany, July 2015.

CHAPTER 16

Dimensionality Reduction of Distributed Vector Word Representations and Emoticon Stemming for Sentiment Analysis

Brian Dickinson, Michael Ganger, Wei Hu

Department of Computer Science, Houghton College, Houghton, NY, USA

ABSTRACT

Social media platforms such as Twitter and the Internet Movie Database (IMDb) contain a vast amount of data which have applications in predictive sentiment analysis for movie sales, stock market fluctuations, brand opinion, or current events. Using a dataset taken from IMDb by Stanford, we identify some of the most significant phrases for identifying sentiment in a wide variety of movie reviews. Data from Twitter are especially attractive due to Twitter's real-time nature through its streaming API. Effectively analyzing this data in a streaming fashion requires efficient models, which may be improved by reducing the dimensionality of input vectors. One way this has been done in the past is by using emoticons; we propose a method for further reducing these features through identifying common structure in emoticons with similar sentiment. We also examine the gender distribution of emoticon usage, finding tendencies towards certain emoticons to be disproportionate between males and females. Despite the roughly equal gender distribution on Twitter, emoticon usage is predominately female. Furthermore, we find that distributed vector representations, such as those produced by Word2Vec, may be reduced through feature selection. This analysis was done on a manually labeled sample of 1000 tweets from a new dataset, the Large Emoticon Corpus, which consisted of about 8.5 million tweets containing emoticons and was collecting over a five day period in May 2015. Additionally, using the common structure of similar emoticons, we are able to characterize positive and negative emoticons using two regular expressions which account for over 90% of emoticon usage in the Large Emoticon Corpus.

Keywords: Natural Language, Emoticon, Twitter, Review

1. INTRODUCTION

In the past few years, the growth of social media platforms has brought with it a wealth of unprocessed data, containing interesting information that may be inferred from the interactions of users. The astronomical amount of information renders manual human analysis unfeasible, and has thus stimulated the development of machine learning algorithms to aid in gleaning information from this data. These algorithms may be running in either a batch processing environment, where a data set is stored and analyzed, or in a streaming environment, responding to changes in the source content. There are many different types of information that may be extracted from social media datasets; one type is that of sentiment. When a user makes a post, it often contains an opinion or feeling that he or she has for something; this opinion may be either explicit or implicit.

One of the most prominent social media platforms for sentiment analysis is Twitter. Over time, Twitter has become a vast source of data for different machine learning analyses, including natural language processing and sentiment classification. Enormous quantities of tweets are processed by Twitter every second, many of which contain users' feelings towards products, brands, events, media, and a wide array of other things. The main difficulty with using these tweets to extract sentiment is that they lack direct labels which convey the overall feeling or leaning of the text. This has led to many different ways to classify the tweets, involving both supervised and unsupervised machine learning. Analyzing Twitter using natural language processing also introduces interesting problems; each tweet is limited to a maximum of 140 characters, meaning that normal grammatical and syntactical conventions are mostly ignored or circumvented. The consequence of this is that models of vocabulary and sentence structure must be rebuilt according to the unique style which has developed on Twitter.

In addition to Twitter, another large source of sentiment information is the Internet Movie Database (IMDb). IMDb allows reviewers with an account to post reviews on different movies, which includes both text and a quantitative rating. The net result is a large data source for sentiment analysis, where each review has a manual label from the review's author according to the overall sentiment of the review; the advantage of such data is that the quality of these labels allows for isolation of prediction inaccuracy, as the labels themselves are already extremely accurate. The type of text in this data source is significantly different from the type in Twitter; the users are not limited to a certain character count, meaning that the text is more syntactically and contextually coherent. The natural consequence of this is that words with similar denotations or connotations generally occur in the same context, which oftentimes will improve the word model built from a dataset.

This paper starts with a discussion of relevant literature on Twitter analysis, and then moves into a discussion of the datasets used in this study, which are derived from data taken from either Twitter or IMDb. After that, the methods utilized to obtain an abstract representation of a body of text as a vector are discussed in detail, as well as operations on this representation. A section dictating the results of these operations is next, followed by concluding remarks and possible avenues of future research.

2. RELATED WORKS

There have been several studies demonstrating the value of data the can be obtained through Twitter. Sentiment analysis of Twitter data has interesting applications such as analyzing movie sales [1] or stock market fluctuations [2] . It may also be used to evaluate public opinion of a brand [3] or current events [4] . The prevalence of public accounts along with incredible volume of tweets and minute-by-minute time resolution through the streaming API has made Twitter a popular choice for mass opinion mining.

Another valuable feature of Twitter data is the popularity of user annotations on their posts. This includes both hashtags (e.g. #smile) and emoticons (e.g. =D). Previous studies have used emoticons as self-labels for labeling tweets based on sentiment [5] [6] . Additionally, many of these studies remove the emoticons after labeling the tweets, leaving the original tweet text without emoticons. However, without the emoticons, the text often does not contain definitive sentiment information. For example, compare the difference between the tweets "heading to be early :)" and "heading to bed early :(". These tweets are both very plausible and understandable and in full express a clear sentiment, however it would be impossible to determine which sentiment the author was feeling once the emoticon is stripped.

As previously stated, there are many benefits to be gained in treating Twitter data as a live stream, including real-time analysis of events, but there are also many challenges. Some of these include time restrictions, unbalanced classes, informal structure, and highly compressed meaning [5] . For these reasons streaming models have to be built to be robust against these challenges. First and most obvious of these is the restriction on processing time from receipt of raw text to classification. This demands that algorithms perform only a single pass over incoming data. Severely unbalanced classes can cause certain algorithms such as SVMs to baseline their predictions with prior probability. This must be avoided to gain any useful insights from algorithms working in raw streams. Finally informal structure and condensed meaning which are common to all problems working with Twitter data can often be accounted for by using built in commonalities such as emoticons and hash tags.

3. DATA

In this study, we used three different datasets, two of which contain tweets (Sentiment Analysis Dataset, and the Large Emoticon Corpus) and another which contains movie reviews (Stanford Movie Review Dataset). The Large Emoticon Corpus is a new dataset introduced with this study. For this reason, we discuss the collection and cleaning methods for this dataset in greater detail below.

3.1. Sentiment Analysis

The Twitter Sentiment Analysis Training Corpus consists of about 1.5 million labeled tweets, and is a compilation of the Sentiment 140 [6] dataset, the Sentiment Classification [7] dataset, and the Twitter Sentiment Corpus [8] . This dataset was labeled using emoticons contained within the tweets, classifying each tweet as either positive or negative. The emoticons were then stripped from the text to avoid circular reasoning (i.e. tweet is labeled positive due to emoticon, emoticon dominates classification). We later removed mentions from the tweets (e.g. @someusertag), as any classification enhancement they provided might be attributed to overfitting.

3.2. Large Emoticon Corpus

The Large Emoticon Corpus, introduced in this paper, is a dataset of tweets which contain one or more emoticons from a unique set of 115 emoticons. It contains roughly 8.5 million tweets, collected over the interval May 13 - 18, 2015 using the Twitter 4J interface to the Twitter Streaming API. For analysis, mentions and retweet tags were removed from the data. A sample of 1000 tweets from this dataset was manually labeled with binary sentiment corresponding to positive and negative emotion. Additionally, a subset of 500 tweets from this sample was manually labeled with user gender.

The tweets in the Large Emoticon Corpus have time and date information associated with them, allowing time-based analytics to be performed. One such inquiry, which we present later, is the frequency of a certain emoticon in tweet text over time. All times are in UTC, so they are not matched with local time, thus preventing time of day analysis. It does, however, open an opportunity for analysis of overall global emoticon usage over time. Additionally, a subset of about 38,000 tweets is tagged with geographic coordinates, which would enable time of day analysis on the subset.

3.3. Stanford Large Movie Review Dataset

The Large Movie Review Dataset from Stanford consists of fifty thousand labeled movie reviews, and an equal number of unlabeled reviews, taken from IMDB. The labeled reviews combine the text of movie reviews with their associated numerical rating. Reviews with a rating of seven or more out of ten are considered positive, while those with a rating of four or lower are considered negative. No more than 30 reviews are allowed for a given movie; additionally, a majority of the reviews are unprofessional. During analysis, stop words (taken from the NLTK Stopwords Corpus [9]) were removed from the text to prevent overfitting.

4. TEXTUAL REPRESENTATION

The problem of representing natural language text in vector form for machine learning analysis has been around for decades, with early models using the Bag-of-Words technique. Textual representation can have a significant impact on the effectiveness of machine learning models as it determines what information is extracted from the original text. The goal is to retain as much pertinent information as possible. For this reason Bag-of-Words models, which lose all of the contextual and structural information of the text, are being replaced by distributed vector models such as Word2Vec and Sent2Vec; these are trained to understand the multifaceted relationships that exist between various words. The following sections highlight the different techniques we used to convert text into vectors.

4.1. Distributed Vector

Word2Vec [10] -[13] is an algorithm which converts individual words to n-dimensional vector representations. For a given text (tweet, sentence, or paragraph), the vectors of each of the words may be combined to produce a single vector which represents the text as a whole. The cosine distance between two vector representations roughly corresponds to the contextual similarity of the two texts [11] , allowing these vectors to be used in various clustering and classification algorithms. To maximize the usefulness of word representations, a model must be built using text similar in structure and context to the target data. To generate models, Word2Vec uses contextual cues to determine the similarity between all of the different words in the training sample, meaning that the size and content of the training sample have a significant influence on the resulting vectors. Another distributed vector model is Sent2Vec [14] -[16] which like Word2Vec creates a vector representation of a text in low dimensional space. It uses the Deep Structured Sematic Model (DSSM) or the DSSM with convolutional-pool- ing structure (CDSSM) to form its vector representations.

4.2. Bag of Words

The Bag-of-Words model of a text relates the frequency of a word or phrase in the text to an element in the resulting representative vector. This leads to sparse vectors in high dimensional space. For this study, phrases with anywhere from 2 to 5 words were used to represent the text in the Stanford Movie Review Dataset. Unlike Word2Vec and Sent2Vec, the Bag-of-Words model does not take into account the similarity in meaning of closely related words. Likewise the n-gram representation of a text is an alternate vector representation which is based on the number occurrences of a "gram" (a sequence of characters). We selected the most commonly occurring grams from the dataset; these were then used to convert each text to a vector by counting the number of occurrences of each gram in the given text. Each position in the resulting vector corresponds to a gram selected from the original text.

4.3. Feature Selection

Feature selection (a type of dimensionality reduction) is a way to reduce the dimensionality of a vector representation of a dataset, removing redundant features and reducing over fitting. Generally, the classification accuracy remains constant or is improved as a result of this process for sparse vectors. The Correlation Based Feature Selection [17] algorithm was used to reduce the size of each of the vectors; this algorithm finds features that are highly correlated to training labels but are not correlated with each other. Another method of feature selection is the Chi2 algorithm [18] . This algorithm takes continuous inputs, and discretizes to determine features. The CFS algorithm was mainly used to perform feature selection; the Chi2 algorithm was used to rank the relative importance of each feature.

4.4. Gender

In addition to manually labeling the sentiment of 1000 tweets, about 500 of the sentiment-labeled tweets were also manually labeled with user gender. A survey of this data is presented in Figure 1and Figure 2. The data indicate that emoticon usage on Twitter is dominated by female users, although surveys suggest total user gender distribution is much more balanced [19] [20] . However, this is a relatively imperfect metric as Twitter does not collect gender information on its users, requiring the use of gender predictive tools to estimate this balance [21] [22] . Additionally, in Figure 2 it is apparent that proportionally, the emoticons utilized by male users are more positive than those utilized by female users.

5. RESULTS

5.1. Stanford IMDB

Representative vectors of the Movie Review dataset were constructed using Word2Vec, Sent2Vec (both CDSSM and DSSM), and Bag-of-Words models. Features were then selected from each of the vector sets to create new, reduced vector sets. Naive Bayes, Random Forest, and SVM algorithms were run on both the full and reduced vectors sets for each model. The results are shown in Figure 3.

Performing feature selection on the Bag-Of-Words model of yielded the phrases most correlated with review sentiment. A selection of the top 20 of these is shown in Table 1. The majority of the phrases only contain 2 words, despite the fact that phrases with 2 to 5 words were used to build the Bag-of-Words; only 2 features from Table 1 contain 3 words. This is likely due to the fact that many higher-complexity phrases (containing a structure of smaller phrases) may have the same meaning, dictated by at least one of these smaller phrases; thus, after feature reduction, the majority of selected phrases have 2 or 3 words.

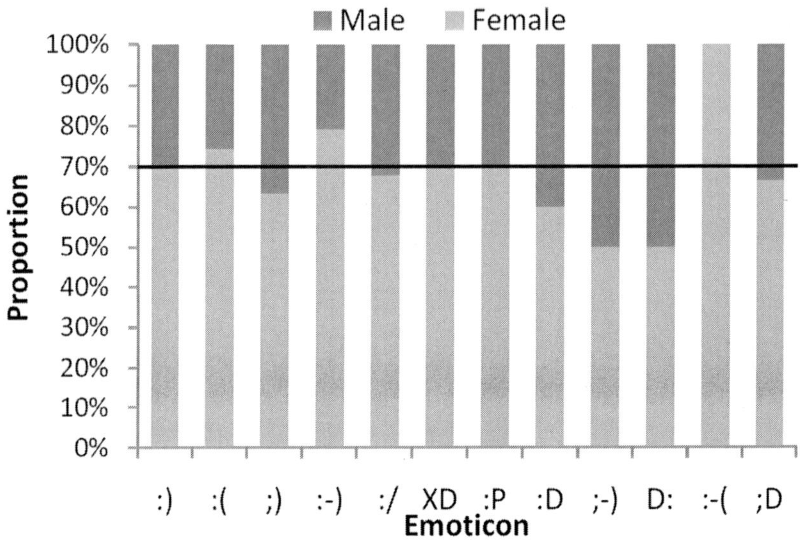

Figure 1. Emoticon usage by gender in a sample of about 500 tweets from the Large Emoticon Corpus manually labeled with user gender. The black line corresponds to the gender weighting of the sample. Note that the first few emoticons are significantly more frequent (Figure 6). While most of these emoticons are female-biased due to the sample, a few are used proportionally more often by males.

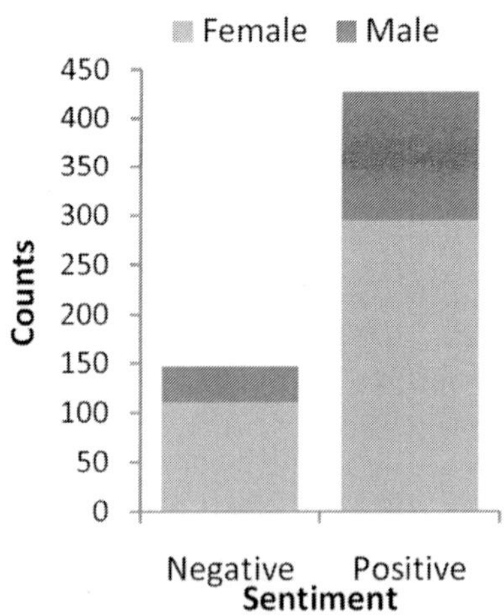

Figure 2. Positive and negative sentiment by user gender. This uses the same data as Figure 1. When males use emoticons, they use proportionally more positive emoticons than females.

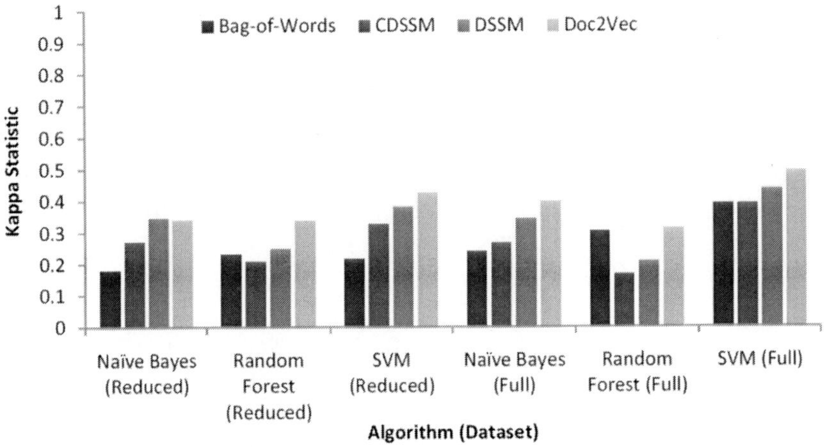

Figure 3. Prediction kappa statistic of different classification algorithms on movie review sentiment with different vector models. In most cases, the Word2Vec model outperformed the other models. Additionally, the Sent2Vec model outperformed the DSSM in all test cases. The Bag-of-Words model appears to be the least consistent of all the models.

5.2. Sentiment Analysis

The analysis of emoticon use over time shows small fluctuations over small periods of time (e.g. minutes or hours) but is relatively constant over longer periods of time (Figure 4). It also shows the frequency of use of the different emoticons. Emoticon use is dominated by the first few most frequent emoticons; the first four most frequent emoticons represent 86% of tweets with emoticons. This indicates that sentiment classification requires relatively few features to achieve reasonable accuracy when emoticons are included in the tweets. To show the significance of keeping emoticons within the tweets, classifying predictions were run on the vectors generated from emoticon-containing text as well as emoticon-absent text (Figure 5). Because of the potential for unbalanced class labels we use the kappa statistic rather than accuracy to evaluate the performance of each algorithm [5] .

As shown in Figure 5, building models with emoticon-containing text and doing predictive analysis with these models is significantly more effective than using emoticon-absent text. In both representations (N-Gram and Word2Vec), the classifying algorithms had superior performance when emoticons were included. Addi- tionally, the utility of using reduced vectors is apparent, as it stabilizes differences in algorithms. Also, the graph indicates that Word2Vec has a higher kappa statistic than N-Grams on the dataset without emoticons, indicating that the predictive ability of the N-Gram representation is highly dependent on the inclusion of emoticons.

Table 1. Ranking of the top 20 features from the Bag-of-Words model. Stop words from the NLTK Stopwords Corpus were removed from the dataset prior to analysis.

Top 1 - 10	Top 11 - 20
worst movie	worst movie ever
one best	worst movies
one worst	highly recommended
highly recommend	bad film
bad movie	well worth
worst film	looks like
bad acting	fast forward
movie bad	one worst movies
really bad	acting bad
must see	piece crap

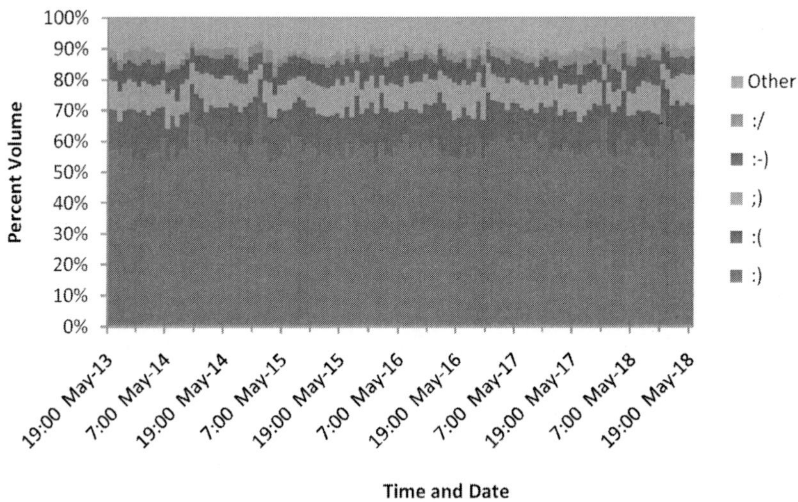

Figure 4. Emoticon usage by hour over a 5 day period from May 13, 2015 to May 18, 2015. The frequencies of each emoticon remain relatively constant over the collection period. The first five emoticons are the most dominate overall emoticon usage in the tweets.

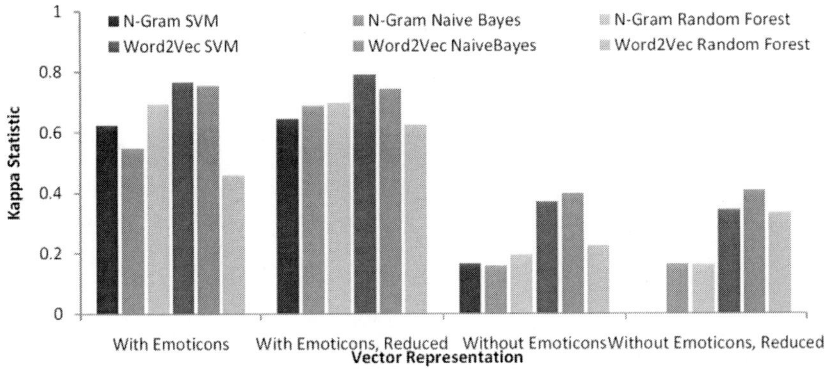

Figure 5. Chart showing comparison of classification algorithm kappa statistic on dataset with and without emoticons. The emoticons significantly affect the predictive ability of models built on the vectors produced from the tweets, especially in the case of N-Grams. Note that the SVM model prediction of the N-Gram reduced vectors had a Kappa Statistic of 0.

In natural language processing, it is common to perform stemming on the input text. This operation usually involves removing the suffixes from different tenses of a word, leaving just the root [23] (e.g. "connect", "connected", and "connecting" all have the root "connect"). A similar process may be applicable to emoticons as well, due to structural similarities in emoticons which convey the same sentiment. For example, the emoticons:), :-), =), ;), and ;) all exhibit the root characteristic ")", while :(, :-(, and = (all have the root characteristic "(" (see Table 2). While it is possible for positive emoticons to be backward (i.e. they contain "()", the frequency of this occurring is orders of magnitude less than the forward emoticons (see Figure 6)).

The structural similarity between emoticons may be used to reduce the amount of processing required to predict the sentiment of a given tweet. For example, the input text may be preprocessed by replacing emoticons with their base element. This reduces the number of unique features which correlate to sentiment, further reducing the computational expense of classifying a given text. An advantage of using this approach is that a regular expression may be used to match positive or negative tweets; this may be computationally more efficient than storing a set of emoticons and comparing each word each word in the set.

5.3. Dimensionality Reduction of Distributed Vectors

In many cases, vectors produced using Word2Vec contain many irrelevant or redundant features, which may increase the computational expense of classification or clustering. They may additionally add noise, which decreases accuracy. One common way to reduce this is by performing feature reduction after calculating representative vectors for text. However, it may be inefficient to generate vectors with excessive dimensionality, only to reduce or discard features.

One way to remedy this is to build a model which minimizes dimensionality without sacrificing predictive quality. To test this theory, we built multiple models based on the Large Emoticon Corpus that cover the spectrum of dimensionality, and tested the predictive ability for sentiment given two datasets. Vectors were obtained using multiple Word2Vec models; labels for this dataset were obtained by using a Random Forest model [24] built on the manually labeled subset of 1000 tweets. One of the datasets contained about 100,000 tweets, with about 90% of the tweets being positive and the other 10% being negative; this was considered the unbalanced set. The balanced set contained a subset of about 20,000 of these tweets, but was chosen such that there was an equal split of positive and negative tweets. Classifying was performed using an SVM model on the unbalanced and balanced datasets; additionally, a Naïve Bayes model was used to classify the unbalanced dataset for comparison. The results are shown in Figure 7.

Table 2. Positive and negative emoticons by root character. The regular expressions matching the positive and negative emoticons are shown in the final row.

<table>
<tr><th>Base</th><th>Positive Emoticons</th><th>Negative Emoticons</th></tr>
<tr><td>)</td><td>:) ;) :-) ;-) =)</td><td></td></tr>
<tr><td>(</td><td>(:</td><td>:(:-(=(</td></tr>
<tr><td>/</td><td></td><td>:/ =/</td></tr>
<tr><td>D</td><td>XD :D ;D =D</td><td></td></tr>
<tr><td>]</td><td>:] =]</td><td></td></tr>
<tr><td>Regular Expression</td><td>([;:\-=X]{1,2}[)\]D])|([(/D][;:\-=X]{1,2})</td><td>[;:\-=X]{1,2}[(/o]</td></tr>
</table>

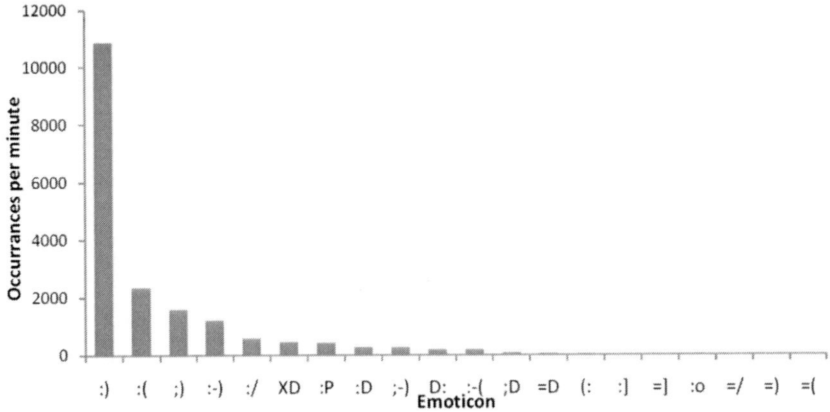

Figure 6. Emoticon usage in the large emoticon corpus. The counts are divided by total collection time in minutes to yield an approximate usage rate. The first few emoticons dominate total emoticon usage; the first emoticon occurs about 16,000 times more frequently than the last emoticon on the chart.

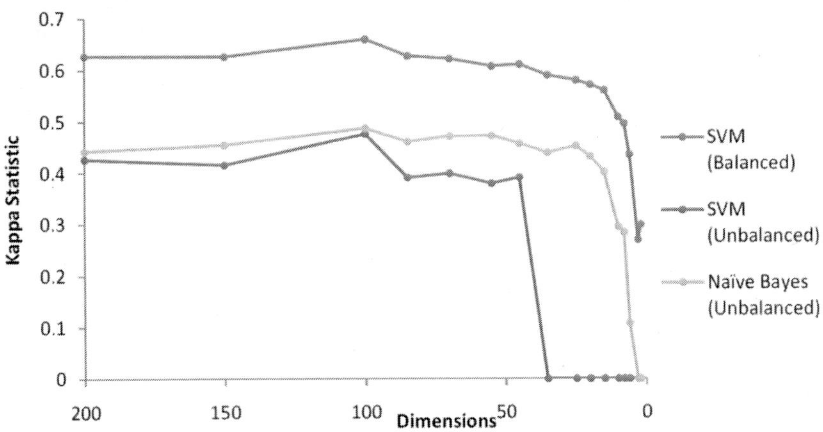

Figure 7. Kappa statistic of sentiment prediction on two samples of the large emotion datasets. The balanced sample has an equal proportion of positive and negative labels, while the unbalanced sample has 90% positive labels.

Depending on the data and model chosen, dimensionality can be significantly reduced without adversely affecting the predictive quality. If the dataset is balanced, it may be possible to reduce the dimensions further than if the dataset is biased toward one label. If the dataset is biased, an

SVM model, which is designed to optimize accuracy, may be unstable with too few dimensions; it may begin to base its predictions entirely on prior probability. In this case, more features may be necessary. This depends on model type, though, as the Naïve Bayes model appears to be much more stable at high bias and low dimensions, for metrics other than accuracy.

6. CONCLUSIONS

This study introduced the Large Emoticon Corpus, which had provided insights into emoticon usage on Twitter, both in the frequency of individual emoticon use and usage based on user gender. We had also shown that positive and negative emoticons might be deconstructed into more basic forms and potentially used to classify positive and negative sentiment by simple regular expressions. Additionally, we found a bias in emoticon usage toward female Twitter users, despite the relatively balanced gender partitioning of the platform as a whole. Also, we ranked the most influential phrases in the Large Movie Review Dataset, finding that the smaller phrases could better represent a set of longer phrases with similar meaning. Finally, we found dimension reduction to be an effective technique on distributed vectors; feature selection might be used to reduce dimensionality of models such as Word2Vec in order to improve computational efficiency when performing real-time analysis with data streams.

A possible area for future research was deeper investigation into gender bias on Twitter with a larger dataset; by using a larger set of labeled gender data, it would be possible to determine to what extent male and female Twitter users utilized different emoticons. Another possible research topic was the optimization of dimension reduction for other labeled datasets, including sets with continuous inputs, compared to the binary inputs used in this study.

ACKNOWLEDGEMENTS

We would like to thank the Summer Research Institute at Houghton College for providing financial support for this study.

REFERENCES

1. Rui, H., Liu, Y. and Whinston, A. (2013) Whose and What Chatter Matters? The Effect of Tweets on Movie Sales. Decision Support Systems, 55, 863-870.

2. Bollen, J., Mao, H. and Zeng, X.-J. (2011) Twitter Mood Predicts. Journal of Computer Science, 2, 1-8.
3. Jansen, B.J., Zhang, M., Sobel, K. and Chowdury, A. (2009) Twitter Power: Tweets as Electronic Word of Mouth. Journal of the American Society for Information Science and Technology, 60, 2169-2188.
4. Wang, H., Can, D., Kazemzadeh, A., Bar, F. and Narayanan, S. (2012) A System for Real-Time Twitter Sentiment Analysis of 2012 U.S. Presidential Election Cycle. Proceedings of the 50th Annual Meeting of the Association for Computational Linguistics, Jeju, 115-120.
5. Bifet, A. and Frank, E. (2010) Sentiment Knowledge Discovery in Twitter Streaming Data. Discovery Science, 1-15.
6. Go, A., Bhayani, R. and Huang, L. (2009) Twitter Sentiment. Stanford Digital Library Technologies Project.
7. University of Michigan (2011) UMICH SI650—Sentiment Classification. https://inclass.kaggle.com/c/si650winter11/data
8. Sanders, N.J. (2011) Sanders-Twitter Sentiment Corpus. Sanders Analytics LLC.
9. Bird, S., Loper, E. and Klein, E. (2009) Natural Language Processing with Python. O'Reilly Media Inc.
10. Le, Q. and Mikolov, T. (2014) Distributed Representations of Sentences and Documents. CoRR, vol. abs/1405.4053.
11. Miklov, T., Chen, K., Corrado, G. and Dean, J. (2013) Efficient Estimation of Word Representations in Vector Space. http://arxiv.org/abs/1301.3781
12. Miklov, T., Sutskever, I., Chen, K., Corrado, G.S. and Dean, J. (2013) Distributed Representations of Words and Phrases and Their Compositionality. In: Advances in Neural Information Processing Systems, Morgan Kaufmann Publishers Inc., San Francisco, 3111-3119.
13. Mikolov, T., Yih, W.-T. and Zweig, G. (2013) Linguistic Regularities in Continuous Space Word Representations. HLT-NAACL, 746-751.
14. Huang, P.S., He, X., Gao, J., Deng, L., Acero, A. and Heck, L. (2013) Learning Deep Structured Semantic Models for Web Search Using Clickthrough Data. Proceedings of the 22nd ACM International Conference on Information & Knowledge Management, San Francisco, 27 October-1 November 2013, 2333-2338. http://dx.doi. org/10.1145/2505515.2505665
15. Shen, Y., He, X., Gao, J., Deng, L. and Mesnil, G. (2014) A Latent Semantic Model with Convolutional-Pooling. Proceedings of the 23rd ACM International Conference on Conference on Information and Knowledge Management, Shanghai, 3-7 November 2014, 101-110.
16. Gao, J., Pantel, P., Gamon, M., He, X., Deng, L. and Shen, Y. (2014) Modeling Interestingness with Deep Neural Networks. Proceedings of the 2014 Conference on Empirical Methods in Natural Language Processing, Doha, 25-29 October 2014. http://dx.doi.org/10.3115/v1/D14-1002
17. Hall, M.A. (1999) Correlation-Based Feature Selection for Machine Learning. PhD Dissertation, University of Waikato, Waikato.
18. Liu, H. and Setiono, R. (1995) Chi2: Feature Selection and Discretization of Numeric Attributes. Proceedings of the IEEE 7th International Conference on Tools with Artificial Intelligence, Herndon, 5-8 November 1995, 388-391.
19. Statistica (2015) Number of Active Twitter Users in the United States from 2010 to 2014, by Gender. http://www.statista.com/statistics/238715/number-of-active-twitter-users-in-the-united-states-by-gender/
20. Beevolve (2012) An Exhaustive Study of Twitter Users across the World. http://www.beevolve.com/twitter-statistics/#a1
21. Miller, Z., Dickinson, B. and Hu, W. (2012) Gender Prediction on Twitter Using Stream Algorithms with N-Gram Character Features. International Journal of Intelligence Science, 2, 143-148. http://dx.doi.org/10.4236/ijis.2012.224019
22. Dietrick, W., Miller, Z., Valyou, B., Dickinson, B., Munson, T. and Hu, W. (2012) Gender Identification on Twitter Using the Modified Balanced Winnow.

Communications and Network, 4, 189-195. http://dx.doi.org/ 10.4236/cn.2012.43023
23. Porter, M.F. (1980) An Algorithm for Suffix Stripping. Program, 14, 130-137. http://dx.doi.org/10.1108/eb046814
24. Pedregosa, F., et al. (2011) Scikit-Learn: Machine Learning in Python. Journal of Machine Learning Research, 12, 2825-2830.

CHAPTER 17

Flexible Composition and Execution of Large Scale Applications on Distributed E-Infrastructures

Stefan J. Zasada, David C.W. Chang[1], Ali N. Haidar[2], Peter V. Coveney*

Centre for Computational Science, University College London, 20 Gordon Street, London WC1H 0AJ, United Kingdom

ABSTRACT

Computer simulation is finding a role in an increasing number of scientific disciplines, concomitant with the rise in available computing power. Marshalling this power facilitates new, more effective and different research than has been hitherto possible. Realizing this inevitably requires access to computational power beyond the desktop, making use of clusters, supercomputers, data repositories, networks and distributed aggregations of these resources. The use of diverse e-infrastructure brings with it the ability to perform distributed multiscale simulations. Accessing one such resource entails a number of usability and security problems; when multiple geographically distributed resources are involved, the difficulty is compounded. In this paper we present a solution, the Application Hosting Environment,[3] which provides a Software as a Service layer on top of distributed e-infrastructure resources. We describe the performance and usability enhancements present in AHE version 3, and show how these have led to a high performance, easy to use gateway for computational scientists working in diverse application domains, from computational physics and chemistry, materials science to biology and biomedicine.

Keywords: E-infrastructure High performance computing Application virtualization Usability

1. INTRODUCTION

Today's computational scientists face a growing number of challenges which affect their ability to fully exploit the computational resources, made available to them via so called *e-infrastructures* (such as PRACE, EGI or EUDAT in Europe, or XSEDE in the USA). Firstly, they have an unprecedented amount of computational power available to them, which will continue to grow inexorably in the future, presenting many opportunities as well as challenges to an increasing number of scientific disciplines that rely on computer based modelling and simulation.

Secondly, the architectures of these large scale high performance computing (HPC) machines point to a growing trend of computers comprised of hybrids of scalar and vector processors [1,2]. This requires application scientists to ensure their code is optimized to take full advantage of the hybrid architecture of a specific machine. Grid computing [3,4] has sought to simplify end user access to and use of HPC resources, but the middleware tools developed to realize the computational grid concept have seldom provided the transparency and ease of use envisaged [5]. The challenges described above are compounded when one attempts to invoke multiple resources, in order to achieve more than just the sum of their individual parts [6].

Alongside grid computing we have witnessed the development of cloud computing. Cloud computing represents a fast growing business model that seeks to commoditize computational infrastructure, and provide access to various distributed resources such as CPU, memory and storage (known as infrastructure services) and applications (software as services). It is a rapidly growing area due to major strategic investments from global software companies such as Microsoft, Amazon, Google and IBM. Cloud storage today is growing in popularity, particularly due to its shared data at low cost capabilities. Nonetheless, there are many security and legal issues in cloud computing that are yet to be resolved.

The Application Hosting Environment [7,8] is a middleware layer designed to simplify the user's ability to exploit computational resources beyond the desktop, greatly facilitating the use of e-infrastructure. It has been deployed in support of a diverse set of projects, including HIV-1 protease modelling [9], immune system simulation [6], and large scale materials modelling [10]. AHE seeks to converge the Software as a Service model of cloud computing with high performance grid computing. In this paper we will discuss the concepts behind AHE, and describe in detail the latest version of the Application Hosting Environment, AHE 3.0, which has been reimplemented using RESTful services [11] rather than WSRF services [12]. We will demonstrate how the work we have done to redesign AHE 3.0 has led to a significant increase in performance compared to AHE 2.0 [8], and we show how this new version of AHE is benefiting various ongoing research projects.

2. SERVICE ORIENTED COMPUTATIONAL SCIENCE

Virtualization is a broad term used in computer science to describe the abstraction of resources. Application virtualization describes a range of technologies designed to separate an application from the operating system that it runs on. In many cases this is achieved by introducing compatibility layers around underlying operating system features and libraries, for example the WINE system used to run Windows applications on UNIX systems [13].

The key aim of virtualization is to abstract away all the details of an underlying hardware or software system from the concern of the user. The benefits are manifold: developers can code to a single virtualized interface or system rather than for a specific hardware implementation; multiple virtual instances of a system can often be run side by side on a single physical system (in machine virtualization for example); and physical resources can be protected.

The growth of virtualization technologies, along with service oriented architectures (SOA), has also driven the development of cloud computing. The use of virtualized interfaces and systems means that the specific details of a cloud's architecture are hidden from consumers of the cloud resources. Several cloud computing models exist; the Infrastructure as a Service (IaaS) cloud paradigm typically takes the form of virtualized servers running on hardware platforms managed by the cloud hosting company, where each user is given access to one or more virtual servers, solely under their control. This also provides a degree of *elasticity*, as the number of virtual machines in a cloud environment can be greater than the number of physical servers available to the hosting entities. The Software as a Service (SaaS) cloud paradigm delivers access to applications centrally hosted on a cloud platform, typically via a web browser.

While virtualization technologies certainly reduce the complexity of using a system, and especially when working across multiple heterogeneous computing environments, they are not widely deployed in high performance computing scenarios. As its name suggest, HPC seeks to obtain maximum performance from computing platforms. Extra software layers impact detrimentally on performance, meaning that in HPC scenarios users typically run the applications as close to the 'bare metal' as possible. In addition to the performance degradation introduced by virtualization technologies, choosing what details to abstract in a virtualized interface is itself very important. Grid and cloud computing support different interaction models. In grid computing, the user interacts with an individual resource (or sometimes a broker) in order to launch jobs into a queuing system. In cloud computing, users interact with a virtual server, in effect putting them in control of their own complete operating system. Both of these interaction models put the onus on the user to understand very specific details of the system that they are dealing with, making life difficult for the end user, typically a scientist who wants to progress his or her scientific investigations without any specific usability hurdles obstructing the pathway.

To address these problems, we have developed a software layer designed to implement the Software as a Service cloud paradigm for scientific applications that rely on high performance computing, mediated by the *Application Interaction Model* which we describe in Section 3, derived from the user requirements also discussed in Section 3. This model is based on the insight that many e-infrastructures impose a steep learning curve on the majority of end users, who do not possess the technical expertise for the most part to compile, optimize, install, debug and finally launch their applications; they simply want to run their applications, obtain results and focus on their scientific endeavours. While an application may consist of a single execution of a computational code, it could also consist of a complex set of operations involving multiple codes, connected as a workflow; AHE enables all kinds of applications to be treated as simple "atomic" units, helping realize the original vision of a grid as "distributed computing performed transparently across multiple administrative domains" [14].

3. USER REQUIREMENTS

For supercomputer class applications, the user generally has to install his/her own application, if that application is not one of the few community codes pre-installed on the machine; it is not possible simply to stage an executable to the target resource as it requires too much bespoke tailoring to the particular hardware setup of the resource. Generally a group of researchers will want to use the same application on a resource. However, many users will not know where a particular application is installed on a target system, nor will they necessarily know the best way to run the application on a particular system. Often, with supercomputer class systems, applications have to be run in specific ways to achieve the best performance. The community's expert users must spend time educating other users on the vagaries of different queuing systems and machines. Typically, the end user will need to stage data to the supercomputer before he/she is able to execute her application. Therefore, the supercomputer must provide accessible interfaces over which data can be staged. In order to launch an application, the users have to prepare a description of the job that they want to run, which is submitted to the queue management system on their target resource, in a format that the queue management system understands and which is potentially incompatible with other instances of the same queue management system running on other resources. Once the job has been submitted, users monitor the progress of their jobs through the queuing system, using interfaces provided by the resources.

Distributed applications can consist of multiple computational codes launched on multiple resources, connected together as workflows of operations, as well as single codes launched on single resources. Applications can get their data from multiple sources, such as online data repositories and databases, and store their output data in similar resources. Typically, users will be given allocations of time on individual grid resources, or the e-infrastructure as a whole, through awards made to their project's principal investigator. These allocations will have a notional associated cost, the cost per CPU hour, derived by the resource operator from their running costs and a projected resource utilization. Such allocation models inhibit the most creative use of and ways of exploiting distributed e-infrastructure.

The scientific end user's primary concern is running their application in a timely fashion, in order to obtain results that further their scientific objectives. All the services and facilities provided by a grid should be subservient to this end. Typically, the user does not even care which machine on the grid their application is run on, as long as results are delivered within a time frame that makes them useful, whether that is the time to publish a scientific paper, or the time to conduct a potentially life-saving medical simulation [15].

A further problem faced by end-users and administrators of computational e-infrastructures arises in connection with the usability of the security mechanisms usually deployed in these environments, in particular identity management. Many of the existing computational grid environments use Public Key Infrastructure (PKI) and X.509 digital certificates as a cornerstone for their security architecture. However, it is well documented that security solutions based on PKI lack user friendliness for both administrators and end-users, which is essential for the uptake of any grid security solution [16,17]. The problems stem from the process of acquiring X.509 digital certificates, which can be a lengthy one, and generating proxy certificates to get access to remote resources as part of the authentication process [17]. As a result, many users engage in practices which weaken the security of the environment, such as the sharing of the private key of a single personal certificate, to get on with their tasks.

From many years of working with high-end computing applications, we have been able to derive a set of requirements for a system designed to make the use of distributed and HPC applications transparent to the end user. Current e-infrastructures focus on submitting jobs to batch schedulers on computational resources, meaning the user has to interact at both job and resource levels. Since users' predominant interest is running their application within a useful time frame, the first requirement of our usability model is that it should promote applications as a first class resource concept. *All user interactions should be with the application, rather than the machine, scheduler and job.*

Current e-infrastructure job submission mechanisms put the onus on the user to manage and curate their application's output data. Our model preserves the full state of each instance of an application, including all parameters and data used to launch the application, and all simulation output. This assists with tracing the provenance of simulation results, and is key to simulation reproducibility. Current grid middleware tools require the user to perform a number of steps in order to launch their code. Our model reduces the number of steps required to the minimum number possible in order to successfully run an application.

Current systems require the user to generate complicated job description documents in order to submit their application. Our model allows the user to launch their application using the simplest set of requirements possible, and takes care of generating whatever job descriptions the middleware on the underlying resource requires [17]. Users of supercomputer class resources may have access to a number of such resources via different computational grids, running different grid middleware stacks, requiring them to learn how to use different middleware tools to submit their jobs [18]. Our model presents a uniform interface to users to access resources running different middleware stacks, allowing them to transparently access not only single grids but also federated resources from multiple grids.

3.1. The Application Interaction Model

Based on our user needs analysis we have derived the Application Interaction Model, designed to allow users to easily control virtualized applications running on remote e-infrastructures. Traditionally, HPC focuses on the concept of 'jobs' to describe distinct workloads submitted to a batch queue. We purposefully focus on the concept of *applications*. An application is a higher level concept than a job; although an application could be realized by a single HPC job, it could equally correspond to a coupled simulation, where two codes (launched as two HPC jobs) pass parameters between themselves, or a steered application which requires steering Web services to be initialized before the code is launched, or a workflow of arbitrary complexity. However the application is composed the user should still interact with a single entity to control the execution of all components of the application.

We define the Application Interaction Model as follows:

(i)The virtualized application is the central entity in the Application Interaction Model.

(ii)An application does not necessarily correspond to a single computational code – it could be composed of multiple computational codes linked together in a workflow, or a computational code and associated steering Web services. However, it is presented to the user as a single application.

(iii)The application encapsulates all of the details of how to launch it, such as where the binaries that constitute the application are located, how to interact with individual resources and so on. These details are shielded from the user, who does not need to know anything about the underlying details.

(iv)Each instance of an application is controlled by a separate application instance, through which it is controlled. The application instance encapsulates all of the state associated with that run of the application, such as the input and output data, the application parameters and so on.

(v)All user interaction occurs through the virtualized application instance, which causes the computational code(s) that constitute the application to be launched on back-end computational resources.

(vi)Operations on the application instance allow the user to stage data associated with the application to the resource where it is needed, launch, monitor and terminate the application. These operations have an effect on the codes running on remote grid resources.

A schematic representation of this interaction model is shown in Fig. 1. The principal motivation behind this approach is to simplify the use of e-infrastructures, by introducing an abstraction layer between the users and the high end computing resources available to them which hides the complexity of the latter, providing an abstract interface to scientific applications deployed on a grid. This abstraction layer takes care of the process of launching the application on one or more HPC resources, and reduces the interaction with an application to those operations most relevant to the user.

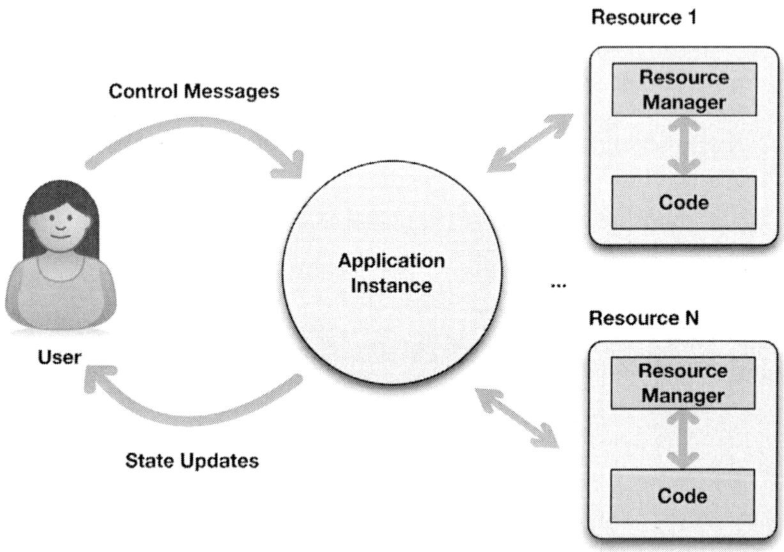

Figure. 1. The Application Interaction Model underpinning AHE. The application instance is the central entity representing each instance of an application that a user launches. All user interaction is mediated via the application instance, which supports operations to launch, monitor and terminate the application, and to manage data sharing.

The Application Interaction Model implies that the task of deploying and configuring an application is taken care of by a system administrator, or a community's 'expert user'. This draws a parallel with many different communities that use applications on high performance computing resources, such as the UK Collaborative Computational Projects (CCPs) [19], where a group of programmers develop a code, which they then distribute to an end user community. Once the expert user has configured AHE to share an application, end users can invoke clients installed on their workstation, tablet or mobile phone to launch and monitor the application across a variety of geographically distributed computational resources.

4. THE APPLICATION HOSTING ENVIRONMENT

The Application Hosting Environment (AHE) is our implementation of the Application Interaction Model. AHE is based on two key concepts to promote usability: application virtualization and community application sharing. Application virtualization allows developers to code against a single virtualized interface instead of the specific underlying software or hardware system, and it also allows multiple virtualized instances to be run side by side on a single physical system. The concept of an application in AHE does not necessarily imply a single computational code executed on a single HPC resource. A virtualized application in AHE can be comprised of more complex workflows, such as coupled simulations where multiple applications are required to pass data to each other, for example coupled quantum and molecular level simulations, made up of separate codes that exchange data via files [20].

AHE is built around the idea of a community model. In this paradigm, an expert user is required to set up and configure AHE with details of a scientific application, the distributed (grid) infrastructure it is deployed on and then uses AHE's RESTful interface to share this scientific application transparently with a group of end users. The resources that a particular application is deployed onto are chosen based on the characteristics and requirements of the application. An end user can then launch and monitor applications through the AHE desktop GUI client, web client or command line client (described in Section 4.2), and any combination of these clients can be used simultaneously. AHE has gone through several software releases, principally AHE 1.0 [7] and AHE 2.0 [8]. AHE 3.0, which we describe in this paper, adds many significant new features to those found in AHE 2.0, with an unremitting focus on usability and reliability.

4.1. The architecture of AHE 3.0

AHE 3.0 is a complete re-implementation of AHE 2.0 in Java (AHE 2.0 and earlier versions were implemented in Perl). AHE 3.0 introduces a new workflow engine based on JBPM [21]allowing complex workflows to be created and integrated into AHE, and accessed by users as single applications. AHE 3.0 also incorporates an object relational mapping framework using Hibernate [22], which simplifies installation and development of AHE by decoupling AHE from the database used to maintain state. A REST ful web service interface based on the Restlet [23] library furnishes a simple and concise HTTP based interface for clients to access AHE services, compared to the WSRF [12] based services used in AHE 2.0.

AHE 3.0 is a departure from AHE 2.0 and earlier releases, having undergone a complete redesign. AHE 3.0 comprises a number of modules which implement the core functionality. AHE runtime controls the start-up and shut-down of the AHE application life cycle; AHE engine implements the core functionalities including the workflow engine as well as facilitating interactions between the different components; AHE connector module implements the functionality required to connect to different types of middleware; AHE security module handles the security component as well as user management of AHE application and grid middleware; AHE interface module provides a REST ful Web service interface as well as command line access to AHE; the file module provides mechanisms to transfer files between different storage resources using Grid FTP. A schematic showing the interaction of these different components is presented in Fig. 2.

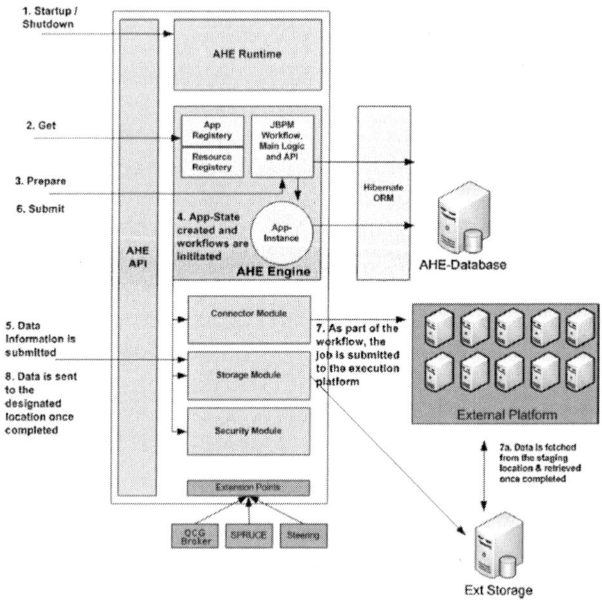

Figure. 2. The architecture of the AHE 3.0 server, showing the relationship between the different software modules.

RESTful Web service provides a simple abstraction of AHE's functionalities to the user by exposing AHE components as resources, each of which is identified by a global identifier (URI). This provides a clean and simple mechanism for end users to access AHE, making client tooling less complicated; it also means that AHE 3.0 can either be deployed via a servlet container such as Tomcat, or as a standalone server. A detailed discussion of the AHE 3.0 server components is presented below.

4.1.1. AHE runtime module

The AHE runtime module is responsible for starting up and shutting down the server in the standalone mode. It also initializes all the components and user configurations as well as the basic registry information relating to users. In standalone mode, AHE uses an embedded Jetty Server to provide web server functionality, including HTTPS with mutual user certificate authentication. In this mode, AHE 3.0 can be started from an executable file. AHE can also be deployed as a Java servlet into a servlet compliant server such as Apache Tomcat [24]. AHE can be deployed with an embedded H2 database [22] or use an external database through the Hibernate framework. More details about the deployment are discussed in Section 6.

4.1.2. AHE engine module

The AHE engine module encapsulates the functionality through which AHE 3.0 virtualizes access to scientific applications. It provides methods to create an Application Instance object, used to represent an instance of a virtualized application. In addition, methods are provided to run and maintain the execution workflow for each virtualized application instance. The workflow describes how the data and computational code(s) associated with this application instance are processed, including details such as the back-end connector (cf. Section 4.1.3) and security mechanism to use.

The AHE engine module also allows higher level workflows to be implemented. These workflows can control multiple application instances to create parameter sweep applications or complex chained application scenarios, in which data created by an application is used as the input for a second one and so on.

4.1.3. AHE connector module

The connector modules provide a set of classes that invoke external Java libraries which allow AHE to act as a client to distributed resource managers (DRMs) such as Globus GRAM. The connector module provides a generic Java interface (using the adapter pattern) which adapters for different DRMs have to implement. This Java interface is used by AHE 3.0 to access external computational resources, providing a loosely-coupled relationship between AHE and external client libraries. Connectors currently exist to allow AHE 3.0 to run applications via Globus 5.0 [25], Unicore 6 [26] and QCG OGSA-BES [27]. Each connector implementation translates the AHE's internal application state model into specific directives to the relevant DRM, such as the number of cores to use. The extensible interface framework means that interfaces to other DRM systems can easily be added as necessary. Each connector module is responsible for trapping errors from the underlying DRM and mapping it to an AHE error state, which is presented to the user.

4.1.4. AHE interface module

This module contains library code used to interface with AHE 3.0. This includes a bridge between the RESTful Web service interface (cf. Section 4.1.7) and the AHE runtime module (cf. Section 4.1.1). The AHE REST Web services exposes the main AHE functionalities and components as resources which can be controlled by performing operations on those resources.

4.1.5. AHE security module

The AHE security module provides a number of important functions, including user management, authorization and authentication control, platform credential management and integration with Audited Credential Delegation [28] (see Section 5). AHE provides a mechanism to delegate security control to ACD; these security functions include user authentication and management as well as virtual organization support and proxy generation for any specified virtual organization. In ACD mode, AHE contacts ACD using RESTful Web service calls to authenticate users as well as to request the generation of proxy credentials on a per user basis. AHE itself is also able to authenticate users via SSL certificates or the more standard username/password credential combination. The security module is able to perform command level authorization, as well as platform credential management such as maintaining private key and certificate information for a user which may be required for him/her to be granted access to particular computational resources and data. Additionally, the security module is able to request updated proxies from a MyProxy server when a certificate is about to expire.

4.1.6. AHE transfer module

In AHE, input data is transferred directly from a location specified by the user to the computational resources where an application is to be run. Once a job is completed, the AHE server takes care of staging any output data back to the user specified location. The AHE Transfer module provides a mechanism to set up the security credential used to authenticate transfers and then initiate transfers between two different storage components through the AHE Java file transfer interface. Different transfer mechanisms have been implemented using this Java interface. Currently, file transfer functionalities make use of the jGlobus and UCC libraries to stage data using the GridFTP [29] and Unicore transfer protocols. The Java interface makes it easy to add new transfer protocols in future should they become necessary.

4.1.7. RESTful web service

One of the main features of AHE 3.0 is the implementation of the RESTful Web services interface, which replaces the WSRF interface used in AHE 2.0

and earlier versions. RESTful Web services expose resources addressable via HTTP and operated on using HTTP operations such as POST and GET. This provides a secure and straightforward universal end point for AHE to provide services to users. AHE 3.0 uses the Java Restlet library for its RESTful implementation [23]. The Restlet library was chosen to underpin the AHE 3.0 server due to the many features it provides, including the ability to develop services that run as standalone services or which can be deployed in a servlet container such as Tomcat (using either the J2SE or the J2EE version of the library), multiple native data representation formats such as XML and JSON, and scalability as well as security support.

The AHE REST command structure is grouped into a number of resources including: User, AppInstance, AppReg (application registry), Resource, PlatformCredential and Cmd (general commands). Each of these resources can be viewed or modified using the GET, POST or DELETE HTTP operations when applied to a suitable AHE resource URI. A typical AHE URI consists of several components; the domain URI followed by the user identifier and the AHE resource that will be operated on. The URI is followed by the command and argument if required.

4.1.8. AHE workflow engine

A key component of AHE 3.0 is the workflow management system built on the JBPM framework. JBPM is a lightweight Java workflow engine, with workflows described using the Business Process Modelling Notation (BPMN) 2.0 specification which calls specific Java classes, scripts or Drool rules to perform arbitrary functions. This allows new complex workflows and scripting functionalities to be introduced quickly to extend AHE. JBPM supports workflow persistence using the Hibernate framework meaning that, in the event of a server crash, the workflow can be recovered quickly and seamlessly. There is also a wide range of tools available to plug in to the JBPM framework, including workflow editors, which eases the integration of JBPM with AHE.

By using a workflow engine, further functionalities and workflows can be introduced into AHE applications. This allows the expert user to tailor customized workflows to complex tasks, such as coupled model applications. It also allows additional features or functionalities to be added, such as fault tolerance, and to integrate AHE with external services such as SPRUCE [30] in order to submit urgent computing jobs, and RealityGrid Steering [31] which allows scientists to interact with running applications.

4.2. AHE client

The AHE server maintains all state information about a particular application instance. This means that client tools need to store no information about individual application runs, and consequently very simple clients containing little configuration data can be created. It also means that clients can interoperate, with one client used to launch an application and another used to monitor it for example.

The simple REST endpoints exposed by AHE 3.0 server mean that in practice any tool which can perform HTTP POST and GET operations (such as the UNIXcurl command) can be used as clients. However, a Java client API has been developed which not only provides methods to call AHE 3.0 server commands, but also provides auxiliary functions such as data staging. This API has been used to produce both graphical and command line clients. It also allows applications hosted in AHE to be accessed from high level tools, and integrated with workflow engines such as Grid Space and Taverna (see Section 8).

In addition, an AHE web client has been developed to provide a simple interface for the end user when interacting with the AHE server via a web browser. The web client interface has been developed using the Google Web Toolkit (GWT) and communicates with the AHE 3.0 server through its RESTful interface. The web client can be deployed on Java servlet compliant servers such as Tomcat or JBoss AS. The web client also allows the user to administer and configure an AHE 3.0 server, providing capabilities to manage users, certificates, applications, target computational resources and the server itself. The web client also allows end users to transfer files, launch and monitor AHE jobs through a web browser.

4.3. AHE application life cycle

AHE manages the whole life cycle of an application when invoking AHE, from input data staging, through job execution, to output data staging; during this process the application transitions through a number of different states. This life cycle is shown in Fig. 3.

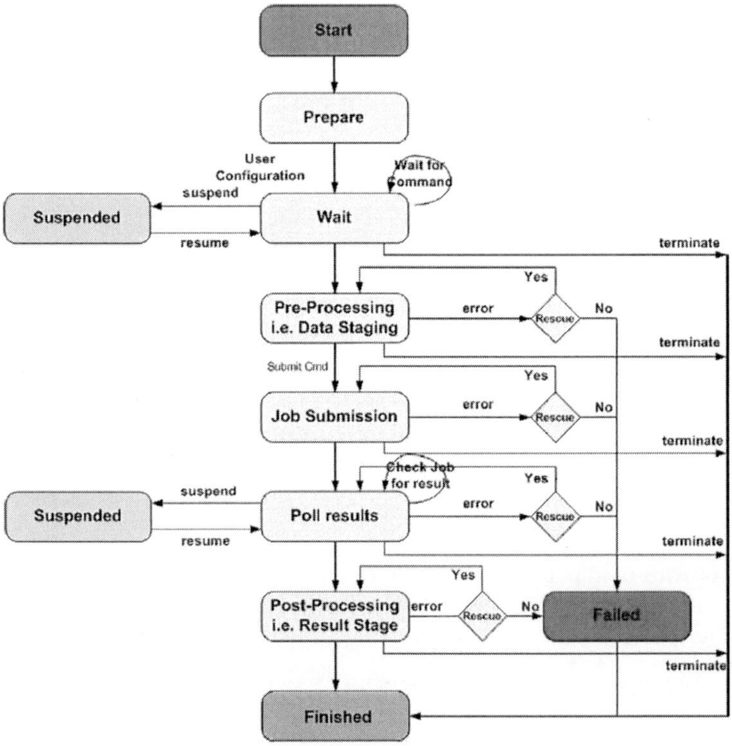

Figure. 3. The application life cycle. AHE server manages the transition of an application instance through a number of states, in order to stage data, execute an application, and handle failures.

The process starts when a prepare command is received by AHE. This puts AHE in a waiting mode, allowing the user to set up additional configuration details required for the application or workflow submission. Once a start command has been submitted, AHE server proceeds to first stage any input data that the user has attached to the application instance; once that is completed, it is then submitted to the execution platform. Once the application has been submitted to an external execution platform, AHE goes into a polling state, checking regularly for the completion of the application. When the job has completed, any output data is staged to the location specified by the user and the job submission process comes to an end.

If errors occur during certain stages of the AHE workflow process, AHE captures the error and allows the user to fix this error and attempt to execute the same step again. This workflow is modelled and executed using the JBoss jBPM workflow library and additional components can be added to the workflow if necessary.

In practice, a user has to go through the following steps in order to run an application:

(i)AHE runtime initializes all components, populates the internal data structures and ensures that the data held in the state database is synchronized with the AHE data structures.

(ii)The user queries the application registry to see what applications are available.

(iii)The Prepare command is issued which tells AHE Engine to create a persistent App-Instance Object that keeps track of the status and state of an executing application, which in turn initiates the AHE workflow process. An App-Instance object is a representation of a virtualized application initiated by the user. This allows AHE to keep track of the state and progress of the virtualized application.

(iv)This App-Instance object is persistent and stored in a local database using the Hibernate Framework, which allows AHE server to be database agnostic. In particular:

(a)the App-Instance object is associated with a user/group and has a unique identifier;

(b)active App-Instance data/objects are held in a registry and checked by AHE Engine to see which processes can be operated on each App-Instance, such as when and how they can be run, when data can be checked or retrieved, and so on.

(v)Input data files required by the application are staged to the target resource. AHE Server records the location and transfer protocol specified for each individual data file and passes that information to the relevant connector module so the job manager knows how to stage the data and retrieve the results if necessary.

(vi)The user next issues the submit command.

(vii)AHE workflow module then schedules the execution of the application using JBPM and quartz scheduler [32]. This allows complex workflows to include asynchronous tasks, as well as multi-thread/concurrency support.

(viii)AHE Engine deals with the security interface requirements and submits tasks to external execution platforms. AHE polls the external execution platform (if it is configured to do so) and retrieves any output data once the application is completed. JBPM allows additional features to be added in order to create more complex workflows incorporating AHE plug-in components. JBPM is persistent so that all events are logged. If the server crashes, the workflow state stored in a database can be retrieved and reinitialized.

(ix)Once the application has completed, the data is retrieved and sent to the scratch disc (temporary file storage) or copied to an external storage resource specified by the user, allowing him/her to access it.

5. SECURING AHE 3.0 WITH AUDITED CREDENTIAL DELEGATION

Efforts to address the usability of e-infrastructures are hampered by existing security mechanisms imposed on users. Typically, these require a user to obtain one or more digital certificates from a certificate authority, as well as to maintain and renew these certificates as necessary. The difficulty in doing this leads to widespread certificate sharing and misuse and a substantial reduction in the number of potential users [16]. In order to remove this barrier, we have coupled AHE 3.0 to Audited Credential Delegation (ACD) [28]. ACD is a usable security system that accommodates the security requirements of both end-users and resource providers, offering facilities to authenticate, authorize and audit all transactions.

The main advantage of ACD is that it entirely removes the use of digital certificates from end-users' experience, minimizing the usability problems caused by such credentials while addressing resource providers' concerns regarding securing access to their shared resources, tracing the users responsible for performing specific tasks on their resources. ACD enables users to invoke security credentials they are familiar with such as their institutional username/password combination (using Shibboleth [33], for example); assuming that they are authenticated it issues a digital certificate to them when necessary in the background.

ACD can be used to set up multiple virtual organizations (VO) to manage dynamic groups of users wishing to access e-infrastructure based resources, and to provide VO administrators with tighter control of users' actions as well as identity management. Existing solutions such as MyProxy [34], Shibboleth, and SARoNGS [35] on their own only provide credential repositories to store short lived X.509 certificates (MyProxy), web based single sign-on (Shibboleth), and web portals to access grid resources using a combination of Shibboleth and VOMS [36] (SARoNGS). None provides a holistic VO controlled security solution in the way ACD does.

The design of ACD is based on the concept of wrappers. A wrapper is a connector between a component and the outside world. It enables controlled access to the functionalities of a component. The ACD security wrapper comprises authentication, authorization and auditing components. Any request by a user to perform an operation using a service secured by ACD is intercepted by the security wrapper to establish the identity of the requester, to check whether or not the user is allowed to perform the task, to record the results of these checks in an audit log, then to perform the task in the distributed environment and, finally, to return results to the user. ACD has been developed in Java and exposes a RESTful interface. This allows its integration with any tool developed in a programming language that is capable of accessing RESTful services. Prior to its implementation, a model of ACD was developed based on formal notation [37], which is used for building safety critical systems, using the recommendations of the Open Web Application Security consortium for developing secure software [38]. This provides rigorous validation of ACD's security features.

The principal features of ACD's architecture are described below.

•*Local authentication service.* The current implementation supports a username-password database specifically for non Shibboleth ACD support. To be authenticated, a user has to provide a username/password pair that matches an entry in the database or use their local institution credentials via Shibboleth. To avoid known vulnerabilities involving usernames and passwords we adopted OWASP best security practices [39] such as storing passwords in encrypted form, rejecting weak passwords chosen by users, forcing the password length to be a minimum of eight characters including special characters, and changing the password on a regular basis. The Shibboleth support in the latest version of ACD provides users with more familiar authentication mechanisms. Shibboleth is currently used by many universities in the UK, EU and beyond to allow students and researchers to access online publishers' resources by invoking their local institutional username/password credentials.

•*Authorization component.* This component controls all actions performed in the VO. It uses the Parametrized Role Based Access Control (PRBAC) model in which permissions are assigned to roles [40]. The VO policy designer associates each user in the VO with the role that best describes his/her job function. The policy is defined at VO set up because it depends on the VO functionalities. The tasks (permissions) assigned to roles are drawn from the VO functionality.

•*Credential repository.* This component is responsible for managing the delegation of identity from the user to ACD via a proxy certificate. It stores the certificates acquired by the VO administrator (known as robot certificates) and their corresponding private keys in order to communicate with the target e-infrastructure. To allow the members of a VO access to computational resources, the VO is assigned a digital certificate, which is used behind the scenes to authenticate requests issued by the VO at the resource provider site. The component also maintains a list of issued proxy certificates (delegated identities), their corresponding private keys and the association between users and proxies in order to trace which proxy was used by which user.

•*Auditing component.* This component records all actions within the VO, including authorized and unauthorized requests to perform tasks within the VO, the username that requested them, the number of login attempts and login times. This allows the VO management to identify those ACD users responsible for having performed any tasks within an e-infrastructure environment.

5.1. Integration with the Application Hosting Environment

When run without ACD, the AHE security model requires each individual user to have a digital certificate, which carries with it the need to go through a lengthy credential acquisition process. To remove the need for such a certificate, we have integrated ACD with AHE. The first step of the integration requires understanding the interaction between AHE and ACD, in other words the functional and administrative tasks that can be performed within the integrated system. The administrative tasks offered by ACD include VO creation, certificate assignment, adding users, resetting user passwords, creating user roles, assigning tasks to roles, and assigning users to roles. The functional tasks offered by AHE include: Prepare Job, Submit Job, Monitor Job, download and Terminate Job. AHE's functional tasks are the same as the tasks permitted for any authorized user on a computational resource that uses the Globus or UNICORE middleware, for example. Therefore, the permissions assignment to the VO is done by the resource owner first, then the VO administrator re-assigns these permissions to the roles in the VO according to the VO authorization requirements.

In the integrated ACD + AHE environment, the authorization requirements determined by the VO administrator are expressed through the introduction of two roles: VO Administrator and Scientist. The former is permitted to perform all the administrative operations above in addition to terminate, monitor and download any job submitted to a resource. The latter is permitted to perform all AHE operations in such a way that a person who submitted a specified job can only perform AHE functional operations on this application. As a result, two users running applications invoking different data will not be able to view the results of each other's activities. In addition, the scientist's role only permits a user to change his/her own password.

The construction of a VO requires that a system administrator goes through the process of acquiring a digital certificate. Once done, the VO administrator creates a VO and assigns the certificate to the named VO using the ACD client. It then becomes possible to add users instantly to the VO and give them genuinely transparent access to e-infrastructure resources. To illustrate how this system works, consider a user named "John Smith" who is a member of a research group in a UK university and would like to use UK National e-Infrastructure Service (NES) [41] resources to run scientific applications using AHE. The user contacts the local VO administrator and requests to join a specific VO. The user can opt to use their local username and password, if their institution is part of the Shibboleth Federation, or requests the creation of a dedicated ACD account. The VO administrator assigns the user to the "Scientist" role described above and assigns the user to a VO that has an access to e-infrastructure resources. The communications between the AHE + ACD client and the wrapped AHE server, as well as between the latter and the grid resources, are protected by the SSL security protocol.

In order to launch an application on a computational resource, the user invokes a request to perform the "Submit Job" task using an AHE client with ACD extension. This request is intercepted by the ACD authentication component that checks if the username and password match an entry in the database or can be authenticated against Shibboleth. The result of the authentication is recorded in the auditing component. The role of the user is picked up from the authorization component, in this case "Scientist". The authorization checks whether the "Submit Job" operation is permitted for the "Scientist" role held by the user. The result of the access control check is recorded in the audit log, and the operation "Submit Job" is invoked on AHE server. Once the request is granted, ACD picks the certificate associated with the VO the user wants to use and checks whether the user is assigned to this VO. If the check is successful ACD generates a proxy certificate from the VO assigned certificate, uploads it to a MyProxy server and records the issued proxies in the credential repository.

ACD then sends the randomly generated username/password pair needed to access MyProxy to the AHE server to download the session proxy. Finally, the AHE server sends the request to the computational resource site along with the proxy. At the target site, the proxy is validated. Certificate authentication succeeds, and the distinguished name on the proxy (VOName) is checked against the resource's authorization system to determine the role of the VOName, which is Scientist. Since this role is allowed to submit an application to resources the task will be invoked. From the e-infrastructure's perspective, it is the VOName that submitted the task, not "John Smith". In order to find out who invoked the "submit job" task on the resource using a specific proxy, the resource administrator passes the public key of the proxy to the VO administrator who can identify the name of the user.

In this way, requests from within the combined ACD/AHE system can be audited. It is thus possible to identify legitimate users and to ensure that only such users are allowed access to resources, in conformance with the policies enforced by the e-infrastructure management. In addition, it is possible to detect unauthorized attempts to access resources from within the VO and to identify persons responsible for such attempts. This form of accountability is an essential requirement for resource providers to be prepared to accept the ACD security model.

6. DEPLOYMENT OF AHE 3.0

AHE can be deployed as a standalone application via the Jetty Server using an embedded database or, in a more complex environment, AHE can be deployed as a Servlet hosted within a Servlet compliant server such as Apache Tomcat and configured to use databases supported by the Hibernate framework.

In the simplest configuration, the standalone mode allows AHE to be executed as an application which launches the Jetty Server with the option of invoking an embedded database or any external database supported by the Hibernate Framework. In this configuration, the user simply downloads the AHE executable, configures the Hibernate configuration file to set up the database connectivity and runs the programme.

With server or network constraints, AHE can be hosted inside a Servlet compliant server such as Apache Tomcat and be configured to use any databases supported by the Hibernate framework. A user should then download the AHE servlet version, deploy it on the Servlet server and configure the database configuration file to ensure AHE runs correctly. Once AHE is running, the system administrator configures user management, hosted applications as well as resources and credentials.

Whichever way the server is deployed, end users can access it either using a web browser, via the web client interface, or using the GUI or command line client tools. The client tools simply require Java to be available on the client machine; after setting an environment variable and running a configuration script these can be easily run.

7. AHE 3.0: COMPARISON WITH AHE 2.0

Our efforts to refactor AHE 3 to expose a restful interface, as well as redesign the AHE server in version 3.0 have not only been done to enhance user experience, but also to improve performance. In order to evaluate the benefits of this work, we ran performance tests comparing the performance of AHE 2.0 against AHE 3.0.

Our experimental set up consisted of a server running both AHE 2.0 and AHE 3.0, with both systems configured to launch applications via the QCG-Computing middleware onto a 96 node cluster within the Centre for Computational Science at UCL. The tests we performed used a workstation to submit batches of applications to AHE 2.0 and AHE 3.0 in turn, measuring the time taken to submit these batches. The application launched was a simple code designed to sort a list of words into alphabetical order, but since we are only interested in the time performance of the AHE server itself, we only measured time taken to submit the application rather than measuring the time the application takes to execute (which would be affected by the cluster load), and the cluster was dedicated to the experiment while the tests were performed. The tests themselves were implemented as JUnit tests calling the AHE client API, while JUnit, executed via the Eclipse development platform, was used to measure the time taken to perform the tests. Each test was repeated three times, and the mean time taken for each test calculated. The results are plotted in Fig. 4, with error bars showing the standard deviation of each result.

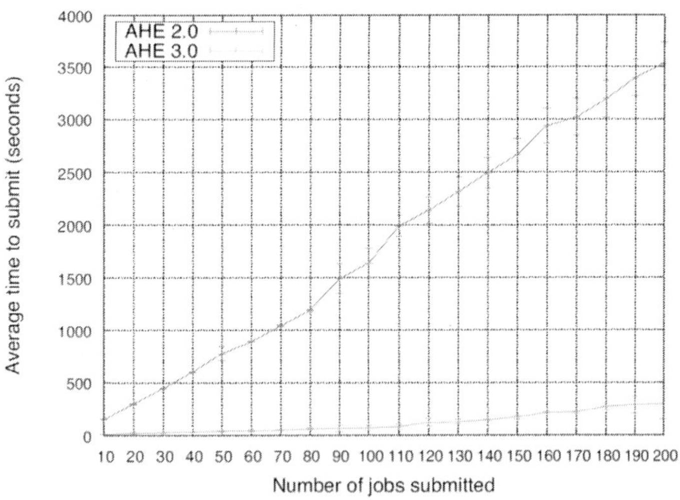

Figure. 4. Comparison of the mean time and standard deviation required to submit using AHE 2.0 and AHE 3.0 for batches of 10–200 jobs.

As Fig. 4 shows, AHE 3.0 performs far faster than AHE 2.0, and the time taken to submit jobs using AHE 2.0 is much more variable, meaning that it is less reliable from a user perspective. This is due to the fact that AHE 3.0 is developed in Java whereas AHE 2.0 was developed with Perl/WSRF::Lite; AHE 3.0 exploits a simple RESTful interface, whereas AHE 2.0 uses the far more complicated WSRF extension to Web services, which increases the complexity of both client and server. Application submission is also faster in AHE 3.0 because the system implements a buffered queuing system between the AHE server and the connector modules, which has the effect of allowing the submission interface to process more simultaneous requests, compared to AHE 2.0. In order for us to better investigate the nature of the performance differences between AHE 2.0 and AHE 3.0, in future work we plan to instrument the server code of both AHE versions with timing routines, and then perform a set of experiments to determine the performance of different parts of the applications launching process, from data staging to job submission.

7.1. Evaluation of the Application Interaction Model

To validate our hypothesis that the Application Interaction Model, by simplifying the process of launching applications on high performance computing resources, is more usable than the traditional grid interaction model, we conducted a rigorous usability study. We have reported this study fully in Zasada et al. [17], and for a comprehensive account of the study methodology we refer readers to that publication.

We compared the AHE command line client with the Globus command line client and the AHE graphical client with the Unicore graphical client. By all of our measurements, the AHE clients were judged to be significantly more usable than either Globus or Unicore [17].

In addition to the usability tests comparing AHE with common middleware tools, we also compared the standard AHE release with the new version of AHE, integrated with the security solution Audited Credential Delegation [37] (discussed in Section 5). Our usability tests clearly established that AHE with ACD is more usable than AHE alone.

8. AHE 3.0 IN ACTION: E-INFRASTRUCTURE BASED MULTISCALE SIMULATION

Like its predecessors, AHE 3.0 is being actively used by several large research projects. AHE provides the principal HPC access tool in the VPH-

Share project [42], a currently funded endeavour within the Virtual Physiological Human (VPH) initiative [43], concerned with patient-specific biomedical modelling and simulation. The aim of this project is to develop a set of intelligent services and supporting network infrastructure that will facilitate the exposure and sharing of data and knowledge. In particular, it is developing a multi-scale framework for the composition of new biomedical workflows to promote collaboration within the VPH community.

As part of this infrastructure, VPH-Share is developing a cloud platform that will allow users to easily access computational as well as data resources. AHE and ACD together constitute the HPC gateway service for VPH-Share, allowing simulations that require more computational power than the VPH-Share cloud infrastructure is able to provide to be seamlessly run on HPC resources. AHE and ACD are deployed based on the Software as a Service (SaaS) model. AHE is responsible for handling the execution life cycle of virtualized applications on computational resources, while ACD bridges the gap between different security infrastructures used by the execution platform and those remote resources. This allows simulation workflows to be deployed which combine resources from a cloud provider such as Amazon in order to execute single core and small scale parallel simulations, but that can switch to high performance computing, accessed via AHE, to run parts of the workflow that require more computational power. The ability of the Taverna workflow system used by VPH-Share to call AHE's RESTful interface allows applications hosted in AHE to be included as components in Taverna workflows.

The system is being used in production runs by VPH-Share scientists to run the Chaste code [44] to model personalized treatments of cardiac arrythmias in patients. AHE allows the researchers to launch Chaste simulations on the HECToR HPC machine in the UK (part of PRACE), marshal input and output data and manage parameter sweeps. It also allows data to be staged in and out of the EUDAT [45] data storage infrastructure as necessary.

Additionally, the AHE client API has been integrated with the GridSpace2 workflow engine [46] within the MAPPER project, and as part of the Virtual Imaging Platform (VIP) [47]. The MAPPER project aims to develop a persistent production infrastructure for distributed multiscale computing [48], making use of resources from multiple European e-infrastructures. AHE provides a key interoperability layer allowing higher level MAPPER services and interface tools to seamlessly access and connect these distinct resources.

MAPPER applications couple codes operating at different temporal and spatial resolutions together. An illustration of such an application is given in Fig. 5, which shows an application from the materials science domain [49]. This application invokes parameters generated at the quantum level (using Car–Parrinello molecular dynamics (CPMD)) to build an atomistic simulation of a material using the Large-scale Atomic/Molecular Massively Parallel Simulator (LAMMPS)), and then uses parameters generated at this level to perform a coarse grained molecular simulation (again using LAMMPS).

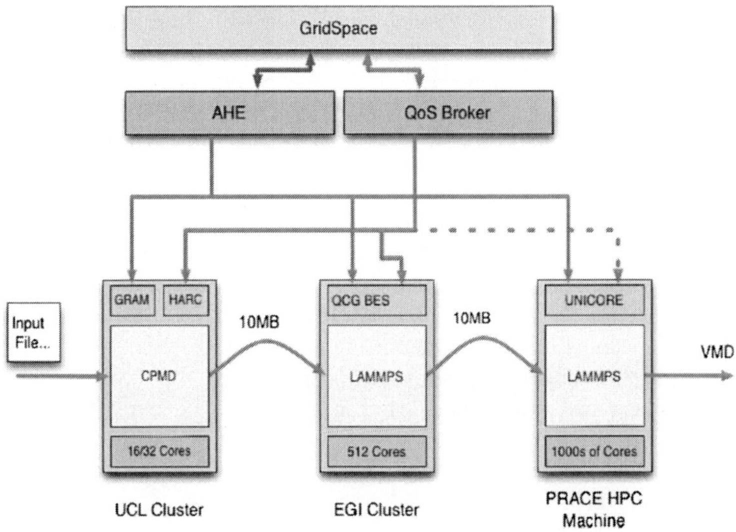

Figure. 5. The architecture of a MAPPER loosely coupled application. AHE provides an interoperability layer to launch codes across a range of platforms.

Each of the MAPPER applications requires access to suitable resources to run, often concurrently or in a particular sequence. AHE's ability to couple with advanced reservation tools such as QCG Computing means that MAPPER applications can be scheduled to run in advance (co-)reservations on HPC resources. This is of course dependent on the machines providing advanced reservation capabilities; it is essential for resource providers to implement such policies in order for these kinds of applications to be run.

The VIP project targets multi-modality, multi-organ and dynamic medical image simulation, integrating proven simulation software to simulate four main imaging modalities. This project builds on the Virtual Imaging Platform, a French ANR (National Agency for Research) project aiming at building a multi-modality simulation platform for the main medical imaging modalities, namely Magnetic Resonance Imaging (MRI), Ultrasound imaging (US), Positron Emission Tomography (PET) and Computed Tomography (CT).

Typically the simulations considered by VIP have been executed in an "embarrassingly parallel" fashion on cluster machines and are managed via the VIP platform. The platform can perform complex simulations, such as whole-body CT scans, in a reasonable time thanks to its connection to EGI, which provides access to a distributed set of clusters. However, some imaging simulation codes can be parallelized and thus benefit substantially from access to high performance computing resources. By integrating calls to AHE, via the AHE client API, with the VIP portal and workflow engine, this requirement has been satisfied. AHE provides a bridge between EGI and the HPC facilities provided by PRACE, so that simulations which require greater power than EGI can provide can instead be run on the PRACE platform.

Common to all of these projects is the need to use AHE's capabilities as an interoperability layer to bridge across disparate e-infrastructures, meaning that AHE provides a single interface to a variety of resources, from PRACE, through EGI to local campus based clusters. AHE's ability to connect to a wide range of different backend middleware tools makes it an ideal candidate to federate resource access from the user's perspective. Integration with ACD means that appropriate security credentials are presented to the target infrastructure by AHE when a user needs to execute a simulation, making the use of multiple e-infrastructure platforms transparent to the user.

9. CONCLUSIONS AND FUTURE WORK

Since its initial release, AHE has been taken up by various user communities, where its usability features have proved extremely important. AHE has been employed to host computational codes from different scientific domains, including widely used codes such as NAMD, CHARMM, LAMMPS, VASP, LB3D and DL_POLY. A key strength of AHE is its flexibility. Since all of its complexity resides on the server side, and all of AHE's functionality is exposed as RESTful Web services, AHE can be used as a building block for systems of much greater complexity.

As our performance tests have shown, the redesign of AHE 3.0 has greatly improved performance over older AHE versions. Our usability results have also confirmed the benefit of the Application Interaction Model in that user interaction is reduced to the most essential components: namely a user interacting with his/her application. Users do not need to worry about the details of particular batch queuing systems, or how to stage data back from particular HPC resources; the specifics of how to launch an application are encapsulated within the Application Interaction Model. The approach virtualizes the HPC resources from a user's point of view. Indeed AHE virtualizes the entirety of a grid's HPC resources, and federates resources stemming from multiple different e-infrastructures.

AHE 3.0 provides a number of capabilities including a workflow engine that allows complex simulations to be created, including coupled simulations where data is automatically transferred from one application to another. ACD provides a security suite that includes support for Shibboleth authentication, as well as user auditing. ACD supports virtual organization management and is able to provide access to grid proxy credentials through RESTful web services.

This combination of usability and performance embedded into a feature-rich platform have led AHE and ACD to become an important cornerstone of many research projects, from materials science through computational physics and chemistry to life and medical sciences. These projects have in common a need for a computational platform to provide access to high performance computational resources and links to cloud computing infrastructures. ACD and AHE allow e-infrastructure to be accessed in a similar manner as an IaaS cloud resource. This is achieved by virtualizing applications using an SaaS model, exposing their functionality as simple RESTful web services, and by abstracting the security mechanism of the e-infrastructure middleware through ACD. Although they may be and often are deployed in combination as part of a project specific e-infrastructure, AHE and ACD are also standalone tools, and can be deployed within any similar e-infrastructure that requires transparent access to high end computing resources.

AHE 3.0 and ACD have been released under the LGPL licence and can be downloaded from: https://sourceforge.net/projects/ahe3/.

ACKNOWLEDGEMENTS

The developments of AHE and ACD reported in this paper have been funded by the EU FP7 VPH-Share (no. 269978), VPH-NoE (no. 223920), MAPPER (no. 261507) and ContraCancrum (no. 223979) projects. Previously they were funded by the EPSRC projects RealityGrid (GR/R67699), Rapid Prototyping of Usable Grid Middleware (GR/T27488/01), User-Friendly Authentication and Authorization for Grid Environments (EP/D051754/1), and also by OMII under the Managed Programme Robust Application Hosting in WSRF::Lite (RAHWL) project. SJZ's PhD studentship is funded by EPSRC.

We would like to thank Maciej Pawlik and Grzegorz Dyk from Cyfronet, Krakow, Poland, and William Romero from CNRS Lyon, France, for their contributions to the code, and also Mariusz Mamonski and Tomasz Piontek from Poznan Super Computing and Networking Center, Poland for useful discussions relating to advance reservation.

REFERENCES

1. D. Turek **HIGH PERFORMANCE COMPUTING AND THE IMPLICATIONS OF MULTI-CORE ARCHITECTURES** CTWatch Quarterly, 3 (2007), pp. 31-33
2. J. Dongarra, D. Gannon, G. Fox, K. Kennedy **THE IMPACT OF MULTICORE ON COMPUTATIONAL SCIENCE SOFTWARE** CTWatch Quarterly, 3 (2007), pp. 1-10
3. P.V. Coveney (Ed.), Scientific Grid Computing, Vol. 363, Philosophical Transactions of the Royal Society A: Mathematical, Physical and Engineering Sciences, 2005
4. Foster, C. Kesselman, S. Tuecke **THE ANATOMY OF THE GRID: ENABLING SCALABLE VIRTUAL ORGANIZATIONS** International Journal of Supercomputer Applications, 15 (2001), 3-23
5. J. Chin, P.V. Coveney **TOWARDS TRACTABLE TOOLKITS FOR THE GRID: A PLEA FOR LIGHTWEIGHT, USEABLE MIDDLEWARE, TECH. REP**(2004) http://nesc.ac.uk/technical_papers/UKeS-2004-01.pdf
6. M. Halling-Brown, D. Moss, C. Sansom, A. Shepherd, **A COMPUTATIONAL GRID FRAMEWORK FOR IMMUNOLOGICAL APPLICATIONS** Philosophical Transactions of the Royal Society A, 367 (2009), pp. 2705-2716
7. P.V. Coveney, R.S. Saksena, S.J. Zasada, M. McKeown, S. Pickles, **THE APPLICATION HOSTING ENVIRONMENT: LIGHTWEIGHT MIDDLEWARE FOR GRID-BASED COMPUTATIONAL SCIENCE**Computer Physics Communications, 176 (6) (2007), pp. 406-418
8. S.J. Zasada, P.V. Coveney, **VIRTUALIZING ACCESS TO SCIENTIFIC APPLICATIONS WITH THE APPLICATION HOSTING ENVIRONMENT** Computer Physics Communications, 180 (12) (2009), pp. 2513-2525
9. S.K. Sadiq, D. Wright, S.J. Watson, S.J. Zasada, I. Stoica, P.V. Coveney, **AUTOMATED MOLECULAR SIMULATION BASED BINDING AFFINITY CALCULATOR FOR LIGAND-BOUND HIV-1 PROTEASES** Journal of Chemical Information and Modeling, 48 (9) (2008), pp. 1909-1919
10. J.L. Suter, P.V. Coveney, H.C. Greenwell, M.-A. Thyveetil, **LARGE-SCALE MOLECULAR DYNAMICS STUDY OF MONTMORILLONITE CLAY: EMERGENCE OF UNDULATORY FLUCTUATIONS AND DETERMINATION OF MATERIAL PROPERTIES** The Journal of Physical Chemistry C, 111 (23) (2007), pp. 8248-8259
11. R.T. Fielding, **ARCHITECTURAL STYLES AND THE DESIGN OF NETWORK-BASED SOFTWARE ARCHITECTURES** (2000) (Ph.D. thesis)
12. S. Graham, A. Karmarkar, J. Mischkinsky, I. Robinson, I. Sedukin, **WEB SERVICES RESOURCE FRAMEWORK, TECH. REP.** (2006) http://docs.oasis-open.org/wsrf/wsrf-ws_resource-1.2-spec-os.pdf WINE, http://www.winehq.org/.
13. P.V. Coveney**SCIENTIFIC GRID COMPUTING** , Philosophical Transactions of the Royal Society A: Mathematical, Physical and Engineering Sciences, 363 (1833) (2005), pp. 1707-1713
14. S. Manos, S.J. Zasada, P.V. Coveney, **LIFE OR DEATH DECISION-MAKING: THE MEDICAL CASE FOR LARGE-SCALE ON-DEMAND GRID COMPUTING** CTWatch Quarterly Journal, 4 (2) (2008), pp. 35-45
15. B. Beckles, V. Welch, J. Basney, **MECHANISMS FOR INCREASING THE USABILITY OF GRID SECURITY** International Journal of Human–Computer Studies, 63 (1/2) (2005), pp. 74-101
16. S.J. Zasada, A.N. Haidar, P.V. Coveney, **ON THE USABILITY OF GRID MIDDLEWARE AND SECURITY MECHANISMS** Philosophical Transactions of the Royal Society A: Mathematical, Physical and Engineering Sciences, 369(1949) (2011), pp. 3413-3428

17. B. Boghosian, P.V. Coveney, S. Dong, L. Finn, S. Jha, G. Karniadakis, N. Karonis, **NEKTAR, SPICE AND VORTONICS: USING FEDERATED GRIDS FOR LARGE SCALE SCIENTIFIC APPLICATIONS** Cluster Computing, 10 (3) (2007), pp. 351-364
18. UK Collaborative Computational Projects, http://www.ccp.ac.uk/
19. S.J. Zasada, M. Mamonski, D. Groen, J. Borgdorff, I. Saverchenko, T. Piontek, K. Kurow ski, P.V. Coveney, **DISTRIBUTED INFRASTRUCTURE FOR MULTISCALE COMPUTING IN** Proceedings of the 2012 IEEE/ACM 16th International Symposium on Distributed Simulation and Real Time Applications, IEEE Computer Society (2012), pp. 65-74
20. JBPM – JBoss Community, http://www.jboss.org/jbpm
21. Hibernate – JBoss Community, http://www.hibernate.org/
22. Restlet – RESTful web services framework for Java, http://www.restlet.org/.
23. The Apache Tomcat Servlet Container, http://tomcat.apache.org.
24. The Globus Project, http://www.globus.org.
25. The UNICORE Project, http://www.unicore.org.
26. K. Kurowski, W. de Back, W. Dubitzky, L. Gulyás, G. Kampis, M. Mamonski, G. Szemes, M. Swain, **COMPLEX SYSTEM SIMULATIONS WITH QOSCOSGRID** Computational Science-ICCS, 2009 (2009), pp. 387-396
27. A.N. Haidar, S.J. Zasada, P.V. Coveney, A.E. Abdallah, B. Beckles, M.A.S. Jones, **AUDITED CREDENTIAL DELEGATION: A USABLE SECURITY SOLUTION FOR THE VIRTUAL PHYSIOLOGICAL HUMAN TOOLKIT** Interface Focus, 1 (3) (2011), pp. 462-473
28. W. Allcock, J. Bester, J. Bresnahan, A. Chervenak, L. Liming, S. Tuecke, GridFTP: Protocol extensions to FTP for the grid, Global Grid Forum GFD-RP 20, http://www.ggf.org/documents/GWD-R/GFD-R.020.pdf
29. P. Beckman, S. Nadella, N. Trebon, I. Beschastnikh, **SPRUCE: A SYSTEM FOR SUPPORTING URGENT HIGH-PERFORMANCE COMPUTING** Grid-Based Problem Solving Environments (2007), pp. 295-311
30. S. Pickles, R. Haines, R. Pinning, A. Porter, **A PRACTICAL TOOLKIT FOR COMPUTATIONAL STEERING** Philosophical Transactions of the Royal Society A: Mathematical, Physical and Engineering Sciences, 363(1833) (2005), pp. 1843-1853
31. C. Cavaness, **QUARTZ JOB SCHEDULING FRAMEWORK: BUILDING OPEN SOURCE ENTERPRISE APPLICATIONS** Prentice Hall, Upper Saddle River, NJ, USA (2006)
32. R.O. Sinnott, J. Jiang, J. Watt, O. Ajayi, **SHIBBOLETH-BASED ACCESS TO AND USAGE OF GRID RESOURCES** IEEE International Conference on Grid Computing, Barcelona, Spain (2006), pp. 28-29
33. J. Novotny, S. Tuecke, V. Welch, **AN ONLINE CREDENTIAL REPOSITORY FOR THE GRID: MYPROXY** 10th IEEE International Symposium on High Performance Distributed Computing, 2001. IEEE Proceedings (2002), pp. 104-111
34. X. Wang, M. Jones, J. Jensen, A. Richards, D. Wallom, T. Ma, R. Frank, D. Spence, S. You ng, C.Devereux, *et al.* , **SHIBBOLETH ACCESS FOR RESOURCES ON THE NATIONAL GRID SERVICE (SARONGS)**
35. Fifth International Conference on Information Assurance and Security, 2009, IAS'09. Vol. 2, IEEE (2009), pp. 338-341
36. R. Alfieri, R. Cecchini, V. Ciaschini, L. dell Agnello, A. Frohner, A. Gianoli, K. Lorentey, F. Spataro, **VOMS, AN AUTHORIZATION SYSTEM FOR VIRTUAL ORGANIZATIONS** Grid Computing, Springer (2004), pp. 33-40
37. A.N. Haidar, P.V. Coveney, A.E. Abdallah, P.Y. Ryan, B. Beckles, J.M. Brooke, M. Jones, **FORMAL MODELLING OF A USABLE IDENTITY MANAGEMENT SOLUTION FOR VIRTUAL ORGANIZATIONS** Proceedings of Formal Aspects of Virtual Organisations (2009), pp. 41-50
38. OWASP top 10: The ten most critical web application security vulnerabilities (2007), http://www.owasp.org/index.php/Top_10_2010-Main.

39. A.Abdallah, E. Khayat, **FORMAL *z* SPECIFICATIONS OF SEVERAL FLAT ROLE-BASED ACCESS CONTROL MODELS**Software Engineering Workshop, 2006, SEW'06. 30th Annual IEEE/NASA, IEEE (2006), pp. 282-292
40. UK National e-Infrastructure Service (NES), http://www.ngs.ac.uk.
41. The VPH-Share Project, http://www.vph-share.eu.
42. P. Hunter, P.V. Coveney, B. de Bono, V. Diaz, J. Fenner, A. Frangi, P. Harris, R. Hose, P. Kohl, P. Lawford,*et al.*, **A VISION AND STRATEGY FOR THE VIRTUAL PHYSIOLOGICAL HUMAN IN 2010 AND BEYOND** Philosophical Transactions of the Royal Society A: Mathematical, Physical and Engineering Sciences, 368(1920) (2010), pp. 2595-2614
43. J. Pitt Francis, P. Pathmanathan, M. Bernabeu, R. Bordas, J. Cooper, A. Fletcher, G. Mirams, P. Murray, J. Osborne, A. Walter, *et al.*, **CHASTE: A TEST-DRIVEN APPROACH TO SOFTWARE DEVELOPMENT FOR BIOLOGICAL MODELING** Computer Physics Communications, 180 (12) (2009), pp. 2452-2471
44. The EUDAT Project, http://www.eudat.eu.
45. M. Malawski, T. Bartynski, M. Bubak, **INVOCATION OF OPERATIONS FROM SCRIPT-BASED GRID APPLICATIONS** Future Generation Computer Systems, 26 (1) (2010), pp. 138-146
46. The VIP Project, http://www.creatis.insa-lyon.fr/vip/.
47. The MAPPER Project, http://www.mapper.eu.
48. J. Suter, D. Groen, L. Kabalan, P.V. Coveney, Distributed multiscale simulations of clay-polymer nanocomposites, MRS Online Proceedings Library 1470 (1) (2012)

Index